Polymer Based Systems on Tissue Engineering, Replacement and Regeneration

NATO Science Series

A Series presenting the results of scientific meetings supported under the NATO Science Programme.

The Series is published by IOS Press, Amsterdam, and Kluwer Academic Publishers in conjunction with the NATO Scientific Affairs Division

Sub-Series

I. **Life and Behavioural Sciences**	IOS Press
II. **Mathematics, Physics and Chemistry**	Kluwer Academic Publishers
III. **Computer and Systems Science**	IOS Press
IV. **Earth and Environmental Sciences**	Kluwer Academic Publishers
V. **Science and Technology Policy**	IOS Press

The NATO Science Series continues the series of books published formerly as the NATO ASI Series.

The NATO Science Programme offers support for collaboration in civil science between scientists of countries of the Euro-Atlantic Partnership Council. The types of scientific meeting generally supported are "Advanced Study Institutes" and "Advanced Research Workshops", although other types of meeting are supported from time to time. The NATO Science Series collects together the results of these meetings. The meetings are co-organized bij scientists from NATO countries and scientists from NATO's Partner countries – countries of the CIS and Central and Eastern Europe.

Advanced Study Institutes are high-level tutorial courses offering in-depth study of latest advances in a field.
Advanced Research Workshops are expert meetings aimed at critical assessment of a field, and identification of directions for future action.

As a consequence of the restructuring of the NATO Science Programme in 1999, the NATO Science Series has been re-organised and there are currently Five Sub-series as noted above. Please consult the following web sites for information on previous volumes published in the Series, as well as details of earlier Sub-series.

http://www.nato.int/science
http://www.wkap.nl
http://www.iospress.nl
http://www.wtv-books.de/nato-pco.htm

Series II: Mathematics, Physics and Chemistry – Vol. 86

Polymer Based Systems on Tissue Engineering, Replacement and Regeneration

edited by

Rui L. Reis

Department of Polymer Engineering,
University of Minho,
Guimarães, Portugal

and

Daniel Cohn

Casali Institute of Applied Chemistry,
The Hebrew University,
Jerusalem, Israel

Kluwer Academic Publishers

Dordrecht / Boston / London

Published in cooperation with NATO Scientific Affairs Division

Proceedings of the NATO Advanced Study Institute on
Polymer Based Systems on Tissue Engineering, Replacement and Regeneration
Alvor, Algarve, Portugal
15–25 October 2001

A C.I.P. Catalogue record for this book is available from the Library of Congress.

ISBN 1-4020-1000-1

Published by Kluwer Academic Publishers,
P.O. Box 17, 3300 AA Dordrecht, The Netherlands.

Sold and distributed in North, Central and South America
by Kluwer Academic Publishers,
101 Philip Drive, Norwell, MA 02061, U.S.A.

In all other countries, sold and distributed
by Kluwer Academic Publishers,
P.O. Box 322, 3300 AH Dordrecht, The Netherlands.

Printed on acid-free paper

TABLE OF CONTENTS

CONTRIBUTORS

ABRAHAM, G.A., *Instituto de Ciencia y Tecnología de Polímeros, CSIC, Juan de la Cierva 3, 28006 Madrid, Spain*

ABRAMSON, S.D., *Department of Chemistry, Rutgers, The State University of New Jersey, New Brunswick, NJ 08903*

AGRAWAL, C. M., *Center for Clinical Bioengineering and Department of Orthopaedics,The University of Texas Health Science Center at San Antonio, San Antonio, TX 28229*

AZEVEDO, M.C., *Department of Polymer Engineering, University of Minho, 4800-058 Guimarães. Portugal*

BAKOŠ, D., *Faculty of Chemical and Food Technology, STU, Radlinského 9, 812 37 Bratislava, Slovak Republic*

BANCROFT, G.N., *Department of Bioengineering, Rice University, 6100 Main, Houston, TX, 77005-1892, U.S.A*

BERNARDO, C.A., *Department of Polymer Engineering, University of Minho, 4800-058 Guimarães. Portugal*

BRENTWOOD, A., *Department of Reconstructive Surgery, University Hospital Basel, Switzerland*

COHN, D., *Casali Institute of Applied Chemistry, The Hebrew University of Jerusalem, Jerusalem 91904, Israel*

CUNHA, A.M., *Department of Polymer Engineering, University of Minho, 4800-058 Guimarães. Portugal*

DE BRUIJN, J.D., *IsoTis NV, Professor Bronkhorstlaan 10, 3723 MB Bilthoven, The Netherlands*

DE GRAAF, L.A., *ATO B.V., P.O. Box 17, 6700-AA Wageningen, the Netherlands*

ELVIRA, C., *Instituto de Ciencia y Tecnología de Polímeros, CSIC, Juan de la Cierva 3, 28006 Madrid, Spain*

FEIJEN, J., *Institute for Biomedical Technology, Polymer Chemistry and Biomaterials Group, Department of Chemical Technology, University of Twente, P.O. Box 217, 7500 AE Enschede, The Netherlands.*

FU, X., *Department of Plastic Surgery, Singapore General Hospital*

GALLARDO, A., *Instituto de Ciencia y Tecnología de Polímeros, CSIC, Juan de la Cierva 3, 28006 Madrid, Spain*

GOMES, M.E., *Department of Polymer Engineering, University of Minho, 4800-058 Guimarães. Portugal*

HUTMACHER, D.W., *Department of Bioengineering & Department of Orthopedic Surgery, Faculty of Medicine, National University of Singapore, 10 Kent Ridge Crescent, Singapore 119260*

IKADA, Y., *Institute for Frontier Medical Sciences, Kyoto University, Shogoin, Sakyo-ku, Kyoto 606-8507, Japan*

KOHN, J., *Department of Chemistry, Rutgers, The State University of New Jersey, New Brunswick, NJ 08903*

KOLLER, J., *Centre for Burns and Reconstructive Surgery, Central Tissue Bank, Ružinov General Hospital, Ružinovská 6, 826 06 Bratislava, Slovak Republic*

LEE, S.T., *Department of Plastic Surgery, Singapore General Hospital*

LEONOR, I.B., *Department of Polymer Engineering, University of Minho, 4800-058 Guimarães. Portugal*

MANO, J.F., *Department of Polymer Engineering, University of Minho, 4800-058 Guimarães. Portugal*

MAROM, G., *Casali Institute of Applied Chemistry, The Hebrew University of Jerusalem, Jerusalem 91904, Israel*

MENDES, S.C., *IsoTis NV, Professor Bronkhorstlaan 10, 3723 MB Bilthoven, The Netherlands*

MIKOS, A.G., *Department of Bioengineering, Rice University, 6100 Main, Houston, TX, 77005-1892, U.S.A*

OLIVEIRA, A.L., *Department of Polymer Engineering, University of Minho, 4800-058 Guimarães. Portugal*

PANGARKAR, N., *Faculty of Business Administration, National University of Singapore FBA2 17, Law Link, Singapore 117591*

PASKULEVA, I., *Department of Polymer Engineering, University of Minho, 4800-058 Guimarães. Portugal*

REIS, R.L., *Department of Polymer Engineering, University of Minho, 4800-058 Guimarães. Portugal*

ROHNER, D., *Department of Plastic Surgery, Singapore General Hospital, National University Singapore*

RYPÁČEK, F. *Institute of Macromolecular Chemistry, Academy of Sciences of the Czech Republic, Heyrovský Sq. 2, 162 06 Prague 6, Czech Republic*

SALGADO, A., *Department of Polymer Engineering, University of Minho, 4800-058 Guimarães. Portugal*

SAN ROMÁN, J., *Instituto de Ciencia y Tecnología de Polímeros, CSIC, Juan de la Cierva 3, 28006 Madrid, Spain*

SCHANTZ, J-T., *Laboratory for Biomedical Engineering, National University Singapore*

SEYDA, A., *Department of Chemistry, Rutgers, The State University of New Jersey, New Brunswick, NJ 08903*

SIT, P.S., *Department of Chemistry, Rutgers, The State University of New Jersey, New Brunswick, NJ 08903*

TAN, B.K., *Department of Plastic Surgery, Singapore General Hospital*

TEMENOFF, J.S., *Department of Bioengineering, Rice University, 6100 Main, Houston, TX, 77005-1892, U.S.A*

VAN BLITTERSWIJK, C.A., *IsoTis NV, Professor Bronkhorstlaan 10, 3723 MB Bilthoven, The Netherlands*

VAZ, C.M., *Department of Polymer Engineering, University of Minho, 4800-058 Guimarães. Portugal*

VÁZQUEZ, B., *Instituto de Ciencia y Tecnología de Polímeros, CSIC, Juan de la Cierva 3, 28006 Madrid, Spain*

WISSINK, M.J.B., *Institute for Biomedical Technology, Polymer Chemistry and Biomaterials Group, Department of Chemical Technology, University of Twente, P.O. Box 217, 7500 AE Enschede, The Netherlands.*

YEOW, V., *Department of Plastic Surgery, Singapore General Hospital*

Welcome Addresses

PREFACE

During the past century, and particularly in the last three decades, conventional materials technology resulted in clear improvements in the field of substitution medicine. The development of artificial hips and knees are only examples of the enormous benefits this technology has had for patients. However, there are still no materials available that can adequately replace several functional tissues, such as bones or large bone segments. Therefore, despite the enormous benefits the contemporary technology has brought, the outer limits have been reached and new breakthroughs can only be expected from a novel hybrid technology that will reduce the shortcomings of the current material technology. Such a combined, biology driven approach is referred to as "tissue engineering", by which biological tissues are engineered through combining material technology and biotechnology. Tissue engineering thus involves the culture of living human cells usually in polymeric scaffold materials, ex vivo, and allows them to develop into a three dimensional tissue. This will be the focus on biomaterials research and substitution medicine in the coming decades. It will also create the need for the education of new scientists and engineers that are also 'hybrid' and can perform multidisciplinary research, combining materials and biotechnology.

Internationally, the combination of materials (namely polymer) technology and biotechnology is seen as the sector in which most major breakthroughs can be expected for medical devices in the coming future. Substantial gains are expected to be obtained both from a medical and economic standpoint as a result of this emerging technology. One of the main difficulties related with performing research in this area is the clearly multidisciplinary approach of the teams. A strong group working on tissue engineering and regeneration must combine the expertises of materials scientists, polymer chemists, engineers, chemists, biologists, biochemists, etc. The problem is not only to join the correct team but also to make people understand all the requirements needed from the polymer and biotechnology side, generating synergies on their daily activities.

Major break-throughs in tissue engineering and regeneration will need to address the following aspects: (i) the development of adequate human cell culture to produce the tissues (cells and matrix) in adequate polymeric scaffold materials, that can subsequently be used as a medical device, (ii) the development of culture technology with which human tissues can be grown ex vivo in three dimensional polymeric scaffold matrices, (iii) the development of a material technology with which degradable, three dimensional polymeric matrices can be produced, being suitable for cell culture (proliferation, differentiation) and that have mechanical properties similar to those of the natural tissue.

3

4

The most important materials that are being used on the development of adequate materials for tissue engineering, replacement and regeneration are based on polymers and its composites reinforced with bioactive ceramics. To design adequate materials for these functions it is necessary:

> to design polymers with the correct chemistry, producing new macromolecules, smart materials, using combinatory chemistry, etc.,
> to understand the available choices among existing polymeric biomaterials that qualify for a certain specific application,
> to use the possibility of local delivery of drugs/growth factors, hormones, etc., to induce tissue regeneration or a certain therapeutic effect, by using adequate carriers,
> to process materials into adequate parts and porous scaffolds, using non-conventional techniques, and to maximize its mechanical properties in order that they can perform their function,
> to design adequate scaffolds and to control their morphology, degradation and surface properties in order to optimise cell adhesion and differention,
> to understand and study the principles of biocompatibility in order to design adequate systems and tailor their properties for the purposed applications,
> to look continuously for new materials that are 'more ideal' from all these perspectives.

Finally the researcher of the future should be able to realize that such a complex technology will in most cases only move to clinical use due to his (or his group) own action, by being able to search for partnerships, protecting the respective intellectual property, moving to industrial spin-offs when needed, etc.

We have decided to organize *a NATO Advanced Study Institute (ASI) on Polymer Based Systems on Tissue Engineering, Replacement and Regeneration*, when we realized that there was a clear need for a course that would address in an integrated way all the above referred to topics. In fact, polymer based systems are playing and will play a key-role on tissue engineering, replacement and regeneration in the near future. This biologically driven materials science is believed to be one of the more appealing and funded research areas in the first decades of the XXI century. No course has addressed before this topic in such an integrated and 'looking forward' perspective. An ASI seemed to be the best forum to educate and brainstorming on this area of such strategic importance.

The *NATO-ASI* was held from the 15th to the 25th of October 2001 in Alvor, Algarve, Portugal. Its structure reflected the integrated and multidisciplinary approach needed in this particular field. The course addressed almost all of the topics listed above, joining together the world-leaders on most of the relevant fields. The Faculty not only gave tutorial lectures, but was always very interactive with the participants trying to open their minds for the future of the field. The lecturers also tried to maximize discussion between themselves and with the participants. The course was organized in several topics and was complemented by short presentations and posters delivered by the participants. The best works presented by the participants have been invited to submit a full manuscript to be considered for publication in a special issue of *Journal of Materials Science: Materials in Medicine.*

Finally, I must say that, as most of you know, nobody can organize a course without the help of hard working people and support from several institutions. I would of course first of all like to thank the NATO Scientific Division for their support that made possible the course and the publication of this book. I would like to acknowledge the many contributions of my co-director and friend Danny Cohn. He was a great support whenever I needed it. The members of the scientific committee and several of the lectures made a lot of useful suggestions. All the invited speakers that accepted our invitation and made the course possible are gratefully acknowledged. Also their contributions to this book made possible to produce a state-of-the-art volume to be used by researchers all around the world in the coming years. But the course, and the program, were also made by the ASI students and their wonderful contributions. All the supporting institutions are gratefully acknowledged. University of Minho and the Department of Polymer Engineering that have supported me and my students in so many ways also deserve a word of appreciation. But I am especially grateful to my group, my colleagues, post-docs, PhD and MSc. students. The outcome of this ASI was mainly the result of their hard work, dedication and ambition. Several of them also supported me on preparing this book. They have put a great number of hours on this enterprise and realized that this was an important organization for all of us. I cannot refer all the names herein, but if you find one of the members of the *3B's Research Group – Biomaterials, Biodegradables and Biomimetics* (that I have the pleasure of directing) in one of the meetings you attend, please just speak with her/him and you will see how fortunate I am for being able to advise such a wonderful group of young and bright researchers!

Please enjoy the science and the lessons contained in this book. We really hope the book will be a useful research and education tool and that it can give the readers the same degree of satisfaction we could experience when preparing it for publication.

Rui L. Reis

30 YEARS OF R&D IN PORTUGAL

CARLOS A.A. BERNARDO
Department of Polymer Engineering, University of Minho
4800-058 Guimarães, Portugal

In 1970 the old dictator Salazar died after a period of more than 40 years of undisputed ruling in Portuguese politics. He left behind a stagnant political regimen that had kept him in power, a legacy of economic underdevelopment, an absence of civic and political liberties, and a war in Africa. The Portuguese research system, reflecting the overall economic situation of the country, was incipient. At the time, the regimen tried to implement some cosmetic changes as a survival tactic, but the situation did not improve significantly until a nearly bloodless revolution in 1974 brought it down.

In spite of this, some changes did occur in the late sixties that had a profound impact in Portuguese science. By that time, a significant number of young Portuguese researchers were sent abroad to obtain doctorate degrees via research. Most of them went to the U.K. (that at the time was an exception in Europe for providing Ph.D. degrees in around 3 years), but many went to other European countries and the United States. When these people returned to Portugal they had many difficulties to continue their research, mostly because of a lack of means, but also due to the sub-critical size of the research groups. It is fair to say that things did not improve substantially after the revolution in 1974. For almost 10 years the country had great difficulties in stabilising democracy and finding its way towards a competitive market-oriented economy. All this affected the growth of the Portuguese R&D system that, in the beginning of the eighties, was minimal, concentrated mostly in Lisbon, with some groups in Porto and Coimbra. There were a few research centres funded and "owned" by the state and a few state laboratories, almost all co-ordinated by the Ministry of Industry. And, of course, there were also the various researchers, almost all working in universities, which had returned with doctorates from abroad. These researchers, although still not significant in R&D output, made a major difference relatively to the previous situation. Many of them were working in the *new universities*, founded in the very last years of the dictatorship, spread all over the country, of which Minho and Aveiro are prime examples.

In 1985 Portugal entered the (at the time) EEC, and things started to change radically and for the better in the Portuguese R&D system. By the early and mid nineties, the state "owned" research centres had almost disappeared. However, new national laboratories and technological infrastructures were operational, co-ordinated or funded by the Ministry of Industry. Some of these, the most successful, were spread all over the country and oriented to a specific industrial sector. Others, mostly in Lisbon, soon become *white elephants*, as they had no focus or target industries to support them. Also, a dramatic change had occurred in the research groups working in the Universities. They had become stronger and larger, their scientific output had increased to near the European average, and they had produced a substantial number of PhDs. This was the result of various factors. First, the universities were themselves stronger and better funded, via the Ministry of Education. Second, the Secretary of State of Science and Technology (at the time there was no Ministry) had launched a programme, called CIÊNCIA, who provided grants for young researchers, funded projects and, above all, financed new scientific infrastructures. The scientific infrastructures branch of the CIÊNCIA programme was probably the major single factor responsible for the development of Portuguese R&D, as it spread state-of-the-art equipment in universities all over the country in a fair and transparent way. The less positive aspect of the policy of science and technology at the time was a big deficit of managers and technicians that did not allow an effective co-ordination of the system. There was no real evaluation of the outputs of the different groups as a pre-condition for funding. There was also no real perception of the utility of this policy by the society, as there were no visible links between scientific results and the national economy. Finally, only about 0.5 to 0.6 % of the GDP were invested in the R&D system.

When the Socialist Party came to power by the end of 1995, the Ministry of Science and Technology (MST) was created and José Mariano Gago became its first minister. He was a professor at Lisbon Technical University, and a disbeliever in the capacity of the universities to support a sustainable scientific growth, mostly due to their administrative inefficiency. He increased the budget allotted to the R&D system to about 0.8% of the GDP and launched some very visible initiatives, like the *Information Society* (computers and Internet for all schools across the country!), in which he truly believes. He, and some of his co-workers, developed the "3 programmes" concept, in which the Portuguese R&D system was to be based:

- A programme for funding research centres, proportionally to the number of doctors working on them and taking into account their ranking, which resulted from an evaluation every 3 years by an international panel of experts;
- A programme to provide grants for young researchers, post-docs and senior scientists;

- A programme for financing research projects, through open tenders whose applications were again evaluated by an international panel of experts.

The co-ordination of these programmes was assigned to another brainchild of Gago, the *Foundation for Science and Technology*, which soon became the central point of the whole R&D system. At the same time, the *Observatory for Science and Technology*, an organism that collects, organises and treats, in a systematic way, statistical data on the system was also created. Additionally, and most important, the Ministry of Science and Technology used an agency, called *Agência de Inovação* (AdI), the Innovation Agency, to evaluate and finance projects with impact in the economy, mostly based in consortia with enterprises. Although the AdI had been created in the last years of the previous government, as a joint venture with the Ministry of Economy, it soon became an important tool of the MST and a most effective vehicle in the diffusion and valorisation of research results.

As a consequence of all this, most people agree that the 6 years of Mariano Gago as Minister of Science and Technology have been globally positive, leading to the stabilisation and strengthening of the Portuguese R&D system, that is now within the mainstream of its European counterparts. On the other hand, its detractors say that he is too biased towards fundamental, expensive, state funded science. Also, that he has and elitist and centralist view on how the system should be managed. Maybe because of that, the MST put recently forward the idea of *associated institutes*. These institutes will be directed to key scientific areas, and staffed by scientists of the highest quality in those areas, recognised by the international community. Although independent from the universities, they can work in association with them. The associated institutes will be generously funded, and consequently their number will necessarily be limited. It is no wonder that they have been the subject of heated discussions between those who support them and those who fear that they will lead to a diminished investment in the remaining part of the R&D system.

Although in the short term the associated institutes will dominate the scientific debate, it is clear that they will not, by themselves, determine the future of Portuguese science. Instead, this future will depend on how successful the government will be in tackling three major problems.

The first is the need to launch a new scientific infrastructures programme, as the equipment acquired under the CIÊNCIA programme is reaching the end of its useful life and needs to be replaced. There is strong evidence that the Ministry of Science and Technology plans to do just that in the near future. The second problem is structural. The majority of the Portuguese scientists work in Universities that depend functionally from the Ministry of Education, and receive their salaries primarily to teach. This means that they are not directly accountable to the MST, which acts only as a complementary source of income, by funding projects and providing grants. But this is not done in a coherent way. For instance, although the MST grants pay researchers for projects, they do not allow hiring technicians to operate scientific infrastructures. All this implies that the two ministries must co-ordinate their policies in what concerns scientific research, which is, after all, one of the main missions of the University. This is not solely the responsibility of the MST, especially because the co-ordination should be extended to other ministries with an interest in R&D, such as the Ministry of Economy, that supervises Portuguese industry. Such co-ordination has not yet happened and is surely a major constraint, with a clear impact in the development of the R&D system.

The third problem, which cuts across Portuguese society, is the need to increase the competitiveness of the economy through innovation. The R&D system is crucial to that end. Indeed this is the only way for research groups, in universities, associated institutes and elsewhere, to assert themselves and grow in a sustainable way. If their existence is perceived as strategic to the survival of the Portuguese enterprises, they will contribute to it, independently of the MST policy in a given moment. In fact, this is already happening in some key sectors, such as shoes, plastics and textiles manufacturing. However, quantitatively, there is still a critical lack of investment in R&D by the private sector in Portugal.

Portugal enters the third millennium with a scientific system that is comparable, in qualitative if not in quantitative terms, to that of the other countries of the European Union. This happens for the first time in its history. Some key factors are still missing, such as more investment in infrastructures, better coordination at governmental level, and higher emphasis and investment by the enterprises in innovation. However, by and large, the Portuguese scientific community is optimistic about its future and ready to fulfil its duty towards society as a whole.

Overview on Polymeric Biomaterials and Biodegradable Polymers

POLYMERIC BIOMATERIALS IN MEDICA L SYSTEMS

Y. IKADA

Institute for Frontier Medical Sciences, Kyoto University, Shogoin, Sakyo-ku, Kyoto 606-8507, Japan

1. Polymers as Biomaterials

Synthetic materials made from polymers, glass ceramics, and metals, as well as materials of biological origin, have been used as biomaterials which can be defined as materials to be used in contact with human living cells. Advantages of polymeric materials over the others are their wide flexibility in physical properties varying from viscous fluid to tough solid. A serious drawback of polymers is their insufficient mechanical strength when used as medical devices in orthopaedic and oral surgeries. In such cases, ceramics and metals are the first choice for the biomaterials, although their modulus is too high in comparison with that of bone and tooth tissues, occasionally resulting in stress shielding to the tissues in contact.

1.1. PREREQUISITES OF BIOMATERIALS

Different kinds of artificial tissues and organs are clinically used or under investigation. However, any materials are not always applicable to the medical systems, because materials to be used as biomaterials should meet several prerequisites that are very different from those for the materials in non-medical use. The prerequisites necessary for biomaterials are given in Fig.1. The most important one is non-toxicity, in other words, safety to the human body. The toxicities which foreign materials will cause include pyrogenicity, hemolysis, sustained inflammation, allergy, and carcinogenesis. It is needless to say that every material for any practical purpose should be effective and durable over the period of their use. Biomaterials are not exceptional; hemodialyser should effectively remove renal toxins from the patient's blood while artificial joints should replace the defective joint of patients with high durability.

13

R.L. Reis and D. Cohn (eds.),
Polymer Based Systems on Tissue Engineering, Replacement and Regeneration, 13–24.
© 2002 *Kluwer Academic Publishers. Printed in the Netherlands.*

1) **Non-toxic (biosafe)**
   Non-pyrogenic, Non-hemolytic, Chronically non-inflammative,
   Non-allergenic, Non-carcinogenic, Non-teratogenic

2) **Effective**
   Functionality, Performance, Durability

3) **Sterilizable**
   Ethylene oxide, γ-Irradiation (electron beams), Autoclave,
   Dry heating

4) **Biocompatible**
   Interfacially, Mechanically, and Biologically

Fig.1 Prerequisites of Polymers as Biomaterials

Sterilizability is also the minimum condition that biomaterials should satisfy. Gamma-irradiation, ethylene oxide gas(EOG) exposure, and autoclaving are the current major sterilization means. In some cases, a small amount of EOG, a toxic gas, remains in the sterilized material even after aeration, while irradiation with high-energy radiation and autoclaving often deteriorate the sterilized biomaterials. It should be kept in mind that water-swollen hydrogels as well as materials with immobilized proteins and cells are not easy to access to the conventional sterilization methods because of their high vulnerability to such harsh conditions.

1.2. CAUSES OF TOXICITY

Very often, even biomaterials scientists do not distinguish non-toxicity from biocompatibility which will be described below. However, clear distinction between non-toxicity and biocompatibility is beneficial in understanding what the biomaterial is.

The toxicity of polymeric biomaterials is mostly caused by leachables from the material surface or inside. Most of them are water-soluble or water-dispersible substances as listed in Fig.2. They include monomers remaining unpolymerized, additives such as catalyst, pigment, UV-absorbent, and surfactant, oligomers, and degradation by-products. Medical devices free of these eventually toxic substances can be fabricated if much care is paid to the manufacturing process. One should be always careful in removing microorganisms from biomaterials, although they will be killed by sterilization, because the dead body of microorganisms might contain pyrogenic fragments.

1. **Impurities**
    Unpolymerized monomer, Catalyst, Initiator fragment,
    Mold-releasing agent, ---

2. **Additives**
    Anti-oxidant, UV absorber, Plasticizer, Pigment, ---

3. **Degradation (corrosion) products**
    Biodegradation fragments, Wear debris, Metal ions, ---

4. **Non-living microorganism**

5. **Materials surface having many positive charges or sharp edges**

Fig.2 Possible Toxic Substances

It is very difficult for us to clearly declare that natural products of biological origin are biosafe even after carefully repeated purification. It is because the purified, naturally-occurring materials possibly still contain antigens which will provoke inflammation and allergy. We should not forget the lesson experienced from the mud-cow disease.

2. Biocompatibility

As biocompatibility involves a large number of influencing factors, it is not easy to give an unambiguous definition to biocompatibility[1]. Generally, when foreign materials come into direct contact with the living structure, they induce so-called foreign-body reactions such as complement activation, blood coagulation, thrombus formation, encapsulation by collagenous tissue, and calcification. Fig.3 illustrates the representative foreign-body reactions. One can say that the biomaterial which evokes less foreign-body reactions is better in biocompatibility. However, as thrombus formation is a preceding event necessary for the neointima formation which is required for large-caliber vascular grafts to become biocompatible, less foreign-body reaction does not always mean to lead to better biocompatibility. It should be stressed that even non-toxic biomaterials without any appreciable leachables induce more or less foreign-body reactions.

Biocompatibility can be divided into two subgroups; interfacial biocompatibility and mechanical biocompatibility.

2.1. INTERFACIAL BIOCOMPATIBILITY

If a material surface has a large effect on the performance of the biomaterial,

16

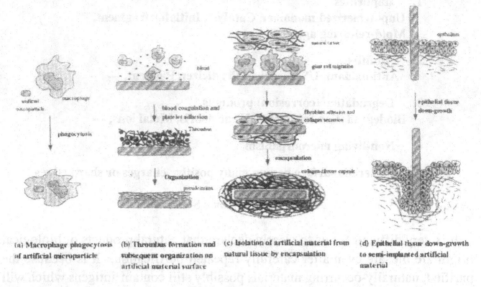

(a) Macrophage phagocytosis of artificial microparticle
(b) Thrombus formation and subsequent organization on artificial material surface
(c) Isolation of artificial material from natural tissue by encapsulation
(d) Epithelial tissue down-growth to semi-implanted artificial material

Fig.3 Representative Foreign-body Reactions against Man-made Materials

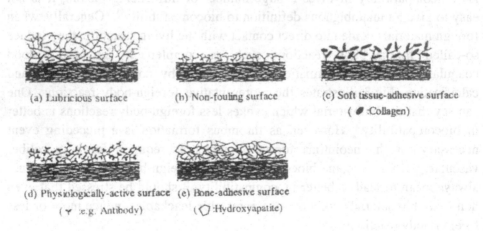

(a) Lubricious surface
(b) Non-fouling surface
(c) Soft tissue-adhesive surface
(● :Collagen)

(d) Physiologically-active surface
(Y :e.g. Antibody)
(e) Bone-adhesive surface
(○ :Hydroxyapatite)

Fig.4 Several Biocompatible Surfaces

we consider its biocompatibility as interfacial. The well-known foreign-body reactions related to interfacial biocompatibility are thrombus formation and encapsulation. As these events are intimately associated with protein adsorption and the subsequent platelet adhesion which occur as the initial biological response to the biomaterial coming into contact with the living body, protein adsorption and cell adhesion onto biomaterials surface have been most extensively studied in the biomaterials science. Depending on the application purpose of biomaterials, either the minimum or the maximum cell adhesion is generally required as highly biocompatible surface.

Several examples of biocompatible surfaces are represented in Fig.4. If a material surface is modified with immobilized, short water-soluble chains of high density, both protein adsorption and cell adhesion will take place to an insignificant extent onto the modified surface, leading to good blood compatibility. On the contrary, if much longer water-soluble chains are immobilized at lower chain density onto a material surface, the biomaterial will evoke much more thrombus formation because of high protein inclusion in the immobilized chain layer, but become very lubricious because of high water uptake into the immobilized layer, giving much less mechanical damage to the biological tissues in contact[2]. It is thought that hydrogels exhibit the least protein and platelet deposition. This is true when the water content ranges from 80 to 90 wt%, as shown in Fig.5[3]. However, even the hydrogels with water contents of this range cannot completely reject platelet adhesion. This is probably due to the presence of a small amount of defects, which will trigger the cascade reaction for thrombus formation, as illustrated in Fig.6[4].

When a biomaterial surface is required to strongly bond to biological tissues, collagen and hydroxyapatite are recommended to be chemically immobilized on the surface for bonding to soft and hard tissues, respectively, as shown in Fig.4.

2.2. MECHANICAL BIOCOMPATIBILITY

Even if a material surface is effectively modified to exhibit good blood compatibility, the material will fail in producing such a vascular graft that is applicable for long-term use. This is because most of man-made materials have much higher rigidity than natural blood vessel walls, when high tensile strength comparable to that of blood vessels is required to give to synthetic materials. As a result, hyperplasia will take place at the anastomotic site, leading to poor patency.

This mismatch of mechanical rigidity or modulus between a biomaterial

18

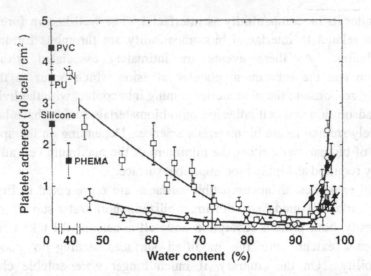

**Fig.5 Effect of Water Content of Hydrogels on Platelet Adhesion: PVA (□),
PAAm (○), methoxy-PEG methacrylate (△), PVP (●), and PEG (▲)**

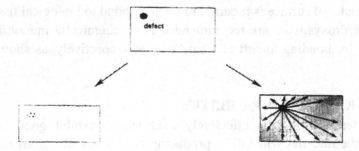

BIOLOGICAL SURFACE

1) No natural defects
2) Self-repairable

ARTIFICIAL SURFACE

1) There are always defects

2) Self-repair is impossible

3) Defects trigger cascade reactions
   (coagulation and complement activation)

**Fig.6   Difference in Trigger of Cascade Reactions between Biological
and Man-made Surface**

and a natural tissue induces deterioration of the tissue bonded to the biomaterial. This is called a stress-shielding effect, especially in metal/bone combination. Therefore, attempts have been made to improve the mechanical biocompatibility of biomaterials by reducing their rigidity or raising their compliance to the level of natural tissues.

3. Clinical Applications of Biomaterials

For any material applications we need both science and technology to get successfully to the final goal. In the medical system, biomaterials research will be the science part while fabrication of medical devices will be the technology part. To contribute to the clinical medicine, scientists of biomaterials should always open the eye to the medical device technology. Basic research only for basic research is no more appreciated.

Medical applications of biomaterials cover a huge field of clinical medicine from short-term to long-term use. Many biomaterials are used extracorporeally while many others are used as implants in the body. Requirements for biomaterials definitely depend on the purpose of their practical application. Some remarks will be given below by classifying biomaterial applications into single-use medical devices, artificial organs, polymer implants, and resorbable medical devices.

3.1. SINGLE-USE MEDICAL DEVICES

Disposable medical devices are an alternative expression for single-use medical devices. They include blood bag, infusion set, tubing, catheter, blood pump, and other devices for extracorporeal blood circulation. Hemodialyser is of single use type in some countries, but its reuse is allowed in USA. Currently, attempts have been made to replace plasticized poly(vinyl chloride)(PVC) because of possible formation of endocrine-disrupting substances, although the plasticized PVC is widely used for bag and tubing. For instance, derivatives of polyolefin and polybutadiene are replacing the plasticized PVC.

Recently, much attention has been directed to sophisticated devices applicable to interventional radiology. It should be also pointed out that disposable soft contact lenses of daily, weekly, and monthly use are now dominating in the contact lens market of advanced countries.

3.2. ARTIFICIAL ORGANS

As mentioned earlier various artificial tissues and organs are currently used or will appear in the clinical medicine including neurosurgery, ophthalmology, plastic surgery, dentistry, cardiovascular surgery, nephrology, orthopaedics and so on. When large part of tissues and organs is damaged or lost, the main therapeutic methods for them are artificial organs and organ transplantation. However, both of the means have problems; poor biocompatibility and insufficient biofunctionality of artificial organs and chronic deficiency of organ donors and severe complications induced by immuno-suppressors for organ transplantation.

However, owing to a remarkable advance in biomaterials research, the durability and performance of some artificial organs have been greatly improved. For instance, artificial kidney of hollow-fiber type has been used for longer than 30-40 years for some patients owing to the development of high-performance hollow fibers. Recent artificial hip joints can be used for aged people without serious problems but not for younger patients because of their loosening. Intraocular lens is also almost free of clinical troubles, but clinically applicable artificial cornea has not yet succeeded in development, probably because of infection occurring around the connection site between the biomaterial and the cornea.

In the cardiovascular surgery any fully-implantable artificial heart of permanent use is not yet developed although huge research money has been funded to this project. Also, small-caliber vascular grafts applicable for coronary artery are not yet successful in clinical application. The major resons for the failure include acute occlusion at the anastomotic line by thrombus formation and late occlusion by hyperplasia. Burst also occurs at the anastomotic site due to inflation of weakened blood vessel or tube material. Artificial heart valves which do not require any anti-coagulant therapy are not yet fabricated from fully synthetic materials. Segmented poly(ether urethane) does not seem to have sufficient durability in the presence of water under a pulsatile condition.

It seems that new break-through technologies are needed for biomaterials to be able to further contribute to reconstructive surgery by replacing defective or lost organs with man-made materials. A very promising break-through is tissue engineering.

3.3. POLYMERS FOR IMPLANTATION
Polymeric biomaterials widely used as implants are limited to a few kinds of polymers, as shown in Fig.7, in comparison with those used outside the body. When a flexible, rubbery material is necessary for long duration in the body, silicone rubber is generally preferred. The examples include hydrocephalus

1. **Poly(dimethyl siloxane) (PDMS, silicone)**
   Hydrocephalus shunts, breat and other prostheses

2. **Poly(ethylene terephthalate) (PET, polyester)**
   Vascular grafts

3. **Polyethylene (ultra-high molecular weight, UHMWPE)**
   Artificial joints

4. **Polytetrafbroroethylene (expanded, e-PTFE)**
   Prostheses for cardiovascular surgery and neurosurgery, blood accesses

5. **Poly(methyl methacrylate) (PMMA)**
   Intraocular lenses, bone cements

6. **Poly(ether urethane)**
   Artificial hearts and ventricular assist devices, pacemaker leads

Fig.7 Polymers Used as Implants

shunt system, mammary implant, intraocular lens, pace-maker lead insulation, and facial prostheses, while woven poly(ethylene terephthalate)(PET) or expanded polytetrafluoroethylene(PTFE) are used when pliable materials with high strength are required, for example, for vascular grafts. Poly(methyl methacrylate)(PMMA) plastic is used as contact lens and intraocular lens because of its excellent optical transparency, while MMA monomer is an important component of bone cements and filling materials for restorative dentistry. Ultrahigh molecular-weight polyethylene(UHMWPE) is indispensable as a sliding component of artificial joints because of its highest resistance against wear among the conventional engineering plastics. The remarkable feature common to all the polymers used as long-term implants is their high resistance against chemical deterioration, in other words, high durability in the aqueous environment. The material property that there is virtually no leachables from these polymers is also an important reason for their selection as implants.

It is interesting to point out that synthetic hydrogels have no wide application as implants, although most of natural soft tissues are composed of hydrogels of water contents ranging from 60 to 80 wt%. This is probably because synthetic hydrogels with high water contents will undergo calcification after long-term implantation as a result of sorption of calcium and phosphate ions. It seems unclear why natural, living hydrogels do not undergo calcification.

TABLE 1 Clinical Applications of biodegradable polymers

Function	Purpose	Examples
Bonding	Suturing	Vascular and intestinal anastomosis
	Fixation	Fractured bone fixation
	Adhesion	Surgical adhesion
Closure	Covering	Wound cover, Local hemostasis
	Occlusion	Vascular embolization
Separation	Isolation	Organ protection
	Contact Inhibition	Adhesion prevention
Scaffold	Cellular proliferation	Skin reconstrution, Blood vessel
	Tissue guide	reconstrution
		Nerve reunion
Capsulation	Controlled drug delivery	Susteined drug release

TABLE 2 Naturally-occurring and Synthetic, Biodegradable polymers

Natural polymer (enzymatic)	Synthetic polymer (non-enzymatic)
1. Proteins	1. Polyesters
collagen, gelatin, fibrin, ----	PGA, PLA, GA-LA cop., poly (ϵ-caprolactone) (PCL), GA-CL cop., La-CL cop., PDS, trimethilene carbonate-GA cop., -----
2. Polysaccharides	2. Polycyanoacrylates
starch, hyalurotane, chitin, chitosan, oxidized cellulose, alginate, ----	isobutyl cyanoacrylate polymer, -----
3. Polyesters	3. Polypeptides [b]
poly (β-hydroxybutyrate) (PHB) [a] , ----	poly(L-glutamic acid), poly(L-lysine),----
4. Nucleic acids	4. Polyanhydres, poly(ortho ester)s, polyphoshazenes

a) non-enzymatic degradation
b) enzymatic degradation

3.4. BIODEGRADABLE POLYMERS

Biodegradable materials have been increasingly attracting much attention in recent decades, as they are completely resorbed into the body after fulfilling their task. Tab.1 summarizes their medical applications. The largest application is as resorbable sutures, while hemostasis and sealing are also their another important application, similar to surgical glue. Application for prevention of tissue adhesion also is recently increasing. Resorbable fixation devices for bone were developed to replace metallic fixation devices, as the latter are necessary to remove from the body by re-operation. Biodegradable polymers have found their application also for slow release of drugs in the area of drug delivery systems(DDS) and tissue engineering.

To pursue the practical purposes listed in Tab.1, a wide variety of biodegradable materials, both synthetic and natural, have been applied. Tab.2 gives the representative biodegradable materials. As can be seen, biodegradable materials include both organic and inorganic substances, but most of them are organic polymers. Only calcium carbonate and phosphates such as β-tricalcium phosphate(β-TCP) are used as resorbables among inorganic biomaterials. A majority of synthetic biodegradable polymers is aliphatic polyesters such as polyglycolide(PGA), polylactide(PLA) and copolymers of glycolide or lactide with other dilactides and lactones[5]. α–Cyanoacrylate polymers are resorbable in the body, but produce formaldehyde as a result of hydrolysis of the main chain.

Natural biodegradable polymers are mostly derived from either proteins (collagen, gelatin, fibrinogen, albumin etc.) and polysaccharides (cellulose, hyaluronate, chitin, alginate etc.) . Biodegradation of natural polymers mostly proceeds through enzymatic hydrolysis, while synthetic polymers degrade through non-enzymatic hydrolysis. An important issue associated with biodegradable materials is the toxicity of biodegradation products, since they are generally water-soluble, low-molecular-weight compounds, which may be capable of passing through cell membranes, in marked contrast with non-biodegradable biomaterials. This indicates that the non-toxicity of biodegradable polymers should be much more carefully examined than that of non-biodegradable materials. In addition, natural biological products might contain antigen, virus, or prion. This possibility demands very severe toxic tests for natural, biodegradable polymers.

4. Concluding Remarks

As demonstrated above, polymeric biomaterials which have been practically applied to patient treatments are restricted to a small number of polymers, although a huge number of polymers have been used in industry. However, this is not surprising because the priority of biomaterials used in the medical system is non-toxicity and hence regulatory authorities as well as medical society become very conservative in using new biomaterials if they have no previous experience in clinical applications.

It should be also mentioned that surface modification and chemical derivatization of polymers are not always welcome by medical device industries, since they consider the cost-effectiveness of biomaterials the most important and want to concentrate their effort on the reduction of device price. This attitude of regulatory, medical, and industrial societies outside the biomaterials science would often discourage young biomaterials scientists.

References

1. Ikada, Y. (1994) Interfacial biocompatibility in Shalaby, S.W., Ikada, Y., Langer, R., and Williams, J.W., (eds.), ACS Symposium Series 540, *Polymers of Biological and Biomedical Significance*, pp.35-48.
2. Ikada, Y. and Uyama, Y. (1993) *Lubricating Polymer Surfaces*, Technomic, Lancaster, PA, USA.
3. Kulik, E. and Ikada, Y. (1996) In vitro platelet adhesion to nonionic and ionic hydrogels with different water contents, *J. Biomed. Mater. Res.*, **30**, 295-304.
4. Ikada, Y.(2001) Super-hydrophilic surfaces for biomedical applications in Morra, M. (ed.), *Water in Biomaterials Surface Science*, John Wiley & Sons, New York, pp.291-305.
5. Ikada, Y. and Tsuji, H. (2000) Biodegradable polyesters for medical and ecological applications, *Macromol. Rapid Commu.*, **21**, 117-132.

BIODEGRADABLE POLYMERS FOR ORTHOPAEDIC APPLICATIONS

C. MAULI AGRAWAL, PH.D., P.E.
Center for Clinical Bioengineering and Department of Orthopaedics
The University of Texas Health Science Center at San Antonio
San Antonio, TX 28229

1. Introduction

Over the past decade there has been an exponential increase in the use of biodegradable polymers in the field of orthopaedics. Such materials are used to fabricate fracture fixation rods, plates, screws, staples, clips, arrows, hooks, suture anchors, sutures and more recently for producing scaffolds for musculoskeletal tissue engineering. Most biodegradable polymeric materials slowly degrade in the body due to hydrolysis or through enzymatic pathways. This renders the need for a second surgery to remove the implant unnecessary. Not only does this reduce healthcare costs but also patient morbidity. Another significant advantage in using biodegradable fixation devices is that such systems can potentially reduce the effects of stress shielding. Bone is a living tissue and remodels in response to the loads it experiences - a phenomenon commonly known as Wolff's Law. In the presence of stiff metal implants, the load on the bone is significantly reduced, and hence, over the long term bone would have a propensity for osteopenia - a phenomenon described as stress-shielding. Fixation devices fabricated from biodegradable polymers can potentially offset this problem because as the fixation device degrades, its mechanical properties deteriorate. Thus, it can support only a decreasing level of load, which results in gradual reloading of the supported or repaired bone until full load bearing is restored.

The use of biodegradable polymers in orthopaedics has been limited in the past because a majority of the applications require some degree of load bearing. In general, polymeric devices have lower strength and stiffness values compared to both cortical bone and the more traditional metal implants. However, with recent technological advancements, significant improvements in these properties have been achieved and this is reflected in the increasing number of biodegradable devices available commercially for orthopaedics.

R.L. Reis and D. Cohn (eds.),
Polymer Based Systems on Tissue Engineering, Replacement and Regeneration, 25–36.
© 2002 *Kluwer Academic Publishers. Printed in the Netherlands.*

2. Common Biodegradable Polymers

2.1. POLYLACTIC ACID AND POLYGLYCOLIC ACID

There are several biodegradable polymers available for medical applications, however, polylactic acid (PLA), polyglycolic acid (PGA), and their copolymers (PLGA) remain the most popular and widely used for orthopaedic applications [1-3]. These are aliphatic polyesters belonging to the α-hydroxy group. They are often synthesized by the ring opening polymerization of cyclic glycolide or lactide diesters. Catalysts such as antimony, zinc, or lead may be used in the polymerization process [4,5].

The primary mechanism of biodegradation for PLA-PGA materials is non-specific hydrolytic scission of their ester linkages, [6,7]. Their degradation can be characterized as bulk degradation. The degradation product eluted by PLA is lactic acid, which is incorporated into the tricarboxylic acid cycle and processed into water and carbon dioxide. PGA, however, can follow two degradation mechanisms - by hydrolysis and by non-specific esterases and carboxypeptidases [8]. PGA degradation products are either excreted in urine or enter the tricarboxylic acid cycle. In general, PLA tends to degrade slower than PGA. Consequently, the degradation rate of the copolymer, PLGA, depends on the exact ratio of PLA and PGA present in the polymer.

PLA-PGA materials are the most common biodegradable polymers used for medical implants in the field of orthopaedics. In addition, in recent years they have been used extensively to fabricate scaffolds for engineering musculoskeletal tissue [9-11].

2.2. POLYANHYDRIDES

Polyanhydrides also degrade by hydrolysis but are different from PLA-PGA in their degradation behavior – they degrade by surface erosion. This characteristic makes them attractive for a variety of controlled delivery systems [12-15]. Studies have suggested that poly(anhydride-co-imide) polymers are similar to PLGA with regard to biocompatibility and can support endosteal and cortical bone regeneration [16,17]. Thus, this material may be a viable candidate for scaffolds used in orthopaedic surgery in weight-bearing applications.

2.3. POLYFUMARATES

In recent years a significant amount of work has been performed on polyfumarates. For instance, Mikos and colleagues have investigated a partially saturated linear polyester based on poly(propylene fumarate) or PPF for use in filling skeletal defects [18]. The mechanical properties of this polymer have been reported to be similar to that of trabecular bone. Additionally, this unsaturated polymer can cure *in vivo*, thereby facilitating the ability to fill skeletal defects of any shape or size with minimal surgical intervention. This material has been also evaluated as a composite with beta-tricalcium phosphate (ß-TCP) [19-21]. Hasiric et al. [22] have produced biodegradable bone plates using semi-interpenetrating networks of crosslinked PPF and PLGA or PLA materials. The use of PPF as an injectable bone cement has also been investigated [23].

2.4. POLYCARBONATES

Poly(BPA-carbonate) is a widely available commercial polycarbonate which exhibits good mechanical properties and stability, and is non-degradable under physiological conditions. However, if the carbonyl oxygen is replaced by an imino group the hydrolytic stability of poly(BPA-carbonate) is significantly decreased and hydrolytically degradable fibers with strength similar to that of poly(BPA-carbonate) are produced [24]. Tyrosine-derived iminocarbonate-amide copolymers have also been produced and tested extensively [25-27]. These materials may exhibit the biocompatibility of amino acids and retain mechanical strength similar to poly(BPA-carbonates). In a canine bone chamber model these materials formed an intimate contact with bone and did not induce a surrounding fibrous layer like that formed about the PLA implants [28,29]. Tyrosine-derived polycarbonates carrying an ethyl ester pendent chain have been found to be a promising orthopaedic implant material [30,31].

2.5. POLYCAPROLACTONES

Polycaprolactones are also polyesters like PLA and PGA with similar biocompatibility [32], however they possess different mechanical properties and exhibit a much slower degradation rate [33]. Fixation of rabbit humeri osteotomies with polycaprolactone versus stainless-steel has been studied [34]. It was found that although polycaprolactone fixation caused less stress shielding than stainless steel, its mechanical strength was not sufficient for load bearing applications. This material may not have use in fracture fixation but may have applications in the field of musculoskeletal tissue engineering.

Although several polymers are being investigated for orthopaedic applications, the most widely used for commercially available implants are PLA, PGA and PLGA materials. Thus, the physical and mechanical properties of these polymers are provided below.

3. Properties of PLA-PGA polymers

The molecular weight is an indication of the average size of the molecular chains comprising the polymer and is related to its intrinsic viscosity. In the case of a biodegradable polymer, its molecular weight is critical to its *in vivo* efficacy because it affects its mechanical properties and degradation behavior [2]. Although the molecular weight is initially determined by the polymerization process, it can be significantly altered by the fabrication and sterilization techniques [35,36]. However, the arrangement of these chains is also important because it determines the crystallinity of a polymer. PGA and L-PLA polymers are semi-crystalline in nature, and contain both crystalline and amorphous regions. The crystallinity of a polymer is influenced by its molecular chain structure, the molecular chemistry, temperature, and the rate of cooling during solidification from a melt. As a result the crystallinity of PLA-PGA polymers can be changed during the fabrication process if heat is used [37]. Crystallinity values for PLA-PGA polymers can range from 46-52% for PGA sutures [4], 15-74% for L-PLA, [4,37,38] and 0% (amorphous) for DL-PLA [4]. The crystallinity of PLA-PGA polymers may change to some extent during the degradation process and this may influence the rate of degradation. It has been postulated that because degradation of these polymers may result in a non-uniform degradation of the amorphous and crystalline phases and the formation of microscopic crystalline debris, absorbable polymers with low crystallinity may be better for medical applications [39].

The mechanical properties of biodegradable polymers vary considerably even within the same family of polymers. These properties may depend on the chemical structure, crystallinity, molecular weight or molecular orientation. For example, a review of the literature indicates that the tensile strength of L-PLA varies between 11.4-82.7 MPa and flexural strength between 45-145 MPa. It has been suggested that for a L-PLA material to function as a load bearing orthopaedic implant, its molecular weight should exceed 100 kDa [5,38]. PGA, which because of its chemical structure degrades faster than PGA [40] exhibits an initial strength between 57-69 MPa [40,41]. To enhance the mechanical behavior of these polymers, fibers may be used as reinforcing elements [42]. Molecular alignment within a polymer can also significantly improve its mechanical properties. Based on this principle, PGA rods fabricated by melting or sintering PGA sutures together can have significantly enhanced properties [43,44]. It should be noted that the tensile strength of fibers can change after exposure to water; to gauge the rate of this change their wet strength half-life can be determined. This wet strength half-life varies from approximately two weeks for PGA to over six months for L-PLA [39].

4. Orthopaedic Applications

As stated earlier there are now a variety of biodegradable implants commercially available for orthopaedic applications. These include PGA pins, poly-P Dioxanone pins, self-reinforced L-PLA anchors, interference screws, screws, pins, and meniscus arrows, and 85:15 (DL-PLA)-PGA suture anchors and interference screws. For instance, Kumta et al. [45] have described the use of PLA pins in fractures of the hand. Others have reported on the use of biodegradable devices for various other types of fixation as reviewed below.

The use of biodegradable devices in the arthroscopic repair of meniscal lesions has seen significant growth in recent years [46]. These are usually a tack or screw type device. Becker et al. [46] described the biomechanical evaluation of menisci repaired with six commonly used biodegradable devices. These were compared to repairs performed with a horizontal mattress suture stitch. In all cases the suture repair exhibited higher ultimate tensile load. The authors recommended that for large lesions it may be advantageous to use a combination of implants and suture techniques. A similar study performed by Barber and Herbert [47] indicated that the various devices had different failure modes. Dervin et al. [48] compared the BIOFIX meniscal Arrow (Bionx, Tampere, Finland) with sutures in an *in vitro* study and demonstrated that the sutures performed better biomechanically. Nevertheless, the ease of use of biodegradable meniscal repair devices has contributed significantly to their popularity, which continues to increase.

Biodegradable devices are also used in surgical procedures to repair anterior cruciate ligaments (ACL). In a 5 year prospective study, Lajtai et al. [49] assessed the local effects of PLGA interference screws used in ACL reconstruction. They reported minimal surgical site edema, minimal reaction to the polymer and complete replacement of the screw by new bone. In a multicenter study the performance of an L-PLA interference screw (Bioscrew; Linvatek, Largo, FL) was compared to metal interference screws in 110 ACL repairs using patellar tendon autografts [50]. The authors reported that the bioabsorbable screws performed well and were comparable to metal screws one year postoperatively. No material related problems were detected.

The use of biodegradable implants can potentially have an important role to play in pediatric fractures because metal implants may interfere with the growth of the child and hence may have to be removed. Several studies have examined the use of PGA based pins to treat pediatric fractures. For example, Svensson et al. [51] used 1.5–2.0 mm PGA pins to repair osteochondral and transphyseal fractures in children. Similarly, Hope et al. [52] treated elbow fractures in children using self-reinforced PGA pins. Repair of humeral fractures of the lateral condyle in 27 children using PGA pins has also been reported [53]. The results indicated that three of the 27 patients developed osteolysis; however, this osteolysis was not accompanied by foreign body reaction or sinus formation, and resolved within six months. There are some indications that there is a lower incidence of complications in children compared to adults; although the exact reasons for this phenomenon are not clear, it may be related to the relatively smaller size of pediatric implants.

Biodegradable screws have also been used extensively in foot surgery. In a large, 14 month study on the fixation of chevron osteotomies of the first metatarsal, 78 osteotomies were fixed with self-reinforced 2 mm diameter PGA pins [54]. Although no immediate postoperative displacement or disturbance of healing was detected, an inflammatory discharge of PGA material and fluid was found approximately 6 weeks postoperatively in two patients. In these cases, the complications did not affect the fracture healing and resolved spontaneously within 21 days. Brunetti et al. [55] used polydioxanone (PDS) pins for stabilizing Austin osteotomies of the first metatarsal. Two pins were placed across each osteotomy in 30 patients in a 6 month follow-up study. No instances of capital fragment displacement, aseptic necrosis, allergy to PDS or dislodgment of the pins were detected. A majority of the patients (63.6%) remained pain free, and the rest (36.4%) had occasional complaints of some mild discomfort.

Bostman and colleagues have reported extensively on the use of biodegradable fixation systems. For example, they used self-reinforced PGA rods to treat ankle (unimalleolar and bimalleolar) fractures in 102 patients [56]. It was demonstrated that these rods held the fragments in place in the majority of cases. In another study Bostman [57] has described the use of L-PLA bolts fractures of the medial malleolus. PLA screws have been investigated for use in fixation of basilar first metatarsal osteotomies [58]. It was reported that no statistically significant differences in load, displacement or stiffness existed between oblique basilar first metatarsal osteotomies fixed with a L-PLA absorbable screw, and osteotomies fixed with a steel screw.

In the past, biodegradable devices fabricated from PGA have exhibited various degrees of tissue reaction. For instance, in a study involving 40 patients, biodegradable PGA pins were used to treat displaced fractures of the distal wrist [59]. However, inflammation was detected at the implantation site up to various time periods post operatively (as long as 145 days in some cases), and a significant percent (22.5%) of the patients required debridement of the inflamed tissue. In another study, intra-articular elbow fractures were treated with PGA rods [60]. Acceptable reduction until union was obtained in 25 out of the 30 cases (83%), although a non-infected sinus formation was noted in four patients. In yet another study on the repair of wrist fractures, biodegradable PGA pins were compared to Kirschner-wires [61]. No significant differences were detected at final follow-up, however, numerous transient complications were reported and hence the use of PGA pins was not recommended.

Others have also reported complications related to PGA devices. In a study of 48 cases of Austin bunionectomies repaired using absorbable PGA (23 cases) or PDS (25 cases) pin fixation the investigators found that there was a 4% incidence of sterile sinus discharge and a 30% incidence of osteolytic changes associated with the use of the PGA devices [62]. No clinical or radiographic complications were noted post operatively with the PDS device.

In summary, in the past there have been reports of adverse tissue reactions although there has been a steady decrease in such reports over time. This may be a factor of better and improved material purity and quality control. Also, adverse tissue reactions may be related to the use of relatively large amounts of degradation products in tissue with low vascularity. This may compromise the ability of the body to flush away these products in an expedient fashion. Although the use of biodegradable polymers as fixation systems in the repair of musculoskeletal tissues has seen a significant increase in recent years there are still limits to their use compared to metal devices. Polymeric devices are not able to bear sustained heavy loads such as those experienced by long bones. The primary reason for this limitation being the mechanical properties of these biodegradable materials which are significantly inferior compared to metals.

5. Tissue Engineering

There has been extensive research on the application of biodegradable polymers as scaffolds for tissue engineering [63-66]. A review of this area for orthopaedics has been provided by Agrawal and Ray [67]. The majority of tissue engineering research in orthopaedics has concentrated on the regeneration of bone and articular cartilage. For example, Klompmaker et al. [68] have described the use of porous polymer implants for meniscal cartilage repair. Freed et al. [64] have reported on culturing chondrocytes on fibrous biodegradable scaffolds which have a porosity of 97%. They recommended high porosity implants (> 90 percent) so that there are minimal diffusion limitations, a high surface area for cell-substrate interactions, and sufficient space for extracellular matrix regeneration.

Mikos and colleagues [65,69] have performed extensive studies on different scaffold types using osetoblasts. They reported that using cells seeded onto biodegradable polymer scaffolds with pore sizes in the range of 150-710 microns, it is possible to grow bone-like tissues *in vitro*. Peter et al. [66] demonstrated that seeding density can have an effect on cell proliferation, although pore size in the range of 150 to 710 microns did not have any significant effects.

Kim et al. [70] have described the generation of cartilage in predetermined shapes using specially configured biodegradable polymer systems and cultured bovine articular chondrocytes. The authors seeded the scaffolds with chondrocytes and then implanted these systems subcutaneously in nude mice. Excised specimens at 12 weeks exhibited the presence of new hyaline cartilage of approximately the same dimensions as the original construct. Others have reported similar efforts [71].

Material or implant related factors that play a significant role in the success of tissue engineering efforts include porosity and permeability of the scaffold, mechanical properties, surface characteristics among others [67,72,73]. Another important parameter may be the cell culture environment. Vunjak-Novakovic and colleagues [74,75] have explored the effects of hydrodynamic conditions in tissue-culture bioreactors. Their work suggests that for chondrocytes grown under static culture conditions, the rate of cell growth is limited by diffusion which is itself influenced by increasing cell numbers and decreasing effective implant porosity due to cartilage matrix regeneration.

6. Sterilization

The sterilization of devices made of biodegradable polymers is a non-trivial issue. These polymers are affected by heat and moisture, as well as by radiation. Materials such as PLA and PGA degrade by hydrolysis and thus heat and moisture are detrimental to their mechanical properties. In fact, moisture results in polymer degradation and decrease in molecular weight which can affect the useful life of implants. Nevertheless, a steam-sterilization program has been reported using temperatures as high as 129°C [76]. The authors reported a significant decrease in molecular weight of L-PLA rods sterilized by this process although they detected only a slight change in mechanical properties. This was a laboratory based study and it is not clear if heating PLA or PGA implants to such high temperatures would result in significant dimensional changes that may affect the *in vivo* performance of devices such as fixation screws.

Radiation sterilization can also attack the molecular chains in the polymer causing bonds to break and in some cases form cross-links. For example, gamma-radiation of PLA has been reported to break ester bonds causing both chain scission and cross-linking [35]. Gamma-radiation (2.5 MRad) of self-reinforced PGA rods can decrease their initial bending strength from 370 to 300 MPa [4]. In a study of PGA sutures sterilized by gamma-radiation, the tensile strength was almost completely lost after ten days of implantation although the initial post-sterilization strength appeared to be unchanged [4]. This may be because the molecular chains had been reduced to a critical threshold size by sterilization - further degradation by hydrolysis *in vivo* resulted in more reduction in chain size and a sudden decrease in mechanical properties. Along this same line of reasoning, in a study of *in vivo* degradation of PGA sutures sterilized by gamma-radiation it was shown that degradation was accelerated compared to non-irradiated specimens [77]. Although gamma-radiation does affect both the molecular weight of biodegradable polymers and their mechanical properties, it is commonly used as the method of choice for sterilization of commercially available polymeric implants. Manufacturers factor the reduction in these parameters into their design process so that the final sterilized product has acceptable properties.

Another common means of sterilizing medical implants is by treatment with gases such as ethylene oxide. In general, this technique should not significantly affect the mechanical properties of the polymer or its molecular weight. However, some studies have indicated otherwise – for example, hydroxyapatite filled L-PLA composites sterilized with ethylene oxide exhibit a decrease in molecular weight and a loss of flexural strength within three weeks [78]. The main issue with ethylene oxide sterilization of biodegradable polymers is potential toxic residues. Thus, the degassing or aeration procedure following treatment is an important step [79]. Studies have shown that vacuum aeration for greater than two weeks may have to be employed to minimize the effects of cytotoxicity [80]. It has been suggested that microwave heating instead of conventional heating can achieve significant improvement in the diffusion of the gas [81].

The wealth of literature on growth factors and tissue engineering suggests that in the future biodegradable devices loaded with growth factors may become commonplace. However, it is not obvious which sterilization techniques would be appropriate for such dual purpose systems because the standard methods described above may affect the growth factors or other biological moieties incorporated in the polymer carrier. For instance, studies have shown that autoclaving, gamma-irradiation and ethylene oxide reduce the osteoinductive nature of human demineralized bone [82,83]. In fact it has been reported that ethylene oxide sterilization can destroy the activity of purified recombinant TGF-ß [82]. Thus, the sterilization process has to be carefully optimized – for example sterilization with ethylene oxide at 29°C for 5 hours has been suggested as a safe method for sterilizing partially purified bone morphogenetic protein [84].

7. Conclusion

The use of biodegradable polymers in the field of orthopaedics is fast becoming mainstream. These polymeric systems offer a host of advantages over the more traditional metal devices in the arena of fixation. However, issues related to inferior mechanical properties are still relevant and preclude the use of such biodegradable devices for the fixation of large bones. In the field of tissue engineering, biodegradable polymers have been used extensively and their role is bound to increase as tissue engineering systems become commercially available.

References:

1. Athanasiou, K.A., Niederauer, G.G. and Agrawal, C.M. (1996) Sterilization, toxicity, biocompatibility, and clinical applications of polylactic acid/polyglycolic acid copolymers, Biomaterials 17(2), 93-102.

2. Agrawal, C.M., Niederauer, G.G. and Athanasiou, K.A. (1995) Fabrication and characterization of PLA-PGA orthopaedic implants, Tissue Engineering 1(3), 241-252.

3. Agrawal, C.M., Niederauer, G.G., Micallef, D.M. and Athanasiou, K.A., Chapter 30: The use of PLA-PGA polymers in orthopaedics, in Encyclopedic Handbook of Biomaterials and Bioengineering. 1995, Marcel Dekker: N.Y. p. 2081-2115.

4. Gilding, D.K. and Reed, A.M. (1979) Biodegradable polymers for use in surgery-polyglycolic/poly (lactic acid) homo-and copolymers: 1., Polymer 20, 1459-1464.

5. Eling, B., Gogolewski, S. and Pennings, A.J. (1982) Biodegradable materials of poly(L-lactic acid): 1. Melt-spun and solution spun fibres, Polymer 23, 1587-1593.

6. Miller, R.A., Brady, J.M. and Cutright, D.E. (1977) Degradation rates of oral resorbable implants (polylactates and polyglycolates): Rate modification with changes in PLA/PGA copolymer ratios, Journal of Biomedical Materials Research 11, 711-719.

7. Lewis, D.H., Controlled release of bioactive agents from lactide/glycolide polymers, in Biodegradable Polymers as Drug Delivery Systems, M. Chasin and R. Langer, Editors. 1990, Marcel Dekker, Inc.: New York. p. 1-41.

8. Williams, D.F., Some observations on the role of cellular enzymes in the in vivo degradation of polymers, in Corrosion and Degradation of Implant Materials, B.C. Syrett and A. Acharya, Editors. 1979. p. 61-75.

9. Hollinger, J.O. and Schmitz, J.P. (1987) Restoration of bone discontinuities in dogs using a biodegradable implant, Journal of Oral and Maxillofacial Surgery 45, 594-600.

10. Heckman, J.D., Boyan, B.D., Aufdemorte, T.B. and Abbott, J.T. (1991) The use of bone morphogenetic protein in the treatment of non-union in a canine model, J Bone Joint Surg (Am) 73(5), 750-764.

11. Agrawal, C.M., Bert, J., Heckman, J.D. and Boyan, B.D. (1995) Protein release kinetics of a biodegradable implant for fracture non-unions., Biomaterials 16(16), 1255-1260.

12. Hamalainen, K.M., Maatta, R., Piirainen, H., Sarkola, M., Vaisanon, A., Ranta, V.P. and Urtti, A. (1998) Roles of acid/base nature and molecular weight in drug release from matrices of gelfoam and monoisopropyl ester of poly(vinyl methyl ether-maleic anhydride), J Controlled Release 56(1-3), 273-283.

13. Hanes, J., Chiba, M. and Langer, R. (1998) Degradation of porous poly(anhydride-co-imide) microspheres and implications for controlled macromolecule delivery, Biomaterials 19(1-3), 163-172.

14. Chiba, M., Hanes, J. and Langer, R. (1997) Controlled protein delivery from biodegradable tyrosine-containing poly(anhydride-co-imide) microspheres, Biomaterials 18(13), 893-901.

15. Chasin, M., Lewis, D. and Langer, R. (1988) Polyanhydrides for controlled drug delivery, Biopharm Mfg. 1(33-46), .

16. Ibim, S.E., Uhrich, K.E., Attawia, M., Shastri, V.R., El-amin, S.F., Bronson, R., Langer, R., and Laurencin, C.T. (1998) Preliminary in vivo report on the osteocompatibility of poly(anhydride-co-imides) evaluated in a tibial model, J Biomed Mater Res 43(4), 374-379.

17. Ibim, S.M., Uhrich, K.E., Bronson, R., El-Amin, S.F., Langer, R.S. and Laurencin, C.T. (1998) Poly(anhydride-co-imides): In vivo biocompatibility in a rat model, Biomaterials 19(10), 941-951.

18. Peter, S.J., Yaszemski, M.J., Suggs, L.J., Payne, R.G., Langer, R., Hayes, W.C., Unroe, M.R., Alemany, L.B., Engel, P.S., and Mikos, A.G. (1997) Characterization of partially saturated poly(propylene fumarate) for orthopaedic application, J Biomater Sci, Polym Ed 8(11), 893-904.

19. Peter, S.J., Miller, S.T., Zhu, G., Yasko, G. and Mikos, A.G. (1998) In vivo degradation of a poly(propylene fumarate)/beta-tricalcium phosphate injectable scaffold, J. Biomed. Mater. Res. 41(1), 1-7.

20. Yaszemski, M.J., Payne, R.G., Hayes, W.C., Langer, R.S., Aufdemorte, T.B. and Mikos, A.G. (1995) The ingrowth of new bone tissue and initial mechanical properties of a degradable polymeric composite scaffold, Tissue Engineering 1(41-52), .

34

21. Peter, S.J., Lu, L., Kim, D.J. and Mikos, A.G. (2000) Marrow Stromal Osteoblast Function on a Poly(propylene Fumarate)/B-Tricalcium Phosphate Biodegradable Orthopaedic Composite. *Biomaterials* 21, 1207-1213.

22. Hasirci, V., Lewandrowski, K.U., Bondre, S.P., Gresser, J.D., Trantolo, D.J. and Wise, D.L. (2000) High strength bioresorbable bone plates: preparation, mechanical properties and int vitro analysis., *Biomed Mater Eng* 10(1), 19-29.

23. Lewandrowski, K.U., Gresser, J.D., Wise, D.L., White, R.L. and Trantolo, D.J. (2000) Osteoconductivity of an injectable and bioresorbable poly(propylene glycol-co-fumaric acid) bone cement., *Biomaterials* 21(3), 293-8.

24. Kohn, J. and Langer, R. (1986) Poly(iminocarbonates) as potential biomaterials, *Biomaterials* 7(3), 176-182.

25. Li, C. and Kohn, J. (1989) Synthesis of poly(iminocarbonates): Degradable polymers with potential applications as disposable plastics and as biomaterials, *Macromolecules* 22, 2029-2036.

26. Pulapura, S., Li, C. and Kohn, J. (1990) Structure-property relationships for the design of polyiminocarbonates, *Biomaterials* 11(9), 666-678.

27. Kohn, J., Pseudo-poly(amino acids), in Biodegradable Polymers as Drug Delivery Systems, R.L.a.M. Chasin, Editor. 1990, Marcel Dekker, Inc.: New York, NY. p. 195-229.

28. Choueka, J., Charvet, J.L., Koval, K.J., Alexander, H., James, K.S., Hooper, K.A. and Kohn, J. (1996) Canine bone response to tyrosine-derived polycarbonates and poly(L-lactic acid), *J Biomed Mater Res* 31(1), 35-41.

29. Ertel, S.I., Kohn, J., Zimmerman, M.C. and Parsons, J.R. (1995) Evaluation of poly(DTH carbonate), a tyrosine-derived degradable polymer, for orthopedic applications, *J Biomed Mater Res* 29(11), 1337-1348.

30. Tangpasuthadol, V., Pendharkar, S.M., Peterson, R.C. and Kohn, J. (2000) Hydrolytic degradation of tyrosine-derived polycarbonates, a class of new biomaterials. Part II: 3-yr study of polymeric devices., *Biomaterials* 21(23), 2379-87.

31. Tangpasuthadol, V., Pendharkar, S.M. and Kohn, J. (2000) Hydrolytic degradation of tyrosine-derived polycarbonates, a class of new biomaterials. PartI: study of model compounds., *Biomaterials* 21(23), 2371-78.

32. Allen, C., Yu, Y., Maysinger, D. and Eisenberg, A. (1998) Polycaprolactone-b-poly(ethylene oxide) block copolymer micelles as a novel drug delivery vehicle for neurotrophic agents FK506 and L-685,818, *Bioconjug Chem* 9(5), 564-572.

33. Pitt, C., Poly-epsilone-caprolactone and its copolymers, in Biodegradable Polymers as Drug Delivery Systems, R.L.a.M. Chasin, Editor. 1990, Marcel Dekker, Inc.: New York, NY. p. 71-120.

34. Lowry, K.J., Hamson, K.R., Bear, L., Peng, Y.B., Calaluce, R., Evans, M.L., Anglen, J.O., and Allen, W.C. (1997) Polycaprolactone/glass bioabsorbable implant in a rabbit humerus fracture model, *J. Biomed. Mater. Res.* 36(4), 536-41.

35. Gupta, M.C. and Deshmukh, V.G. (1983) Radiation effects on poly(lactic acid)., *Polymer* 24, 827-830.

36. Matsusue, Y., Yamamuro, T., Oka, M., Shikinami, Y., Hyon, S.-H. and Ikada, Y. (1992) In vitro and in vivo studies on bioabsorbable ultra-high-strength poly (L-lactide) rods, *Journal of Biomedical Materials Research* 26, 1553-1567.

37. Suuronen, R., Pohjonen, T., Taurio, R., Törmälä, P., Wessman, L., Rönkkö, K. and Vainionpää, S. (1992) Strength retention of self-reinforced poly-L-lactide screws and plates: an in vivo and in vitro study, *Journal of Materials Science: Materials in Medicine* 3, 426-431.

38. Engelberg, I. and Kohn, J. (1991) Physico-mechanical properties of degradable polymers used in medical applications: A comparative study, *Biomaterials* 12(3), 292-304.

39. Andriano, K.P., Pohjonen, T. and Tormala, P. (1994) Processing and characterization of absorbable polylactide polymers for use in surgical implants, *Journal of Applied Biomaterials* 5, 133-140.

40. Christel, P., Chabot, F., Leray, J.L., Morin, C. and Vert, M., Biodegradable composites for internal fixation, in Biomaterials 1980, G.L. Winters, D.F. Gibbons, and H. Plenk, Editors. 1982, John Wiley & Sons. p. 271-280.

41. Birmingham Polymers, I., Properties of biodegradable polymers. . 1993.

42. Daniels, A.U., Chang, M.K.O. and Andriano, K.P. (1990) Mechanical properties of biodegradable polymers and composites proposed for internal fixation of bone., *Journal of Applied Biomaterials* 1, 57-78.

43. Törmälä, P., Vasenius, J., Vainiompää, S., Pohjonen, T., Rokkanen, P. and Laiho, J. (1991) Ultra high strength absorbable self-reinforced polyglycolide (SR-PGA) composite rods for internal fixation of bone fractures: *In vitro* and *in vivo* study, *Journal of Biomedical Materials Research* 25, 1-22.

44. Vainionpää, S., Kilpikari, J., Laiho, J., Helevirta, P., Rokkanen, P. and Törmälä, P. (1987) Strength and strength retention in vitro, of absorbable, self-reinforced polyglycolide (PGA) rods for fracture fixation, *Biomaterials* 8(January), 46-48.

45. Kumta, S.M. and Leung, P.C. (1998) The technique and indications for the use of biodegradable implants in fractures of the hand., *Techniques in Orthopaedics* 13(2), 160-163.

46. Becker, R., Schroder, M., Starke, C., Urbach, D. and Nebelung, W. (2001) Biomechanical investigations of different meniscal repair implants in comparison with horizontal sutures on human menicus., *Arthroscopy* 17(5), 439-44.

47. Barber, F.A. and Herbert, M.A. (2000) Meniscal repair devices., *Arthroscopy* 16(6), 613-8.

48. Dervin, G.F., Downing, K.J., Keene, G.D. and McBride, D.G. (1997) Failure strengths of suture versus biodegradable arrow for meniscal repair: an in vitro study., *Arthroscopy* 13(3), 296-300.

49. Lajtai, G., Schmiedhuber, G., Unger, F., Aitzetmuller, G., Klein, M., Noszian, I. and Orthner, E. (2001) Bone tunnel remodeling at the site of biodegradable interference screws used for anterior cruciate ligament reconstruction: 5-year follow-up., *Arthroscopy* 17(6), 597-602.

50. Barber, F.A., Elrod, B.F., McGuire, D.A. and Paulos, L.E. (1995) Preliminary results of an absorbable interference screw, *Arthroscopy* 11(5), 537-548.

51. Svensson, P., Janary, P. and Hirsch, G. (1994) Internal fixation with biodegradable rods in pediatric fractures: one-year follow-up of fifty patients., *J Pediatr Orthop* 14, 220-4.

52. Hope, P.G., Williamson, D.M., Coates, C.J. and Cole, W.G. (1991) Biodegradable pin fixations of elbow fractures in children. A randomized trial., *J Bone Joint Surg [Br]* 73, 965-8.

53. Fraser, R.K. and Cole, W.G. (1992) Osteolysis after biodegradable pin fixation of fractures in children, *Journal of Bone and Joint Surgery* 74-B(November), 929-930.

54. Hirvensalo, E., Bostman, O., Tormala, P., Vainoonpaa, S. and Rokkanen, P. (1991) Chevron osteotomy fixed with absorbable polyglycolide pins, *Foot & Ankle Journal* 11, 212-218.

55. Brunetti, V.A., Trepal, M.J. and Jules, K.T. (1991) Fixation of the austin osteotomy with bioresorbable pins, *The Journal of Foot Surgery* 30(1), 56-65.

56. Böstman, O., Hirvensalo, E., Vainionpää, S., Mäkelä, A., Vihtonen, K., Törmälä, P. and Rokkanen, R. (1989) Ankle fractures treated using biodegradable internal fixation, *Clinical Orthopaedics and Related Research* 238(1), 195-203.

57. Pihlajamäki, H.K. and Böstman, O.M. (1998) Biodegradable expansion bolt for fractures of the medial malleolus., *Techniques in Orthopaedics* 13(2), 177-9.

58. Lavery, L.A., Higgins, K.R., Ashry, H.R. and Athanasiou, K.A. (1994) Mechanical characteristics of poly-L-lactic acid absorbable screws and stainless steel screws in basilar osteotomies of the first metatarsal, *The Journal of Foot and Ankle Surgery* 33(3), 249-254.

59. Hoffmann, R., Krettek, C., Haas, N. and Tscherne, H. (1989) Die distale Radiusfraktur. Frakturstabilisierung mit biodegradablen Osteosynthes Stiften (Biofix), *Experimentelle Untersuchungen und erste klinische Erfahrungen.* 92, 430-434.

60. Hirvensalo, E., Böstman, O., Vainionpää, S., Törmälä, P. and Rokkanen, P. (1988) Biodegradable fixation in intraarticulate fractures of the elbow joint., *Acta Orthop. Scandinavica Supplementum*, 227, 78-79.

61. Casteleyn, P.P., Handelberg, F. and Haentjens, P. (1992) Biodegradable rods versus Kirschner wire fixation of wrist fractures, *Journal of Bone and Joint Surgery* 74B, 858-861.

62. Gerbert, J. (1992) Effectiveness of absorbable fixation devices in Austin bunionectomies, *Journal of the American Podiatric Medical Association* 82(4), 189-195.

63. Agrawal, C.M., McKinney, J.S., Huang, D. and Athanasiou, K.A., The use of the vibrating particle technique to fabricate highly permeable biodegradable scaffolds, in STP 1396: Synthetic Bioabsorbable Polymers for Implants, C.M. Agrawal, J. Parr, and S. Lin, Editors. 2000, ASTM, 99-114.

36

64. Freed, L.E., Vunjak-Novakovic, G., Biron, R.J., Eagles, D.B., Lesnoy, D.C., Barlow, S.K. and Langer, R. (1994) Biodegradable polymer scaffolds for tissue engineering. *Biotechnology (NY)* 12(7), 689-693.

65. Ishaug-Riley, S. (1997) Bone formation by three-dimensional stromal osteoblast culture in biodegradable polymer scaffolds, *J of Biomed Mater Res* 36(1), 17-28.

66. Peter, S.J., Miller, M.J., Yasko, A.W., Yaszemski, M.J. and Mikos, A.G. (1998) Polymer concepts in tissue engineering, *J Biomed Mater Res* 43(4), 422-427.

67. Agrawal, C.M. and Ray, R.B. (2001) Biodegradable polymeric scaffolds for musculoskeletal tissue engineering., *J Biomed Mater Res* 55(2), 141-50.

68. Klompmaker (1991) Porous polymer implant for repair of meniscal lesions: A preliminary study in dogs, *Biomaterials* 12(9), 810-816.

69. Ishaug-Riley, S.L., Crane-Kruger, G.M., Yaszemski, M.J. and Mikos, A.G. (1998) Three-dimensional culture of rat calvarial osteoblasts in porous biodegradable polymers, *Biomaterials* 19(15), 1405-1412.

70. Kim, W.S., Vacanti, J.P., Cima, L., Mooney, D., Upton, J., Puslacher, W.C. and Vacanti, C.A. (1994) Cartilage engineered in predetermined shapes employing cell transplantation on synthetic biodegradable polymers, *Plast Reconstr Surg* 94(2), 233-240.

71. Vacanti, C.A. and Upton, J. (1994) Tissue-engineered morphogenesis of cartilage and bone by means of cell transplantation using synthetic biodegradable polymer matrices, *Clin Plast Surg* 21(3), 445-462.

72. Agrawal, C.M., McKinney, J. and Athanasiou, K.A. (2000) Effects of flow on the in vitro degradation kinetics of biodegradable scaffolds for tissue engineering., *Biomaterials* 21(23), 2443-2452.

73. Athanasiou, K.A., Schmitz, J.P. and Agrawal, C.M. (1998) The effects of porosity on degradation of PLA-PGA implants, *Tissue Engineering* 4, 53-63.

74. Freed, L.E., Vunjak-Novakovic, G. and Langer, R. (1993) Cultivation of cell-polymer cartilage implants in bioreactors, *J Cell Biochem* 51(3), 257-264.

75. Vunjak-Novakovic, G., Martin, I., Obradovic, B., Treppo, S., Grodzinsky, A.J., Langer, R. and Freed, L.E. (1999) Bioreactor cultivation conditions modulate the composition and mechanical properties of tissue-engineering cartilage, *J Orthop Res* 17(1), 130-138.

76. Rozema, F.R., Bos, R.R.M., Boering, G., Asten, J.A.A.M.v., Nijenhuls, A.J. and Pennings, A.J. (1991) The effects of different steam-sterilization programs on material properties of poly (L-lactide)., *Journal of Applied Biomaterials* 2, 23-28.

77. Chu, C.C. and Williams, D.F. (1983) The effect of gamma irradiation on the enzymatic degradation of polyglycolic acid absorbable sutures., *Journal of Biomedical Materials Research* 17, 1029-1040.

78. Verheyen, C.C.P.M., Wijn, J.R.d., Blitterswijk, C.A.v. and Groot, K.d. (1992) Evaluation of hydroxylapatite/poly(L-lactide) composites: Mechanical behavior, *Journal of Biomedical Materials Research* 26, 1277-1296.

79. Vink, P. and Pleijsier, K. (1986) Aeration of ethylene oxide-sterilized polymers, *Journal of Biomaterials* 7, 225-230.

80. Zislis, T., Martin, S.A., Cerbas, E., Heath, J.R., Mansfield, J.L. and Hollinger, J.O. (1989) A scanning electron microscopic study *in vitro* toxicity of ethylene-oxide-sterilized bone repair materials, *Journal of Oral Implantology* 25(1), 41-46.

81. Matthews, I.P., Gibson, C. and Samuel, A.H. (1989) Enhancement of the kinetics of the aeration of ethylene oxide sterilized polymers using microwave radiation, *Journal of Biomedical Materials Research* 23, 143-156.

82. Puolakkainen, P.A., Ranchalis, J.E., Strong, D.M. and Twardzik, D.R. (1993) The effect of sterilization on transforming growth factor *B* isolated from demineralized human bone, *Transfusion* 33, 679-685.

83. Doherty, M.J., Mollan, R.A.B. and Wilson, D.J. (1993) Effect of ethylene oxide sterilization on human demineralized bone, *Journal of Biomaterials* 14(13), 994-998.

84. Ijiri, S., Yamamuro, T., Nakamura, T., Kotani, S. and Notoya, K. (1994) Effect of sterilization on bone morphogenetic protein, *Journal of Orthopaedic Research* 12(5), 628-636.

POLYMERIC MATRICES FOR RELEASE OF GROWTH FACTORS, HORMONES AND OTHER BIOACTIVE AGENTS

A. GALLARDO, G.A. ABRAHAM, C. ELVIRA, B. VÁZQUEZ, AND J. SAN ROMÁN

Instituto de Ciencia y Tecnología de Polímeros, CSIC

Juan de la Cierva 3, 28006 Madrid, Spain

1. Introduction

The treatment, repair and regeneration of damaged tissues, which is the aim of modern tissue engineering, can be tailored to provide the optimum environment incorporating controlled release of the appropriate active compounds. In this sense, a big challenge is to optimize the delivery of immunosuppressors, hormones, growth factors, vitamins and other active agents. Great efforts are nowadays directed towards the preparation of matrices for the controlled delivery of this kind of compounds.

Two basic types of delivery systems have to be taken into account: polymer-active covalent conjugates, and matrices that physically incorporate the drug. In both cases and for these particular applications, biodegradability (and/or resorbability), and biocompatibility are indispensable requirements since most of the times the biomaterial will be implanted or parenterally administered.

Examples of polymeric supports for chemically-linked bioactive compounds and physically-loaded systems are presented. A critical discussion of their advantages and practical problems is also included in this chapter.

37

R.L. Reis and D. Cohn (eds.),
Polymer Based Systems on Tissue Engineering, Replacement and Regeneration, 37–52.
© 2002 *Kluwer Academic Publishers. Printed in the Netherlands.*

2. Covalent Conjugates

Synthetic or natural polymers may be conjugated covalently by weak hydrolytically sensitive bonds with a great number of bioactive compounds including drugs, peptides, proteins, growth factors, hormones, enzymes, etc. These possibilities make the polymeric conjugated systems very useful for applications not only in medication, but also in tissue engineering, biosensors, affinity separations, enzymatic processes, cell culture, etc. One of the most attractive advantages of synthetic polymers for the use in the biomedical field is the great diversity in composition, molecular structure, molecular weight and molecular weight distribution. In addition, it is possible to control the hydrophilic or hydrophobic character of the conjugate by copolymerization reactions, which makes the system very suitable for specific applications including the design of targeting. Also there is the possibility to use reactive conjugation sites at one or both ends of a macromolecule, or distributed along the polymer chain as pendant side groups. All these concepts are described in an excellent chapter published recently by Hoffman [1].

According to these characteristics, an elegant rationalized model of the drug-polymer conjugation was proposed by Ringsdorf in 1975 [2] with three components incorporated to a macromolecular chain: a component chemically linked to the drug (may be through a spacer), a targeting unit, and a molecule to control the solubility or the hydrophilicity. Since then, numerous conjugated systems have been designed and investigated. Particularly relevant are the hydroxypropyl methacrylamide (HPMA) copolymer conjugates, developed and optimised during the 1980's [3,4] and the styrene-co-maleic acid/anhydride neocarcinostatin (SMANCS) based on the work of Maeda *et al.* [5], both currently used in clinic. So far, five compounds based on HPMA copolymers have entered early clinical trials as anticancer agents. A molecular weight of approximately 30,000 Da was chosen to ensure that this non degradable polymer would be cleared from the body through metabolic pathways of the kidney. We have been interested in the preparation and application of families of polymer-drug conjugates based on acrylic derivatives of several compounds with pharmacological and medical interest (Figure 1) for the treatment of different tissue dysfunctions: vitamin E [6], antiaggregating pharmacons as Triflusal [7], analgesic compounds such as salicylic acid or paracetamol [8], and other non steroidal antiinflammatory agents (NSAIDs) based on derivatives of phenyl acetic or propionic acids [9,10,11].

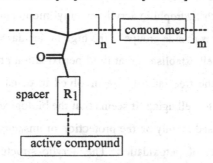

Active Compound	Spacer, R1	Comonomer
Vitamine E	-	HEMA
Vitamine E	-	DMA
Vitamine E	-	VP
Salicylic acid	-	HEMA
Salicylic acid	-	VP
Triflusal	—O—CH₂–CH₂–O—	-
Triflusal	—O—CH₂–CH₂–O—	DMA
Ketoprofen	—O—CH₂–CH₂–O—	HEMA
Ibuprofen	—O—CH₂–CH₂–O—	HEMA
Ketoprofen	—NH—⬡—O—	HEMA
Ibuprofen	—NH—⬡—O—	HEMA
Ketoprofen	—NH—⬡—O—CH₂–CH₂–O—	HEMA
Ibuprofen	—NH—⬡—O—CH₂–CH₂–O—	HEMA

Figure 1: Chemical structures of the polymer-active compound conjugates synthesized in our laboratory.

2.1. VITAMIN E CONJUGATES

We have reported recently the synthesis of a new acrylic derivative of vitamin E that can be copolymerized with hydrophilic acrylic or vinyl monomers to obtain hydrogels bearing vitamin E structures as side groups of the high molecular weight polymer chains [12]. In this sense, it is well established that lipid peroxidation proceeds through a free radical mechanism, and the free radicals are involved in damaging processes of cell membranes as well as in the cell aging. It seems that the biological antioxidant function of vitamin E *in vivo* is based mainly on the protection of unsaturated lipids of the cells from the damaging effect of peroxidation. The acrylic structure of this vitamin E derivative offers a new route for the design and preparation of biomaterials with specific properties. Moreover, recently has been postulated that the presence of vitamin E-derivatives plays a crucial function in the levels of triglycerids, LDH and HDH, related to the presence of cholesterol is blood vessels. The copolymerization of the acrylic derivative with comonomers like 2-hydroxyethyl methacrylate, HEMA, dimethylacrylamide, DMA, or vinyl pyrrolidone, VP, provides biocompatible polymeric drugs with controlled hydrophilic character according to the average composition of copolymer chains. The physicochemical properties of these systems are excellent for the application of these copolymers as powder that after hydration forms a pharmacologically active hydrogel. Particularly we have analysed the excellent behaviour of hydrogels prepared from copolymers of the acrylic derivative of vitamin E and HEMA in the healing process of Achilles tendon of rabbits (see Figure 2 for the experimental model). The regeneration of the tendon is clearly favoured by the presence of the active polymeric system, with a good reorientation of the fibrillar collagen in the longitudinal direction. The histological study of the regenerated tissue demonstrated that the polymeric derivative of vitamin E stimulates the regenerative process as a consequence of the antiaging effect in the local area of application.

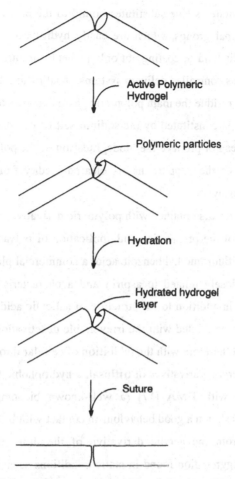

Figure 2: Schematic design of the experimental model applied to the Achilles tendon regeneration using the vitamin E derivative.

2.2 ANTITHROMBOGENIC CONJUGATES

We have demonstrated that the application of coatings polymeric derivatives with salicylic acid on the inner surface of small diameter Dacron or Goretex vascular grafts, improved the prevention of adhesion and aggregation of platelets on the surface of the vascular graft under dynamic conditions [13,14]. The derivatives of salicylic acid studied were constituted by high molecular weight polyacrylic chains ($M_n > 40,000$ Da) bearing the salicylic residue as side substitutes bound to the polymeric chain by weak carboxylic ester functional groups, which are easily hydrolyzed in the physiological medium. In this way, this kind of coating not only presents an intrinsic antiaggregating character, but also acts as controlled delivery systems of salicylic acid. In addition, after the release of the active residue the main polymeric chain becomes totally soluble in the physiological fluids, as is constituted by the sodium salt of polymethacrylic acid. The hydrolytical process does not produce the biodegradation of the polyacrylic chains, but changes the solubility of the support that is cleared readily from the body by the classical metabolic pathway.

In view of the previous results obtained with polymeric derivatives of salicylic acid, we considered the interest of the preparation and application of polyacrylic derivatives of triflusal (2-acetoxy-4-trifluoromethyl benzoic acid, a commercial platelet inhibitor with a chemical structure closely related to aspirin and a characteristic pharmacological profile) [15,16], which in addition to the structure of salicylic acid, has acetyl groups, which are considered to be related with the irreversible deactivation of platelets in the aggregation process and therefore with the inhibition of cellular thrombus.

Two types of polyacrylic derivatives of triflusal, a hydrophobic homopolymer and a hydrophilic copolymer with DMA [17] (a well-known biocompatible hydrophilic component that also has shown a good behaviour in contact with blood) [18], have been synthesized, starting from the acrylic derivative of the drug (see Figure 1). The preliminary results of aggregation found in static conditions seem to indicate that the coating of the surface of vascular grafts of Goretex improves the antithrombogenic character of the prostheses and provides a resorbable systems, which allows the reendothelization of the prosthesis after implantation in a moderate interval of time. In addition, the new polyacrylic systems derived from triflusal are truly controlled release systems of the antithrombogenic drug triflusal, leading to constant *in vitro* release during several weeks [7]. There is a clear dependence on the hydrophilia of the system, being the release rate of the copolymer much higher than the exhibited by the hydrophobic homopolymer.

2.3 IBUPROFEN AND KETOPROFEN CONJUGATES

Ibuprofen and ketoprofen are two well-known NSAIDs widely used in anti-inflammatory therapy. Their main disadvantages are the relatively short half-life in plasma and a significant gut- and nephro-toxicity. Therefore, a controlled release system with hydrophilic character would be useful especially in chronic diseases as rheumatoid arthritis. In this sense, several research groups have devoted attention to the preparation of polymeric drugs bearing ibuprofen and comonomeric hydrophilic components [19,20,21,22]. These systems seem quite promising for local, long time (intra-articular) applications. In addition, these drug delivery systems are expected to be clearable after hydrolysis because of the formation of the corresponding sodium salts, soluble in the physiological fluids.

The *in vitro release* profiles of these systems, which are copolymers with HEMA (see Figure 1), have been tailored at some extent during the synthetic procedure. The control on the composition, hydrophilic/hydrophobic character, spacer nature and microstructure leads to a control in the hydrolysis rate [11].

3. Physically-loaded Matrices and Scaffolds

The design of delivery systems based on polymeric matrices with homogeneously distributed drugs, which can be released by diffusion or/and erosion or involving dissolution, swelling and osmosis processes, is probably one of the most important topics of a great number of researchers in the field of controlled release technologies. Excellent books and reviews on this topic can be found in the literature [23,24]. In this context, the design of biodegradable scaffolds in tissue engineering should accomplish some requirements to get the desired tissue growth, and this includes for example the incorporation and controlled delivery of growth factors into the polymer to stimulate cell differentiation and growth [25].

Tissue remodeling is an essential component in the process of wound healing and thus in the normal maintenance and survival of all organisms. Growth factors are proteins that bind to cell surface receptors, activating cellular proliferation and/or differentiation. They include platelet-derived growth factor (PDGF), epithelial growth factor (EGF), tissue growth factor (TGF)-a, TGF-b, fibroblast growth factor (FGF), nerve growth factor (NGF), erythropoietin, insulin-like growth factor (IGF)-I and IGF-II, bone morphogenetic proteins (BMP), among others. Numerous potential applications are being investigated such as bone growth, angiogenesis and nerve regeneration. The controlled delivery of protein/cytokine growth factors is an important aspect of tissue engineering technology.

Growth factors can stimulate or inhibit cell division, differentiation and migration [26].They up- or down-regulate cellular processes such as gene expression, DNA and protein synthesis and autocrine and paracrine factor release. Therefore, they initiate, sustain and propagate a number of the normal processes required for the orderly progression of inflammation, tissue remodeling, and repair [27] Various approaches to the use of growth factors in tissue engineering have been proposed [28,29]. Growth factors may be added to cell culture media, thus acting on growing tissues *in vitro*. Controlled delivery of growth factors *in vivo* also has been proposed [30].

On the other hand, gene therapy may also has potential applications in combination with devices. Genes have been studied in the treatment of several spinal disorders that require disc regeneration and spinal fusion, restenosis or other diseases. The transfer of genes to cells enables recipient cells to ultimately synthesize the proteins these genes encode and exert a continual localized influence on cell growth to treat acquired diseases. Therefore, gene transfer systems have also been used to treat vascular tissue, chondrocytes, tendons, ligaments, meniscus, muscles and bones.

Although growth factor delivery included in the first stages non-degradable polymeric carriers, such as poly(ethylene vinyl acetate) and poly(vinyl alcohol) sponges, the growth of seeded cells in a scaffold in the presence of GF delivery system was well contrasted with control experiment without the delivery system. The results indicate that the release from biodegradable polymer microspheres will be beneficial in animal cell culture systems. Thus, controlled-release devices made from biodegradable polymers or gels which incorporate growth factors may be used *in vivo* in order to aid tissue regrowth or regeneration. These devices can be implanted either alone or in conjunction with another structure, such as a polymeric scaffold supporting the regenerating tissue.

Bone tissue regeneration requires the use of growth factors to markedly increase scaffold effectiveness. Bone morphogenetic proteins (BMP) are of particular interest to the field of orthopedics because the critical role that they have been shown to play in embryological bone formation, osteoinduction, and bone repair as well as the possibility that they may assist in the replacement of the standard autogeneous bone graft [31]. Delivery systems have included demineralized bone matrix, collagen composites, fibrin, calcium phosphate, polylactide, poly(lactide-co-glycolide), polylactide-polyehtylene glycol, hydroxyapatite and titanium. An ideal delivery system would allow a slow release of the BMPs, be biologically and immunologically inert, quickly absorbed, and supportive of cell proliferation and angiogenesis. It would also possess enough rigidity to withstand deforming forces until absorbed.

We have studied recently some applications of non-crosslinked hydrogels of controlled hydrophilic/hydrophobic character based on copolymers of vinyl pyrrolidone VP, with 2-hydroxyethyl methacrylate, HEMA, prepared by free radical polymerization [32]. In these supports, the release is controlled by the solubilization rate of the matrix after a relatively fast swelling process; therefore, as defined by Peppas *et al.* [33], these systems can be considered as dissolution controlled systems. Poly(VP-*co*-HEMA) is a well known biocompatible hydrogel with broad applications in the biomedical area [34,35], mainly as crosslinked networks. Because of its biocompatibility and hydrophilic nature, it has been investigated as carrier for drug delivery [36] However, crosslinked matrices are not resorbable. In this sense, we have proposed the use of non-crosslinked VP-HEMA copolymers as entirely resorbable delivery systems.

As a consequence of its relatively high hydrophilia, this copolymer has shown to be successful for the release of immunosuppressors like cyclosporine with very low solubility in water [37] (and associated to serious toxic side effects, primarily renal and hepatic) [38], and for the release of macromolecules like Growth Hormone (GH) [39]. There is a clear correlation between the solubilization and the release rate [37][39].

Further *in vivo* experiments using rat as animal model exhibited a very good correlation with this previous *in vitro* data for the cyclosporine-loaded material [40] as it is shown in Figure 3. The most hydrophobic implant reverts a provoked immune response slower (2-4 weeks) than the most hydrophilic one (1-2 weeks), being in agreement with the release time-scales obtained from the *in vitro* experiments.

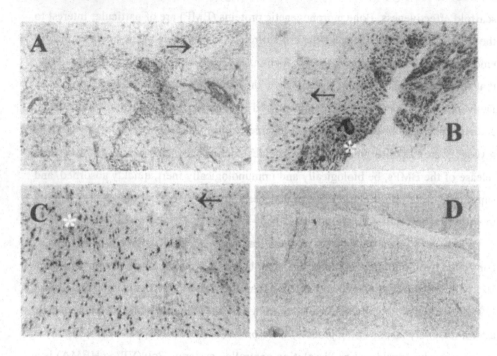

Figure 3. Microphotographs of the surrounding tissue of the control group at 2 (A) and 14 (B). (→) lymphocytes and inflammation; (←) edema; (*) granulation tissue, (original magnification ×16, hematoxylin-eosin staining, H.E.). (C) Microphotograph corresponding to the VP-HEMA 60-40 (wt %) implant at 14 days. (*) incipient cicatrizial fibrosis; (←) edema, (×40, H.E.). (D) Microphotographs corresponding to the VP-HEMA 40-60 (wt %) implant at 28 days, (×16, H.E.).

We also propose the use of the VP-HEMA copolymer as dissolution-controlled system for the release of macromolecules as GH. Recombinant Growth Hormone (GH) controlled delivery systems might be very helpful in therapeutic treatments for tissue regeneration associated to some pathologies. Typical examples are the regeneration of bone tissue or the healing of persistent ulcers in diabetic patients. The controlled release of macromolecules (protein is the best example), is not an easy issue. The first in vitro and in vivo demonstrations of macromolecular drugs loading and release from non-degradable polymeric matrices, such as poly(2-hydroxyethyl methacrylate), poly(vinyl alcohol) and copolymers of ethylene vinyl acetate, were made by Langer *et al.* [41] Polyurethanes and polydimethylsiloxane have been also used since the development of polymeric drug delivery systems. Although these types of polymers deliver drugs effectively, they have found limited clinical use because they remain as space-occupying foreign bodies. Furthermore, they typically do not deliver drugs with zero-order kinetics. A wide variety of macromolecules that include lysozyme, alkaline phosphatase, tumor angiogenesis factor, insulin, bovine serum albumin, and methyl β-lactoglobulin A, among others, were used. The release pattern of these macromolecules was characterized by an initial high burst effect followed by sustained release. By improving method and developing a sintering technique, a more controlled and reproducible release pattern was obtained. Thus, a zero-order release of insulin was obtained. This kind of systems requires surgical removal of the exhausted device after a determined period of time, due to the non degradability of the polymeric support.

In our case, due to the low stability of GH in physiological media (its half life is about a few hours), a polymeric depot loaded with the hormone can be used not only as controlled delivery system of the protein, but also as a protective matrix. Figure 4 exhibits the release profiles from different GH loaded matrices, together with the dissolution rates, showing again the dependence of both processes and the possibilities of these dissolution-controlled systems for the release of macromolecules. The solubilization properties of this VP-*co*-HEMA copolymer depend on the hydrophilic/hydrophobic balance associated to the VP/HEMA composition, respectively. This balance can be tailored in the synthetic procedure (radical polymerization) profiting from the particular characteristic of this polymerization process [32]. Basically, HEMA is much more reactive than VP, and therefore the copolymer microstructure formed in the course of the reaction can be considered as a blend of: 1) a HEMA-rich copolymer with an average composition regulated by the initial molar ratio in the feed, and 2) a VP-rich copolymer formed in the last steps of the polymerization when most of the more reactive HEMA has been consumed. The control

48

on the feed and the conversion leads to a control on the species ratio and on the properties of the final materials. In this way, we have been able to obtain solubilization rates *(in vitro)* ranging from few hours to several months.

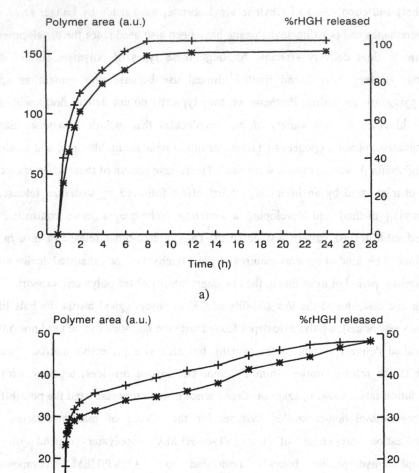

a)

b)

Figure 4. Polymer area vs. release time (*) and GH vs. release time (+) for a) VP-HEMA 70-30 (wt %) and b) VP-HEMA 40-60 (wt %) copolymer systems.

According to the microstructure of copolymer chains, the poly(VP-*co*-HEMA) systems prepared at high conversion could be considered as interpenetrating physical networks constituted by chains very rich in VP, which are very soluble in water, entangled into chains of random copolymers of VP-HEMA which are less soluble but highly hydrated in water or physiological fluids. After the hydration process the VP-rich chains become very soluble in the hydrated medium being incorporated to the solution and giving rise to the formation of microchannels which favor the release of the drug or the bioactive component loaded into the polymer matrix. These structures formed during the free radical polymerization of the VP and HEMA comonomers modulate the release of the bioactive component and the resorption of the hydrogel within periods of time up to four or six weeks, showing a very good biocompatibility as it was tested by implantation of discs in the dorsal muscle of rats.

The preparation of this kind of controlled delivery systems is very easy because of the solubility of the copolymeric matrix in common organic solvents compatible with the bioactive compound. It is possible to prepare powder by liophilization, or films by casting from water/dioxane solutions. In any case, when the system is in contact with the hydrated medium, swells very fast to become a mucoadhesive gel, which releases the drug and becomes resorbable simultaneously.

On this basis, we have developed a new model for the regeneration of peripheral nerve gaps based on the application of hollow fibers oriented longitudinally to the direction of the original nerve fibers [42]. These fibers were filled with poly(VP-*co*-HEMA) blended with growth hormone Norditropin. The histological results after four weeks of implantation were positive, with a good tissue regeneration and electrophysiological response. The application of the polymeric component in conjunction with GH provided, after hydration, an adequate medium for the growing of the nervous fibers, being the polymer resorbed in the applied defect without any foreign response. Figure 5 shows an schematic representation of this system.

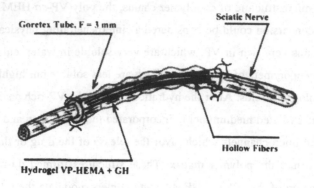

Figure 5. Experimental model used for the sciatic nerve regeneration.

4. References

1. Hoffman, A.S. (1998) A commentary on the advantages and limitations of synthetic polymer-biomolecule conjugates, in T. Okano (ed.), *Biorelated polymer and gels*, Academic Press, San Diego, pp. 231-248.
2. Ringsdorf, H. (1975) Structure and properties of pharmacologically active polymers, *J. Polym. Sci. Polym. Symp.* **51**, 35–53.
3. Duncan, R. (1992) Drug-polymer conjugates: potential for improved chemotherapy, *Anti-Cancer Drugs* **3**, 175–210.
4. Putnam, D. and Kopecek J. (1995) Polymer conjugates with anticancer activity, *Adv. in Polym. Sci.* **122**, 55–123.
5. Maeda, H. (1991) SMANCS and polymer-conjugated macromolecular drugs: advantages in cancer chemotherapy, *Adv. Drug Delivery Review* **6**, 181–202.
6. Ortiz, C., Vázquez, B. and San Román, J. (1998) Synthesis, characterization and properties of polyacrylic systems derived from vitamin E, *Polymer* **39**, 4107–4114.
7. Rodríguez, G., Gallardo, A., San Román, J., Rebuelta, M., Bermejo, P., Buján, J., Bellón, J.M., Honduvilla, N.G. and Escudero, C. (1999) New resorbable polymeric systems with antithrombogenic activity, *J. Mater. Sci. Mater. in Med.* **10**, 873–878.
8. Elvira, C. and San Román, J. (1997) Complexation of polymeric drugs based on polyacrylic chains with aminosalicylic acid side groups, *J. Mat. Sci.: Mat. in Med.* **8**, 743–746.
9. Liso, P.A., Rebuelta, M., San Román, J., Gallardo, A. and Villar A.M. (1995) Antinociceptive and antipyretic properties of a new conjugated ibuprofen-methacrylic polymeric controlled delivery system, *J. Controlled Release* **33**, 429–436.
10. Gallardo, A. and San Román, J. (1993) Synthesis and characterization of a new poly(methacrylamide) bearing side groups of biomedical interest, *Polymer* **34**, 394–400.
11. Gallardo, A., Parejo, C. and San Román, J. (2001) NSAIDs bound to methacrylic carriers: microstructural characterization and in vitro release analysis, *J. Controlled Release* **71**, 127–140.
12. Vázquez, B., Ortiz, C., San Román, J., Plasencia, M.A. and López-Bravo, A. (2000) Hydrophilic polymers derived from vitamin E, *J. Biomat. Appl.* **15**, 118–139.

13. San Román, J., Escudero, M.C., Gallardo, A., Santa Cruz, R., Jorge, E., de Haro, J., Álvarez, L., Millán, J., Buján, J., Bellón, J.M. and Castillo-Olivares, J.C. (1994) Application of new coating for vascular grafts based on polyacrylic systems with antiaggregating activity, *Biomaterials* **15**, 759–765.
14. San Román, J., Buján, J., Bellón, J.M., Gallardo, A., Escudero, M.C., Jorge, E., de Haro, J., Álvarez, L. and Castillo-Olivares, J.C. (1996) Experimental study of the antithrombogenic behavior of dacron vascular grafts coated with hydrophilic acrylic copolymers bearing salicylic acid residues, *J. Biomed. Mater. Res.* **32**, 19–27.
15. Guiteras, P., Altimiras, J., Aris, A., Auge, J.M., Bassons, T., Bonal, J., Casalps, J.M., Castellarnau, C., Crexells, C., Masotti, M., Oriol, A., Padró, J.M. and Rutllant, M. (1989) Prevention of aortocoronary vein-graft attrition with low-dose aspirin and triflusal, both associated with dipyridamole: a randomized, double-blind, placebo-controlled trial, *Eur. Heart J.* **10**, 159–167.
16. de la Cruz, J.P., Villalobos, M.A., García, P.J., Smith-Agreda, J.M. and Sánchez de la Cuesta, F. (1995) Effects of triflusal and its main metabolite HTB on platelet interacting with subendothelium in healthy volunteers, *Eur. J. Clin. Pharmacol.* **47**, 497–502.
17. Rodríguez, G., Gallardo, A., Rebuelta, M., Bermejo, P., Buján, J., Bellón, J.M., Honduvilla, N.G., Escudero, C. and San Román, J. (manuscript submitted) Polymeric hydrophilic systems derived from aine's : characterization and in vitro release of polyacrylic systems of triflusal, *J. Biomat. Sci. Polym. Edn.*
18. de Queiroz, A.A.A., Gallardo, A. and San Román, J. (2000) Vinyl pyrrolidone-N,N-dimethyl acrylamide water soluble copolymers: synthesis, physical-chemical properties and protein interactions, *Biomaterials* **21**, 1631–1643.
19. Davaran, S. and Entezami, A.A. (1998) Hydrophilic copolymers prepared from acrylic type derivatives of ibuprofen containing hydrolyzable thioester bond, *Eur. Polym. J.* **34**, 187–192.
20. Davaran, S. and Entezami, A.A. (1997) Acrylic type polymers containing ibuprofen and indomethacin with difunctional spacer group: synthesis and hydrolysis, *J. Control. Rel.* **47**, 41–49.
21. Cecchi, R., Rusconi, L., Tauzi, H.C., Danusso, F. and Ferruti, P. (1981) Synthesis and pharmacological evaluation of poly(oxyethylene) derivatives of 4-isobutylphenyl-2-propionic acid (ibuprofen), *J. Med. Chem.* **24**, 622–625.
22. Larsen, C. and Johansen, M. (1989) Incorporation of acrylic salicylic derivatives to hydrophilic copolymer systems with biomedical applications, *Acta Pharm. Nordica* **2**, 57–66.
23. Kydonieus, A. (ed.) (1991) *Treatise on controlled drug delivery*, Marcel Deker, New York.
24. Mathiowitz, E. (ed.) (1999) *Encyclopedia of Controlled Drug Delivery*, Wiley, New York.
25. Chapekar, M.S. (2000) Tissue engineering: challenges and opportunities, *J. Biomed. Mat. Res. (Applied Biomaterials)* **53**, 617–620.
26. Gooch, K., Blunk, T., Vunjak-Novakovic, G., Langer, R., Freed, L. and Tennant, C.J. (1998) Mechanical forces and growth factors utilized in tissue engineering, in C.W. Patric Jr., A. Mikos, L. V. McIntire (eds.), *Frontiers tissue engineering*, Pergamon, Oxford, Chapter II.3
27. Deuel, T.F. and Zhang, N. (2000) Growth factors, in R.P. Lanza, R. Langer, J. Vacanti (eds.) *Principles of tissue engineering*, Academic Press, San Diego, chapter 12.
28. Langer, R. and Vacanti, J.P. (1993) Tissue engineering, *Science* **260**, 920–926.
29. Reddi, A.H. (1994) Symbiosis of biotechnology and biomaterials: Applications in tissue engineering of bone and cartilage, *J. cellular biochemistry* **56**: 192–195.
30. Saltzman, W.M. (1996) Growth-factor delivery in tissue engineering, *MRS bull.* nov, 62–65.
31. Burg, K.J.L., Porter, S. and Kellam J.F. (2000) Biomaterial development for bone tissue engineering, *Biomaterials* **21**, 2347–2359.
32. Gallardo, A., Lemus, A.R., San Román, J., Cifuentes, A. and Díez-Masa, J.C. (1999) Micellar electrokinetic chromatography applied to copolymer systems with heterogeneous distribution, *Macromolecules* **32**, 610–617.
33. Narasimhan, B. and Peppas, N.A. (1997) Molecular analysis of drug delivery systems controlled by dissolution of the polymeric carrier, *J. Pharm. Sci.* **86**, 297–304.
34. Bell, C.L. and Peppas, N.A. (1995) Biomedical membranes from hydrogels and interpolymer complexes, *Adv. Polym. Sci.* **122**, 125–176.
35. Laporte, R.J. (1997) Hydrophilic polymer coatings for medical devices. Technomic Pbl., Lancaster, p. 58.
36. Blanco, M.D., Trigo, R.M., García, O. and Teijón, J.M. (1997) Controlled release of cytarabine from poly(2-hydroxyethyl methacrylate-co-N-vinyl-2-pyrrolidone) hydrogels, *J. Biomater. Sci. Polym. Edn.* **8**, 709–719.

37. Gallardo, A., Fernández, F., Bermejo, P., Rebuelta, M., Cifuentes, A., Díez-Masa, J.C. and San Román, J. (2000) Controlled release of cyclosporine from VP-HEMA copolymer systems of adjustable resorption monitorized by MEKC, *Biomaterials* **21**, 915–921.
38. Hassan, M.M.A., Al-Yahya, M.A. (1987) Cyclosporine, in K. Florey (ed.) *Analytical Profiles of Drug Substances*, Academic Press, London, **16**, pp. 146–206.
39. Cifuentes, A., Diez-Masa, J.C., Montenegro, C., Rebuelta, M., Gallardo, A., Elvira, C. and San Román, J. (2000) Recombinant growth hormone delivery systems based on vinylpyrrolidone-hydroxyethyl methacrylate copolymer matrices: Monitoring optimization by capillary zone electrophoresis, *J. Biomater. Sci. Polymer Edn.* **11**, 993–1005.
40. Gallardo, A., Fernández, F., Cifuentes, A., Diez-Masa, J.C., Bermejo, P., Rebuelta, M., López-Bravo, A., and San Román J. (2001) Modulated release of cyclosporine from soluble vinyl-pyrrolidone-hydroxyethyl methacrylate copolymer hydrogels. A correlation of "in vitro" and "in vivo" experiments, *J. Controlled Release* **72**, 1–11.
41. Folkman, J. and Langer, R. (1976) Polymers for sustained release of proteins and other macromolecules, *Nature* **263**, 797–800.
42. Tomás, D., de Pedro, J.A., López Bravo, A., Lemus-Q, R., Gallardo, A. and San Román, J. (1998) Regeneration of peripheral nerve gaps stimulated by resorbable hydrogels, growth homone and guided by hollow fibres, Proceeding of 14th ESB Conference, The Hague, p. 184.

Novel Systems, Hydrogels and Bone Cements

MEMBRANES AND HYDROGELS IN RECONSTRUCTIVE SURGERY

D. BAKOŠ
Faculty of Chemical and Food Technology, STU
Radlinského 9, 812 37 Bratislava, Slovak Republic
J. KOLLER
Centre for Burns and Reconstructive Surgery, Central Tissue Bank,
Ružinov General Hospital
Ružinovská 6, 826 06 Bratislava, Slovak Republic

1. Introduction

The clinical requirement for artificial graft materials to promote effective wound repair is large. Examples of chronic or extensive wounds include burn injuries, post traumatic skin and soft tissue defects, pressure sores (decubitus ulcers), diabetic skin ulcers, venous stasis ulcers, and defects arising following tumor excision. Especially, progress in burn care is dependent much on a suitable sophisticated skin substitute. The successful development of a permanent skin substitute will have a very strong impact on care of patients with serious burns.

With regard to woundcare, the desirable final goal is that the engineered tissue becomes integrated within the patient, affording a potentially permanent and specific cure [1]. Engineering of skin substitutes provides a prospective source of advanced therapies for treatment of acute and chronic wounds. The most common tissue engineering strategies involve the use of isolated cells or cell substitutes, tissue-inducing substances, and cells seeded on or within polymer matrices with a three-dimensional architecture as well as appropriate biological and mechanical properties. By design and incorporation of specific therapeutic properties in skin substitutes, reduction of morbidity and mortality from full-thickness wounds can be facilitated [2].

Several companies are investigating the use of biodegradable polymers – natural and/or synthetic as a matrix for living cells. In this regard, important properties of the materials include porosity for cell in-growth, an appropriate surface for cellular attachment, mechanical properties compatible with those of the natural tissues, and degradation rate with by-products. In case of skin substitute in the form of a membrane, the polymer matrix, may represent the device itself, or can be a scaffold for cell growth in vitro that is degraded by the growing cells prior to implantation.

Skin substitutes composed of cultured cells and biopolymers provide alternative materials for study of skin biology and pathology, treatment of skin wounds, safety testing of consumer products, and therapeutic delivery of gene products [3]. Each of

R.L. Reis and D. Cohn (eds.),
Polymer Based Systems on Tissue Engineering, Replacement and Regeneration, 55–67.
© 2002 *Kluwer Academic Publishers. Printed in the Netherlands.*

these prospective applications may have distinct requirements for validation of skin substitutes. However, this paper is focused on considerations with respect to treatment and closure of skin wounds.

2. Objectives of Skin Substitutes

One of the most important requirements for a skin substitute is to reestablish the epidermal barrier to fluid loss and microorganisms [4]. At present, extensive full thickness skin wounds are best closed as quickly as possible by split thickness autografts or temporary substitutes. In full-thickness skin loss, replacement of both the epidermis and dermis must also mininize scar formation and restore acceptable function and cosmetics [5]. In these cases, split thickness autograft is an imperfect replacement for full thickness skin (also limited in quantity and associated with donor site morbidity).

Different techniques of skin substitution have been introduced in the last decades to minimise scar formation and to accelerate healing time. Currently available wound covers fall into two categories: permanent closure represented by autograft and by developing permanent skin substitutes, and temporary coverings such as allografts, xenografts (porcine) and all kinds of temporary synthetic wound coverings.

At present, the effort is focused on development of improved temporary skin substitutes more stimulating native wound healing and realization of a permanent composite skin replacement. The solution is very complex and requires an interdisciplinary approach, as it results from the characteristics of the ideal skin substitute (Table 1) [6].

TABLE 1 Characteristics of the ideal skin substitute

1. Inexpensive
2. Long shelf life
3. Used off the shelf
4. Non-antigenic
5. Durable
6. Flexible
7. Prevents water loss
8. Barrier to bacteria
9. Conforms to irregular wound surfaces
10. Easy to secure
11. Grows with child
12. Applied in one operation
13. Does not become hypertrophic
14. Does not exist at the present time

Biologic dressings remain the gold standard for temporary wound coverage and closure and as a skin substitute. However, increasing interest has resulted in construction of numerous synthetic wound dressings and skin substitutes imitating biological dressings and substitutes. On the other hand, the increase in availability of skin grafts based on in vitro cell cultures for wound closure drives the research and development of cultured skin substitutes.

3. Synthetic membranes as wound dressings

As temporary dressings, there exist a number of different kinds of membranes of several proprietary types that are designed to provide a vapor and bacterial barrier and control pain while the underlying superficial wound reepithelializes. Biodegradable polymers used for wound dressings may be natural or synthetic. Collagen is commonly used in medical devices as an implant or a coating onto which cells can attach and migrate during healing [7]. Other natural biopolymers of increasing importance are glycosaminoglycans like hyaluronic acid (usually associated with scarless wound healing in the fetus) and chondroitin-6-sulphate, basement membrane proteins, fibrin, fibronectin, chitin/chitosan, etc. By comparison, synthetic polymers (i.e. polylactic acid – PLA and polyglycolic acid – PGA) became standards for different medical applications, and also for delivery of skin cells [8].

The wound dressings can be more complex in the structure. In 1979 Biobrane was successfully introduced for the treatment of burn wounds and donor sites. Biobrane is a biosynthetic wound dressing constructed of a silicone film with a nylon fabric partially embedded into the film. The fabric presents to the wound bed a complex three-dimensional structure of trifilament threads. Porcine collagen is incorporated in both silicone and nylon components by being chemically crosslinked. In freshly excised wounds covered by Biobrane, blood/plasma can form fibrin clots in the nylon matrix, thereby firmly adhering the dressing to the wound. In partial thickness burns Biobrane becomes attached to the wound by fibroelastic bond until epithelialization occurs [9].

The natural growth matrix for mammalian cells is collagen - a fibrous protein. The matrix is modified by various glycoamino- glycans to provide an aqueous medium into which cells may grow and proliferate. Simplified, a protein is a polymer with various hydrophobic and hydrophilic pendant side groups.

Synthetic polymers that exhibit cell growth characteristics would be expected to have a carbon-carbon backbone with similar types of pendant groups as proteins. As example, the study of simplified synthetic acrylic co, ter and tetra polymers [9] led to the conclusions that positive charges in low surface density encouraged cell attachment but did not favour proliferation particularly when the density was sufficiently high as to cause cell membranes to rupture on contact. On the other hand, negative charged carboxyl groups were found not only to promote cell attachment, but also to allow uninhibited cell growth up to certain levels, above which surface hydration impedes growth. Counter cations were shown to be important in the type of cellular reaction at the interface - *in vivo*. Generally, hydrophilicity and surface change should play some part in controlling adhesion, especially when hydrogen bonding and electrostatic forces are involved [10].

A number of single layer semi-permeable synthetic membranes exist that provide a mechanical barrier to bacteria and have physiologic vapor transmission characteristics.

Transparent adhesive films are typically semi-permeable membrane dressings that are waterproof, yet permeable to oxygen and water vapors. These semipermeable membranes can be used with good effect in clean superficial wounds. Synthetic polymers have several advantages such as ability to adhere to the wound edges, ability to drape to the wound contour, and ease of use. The major disadvantage is the lack of biological properties such as enhancing wound healing via attraction of cells involved in the healing process [11].

Hydrocolloid dressings, hydrogels and hydrogel sheets are an important group of dressing material which create a moist wound environment while absorbing wound exudate. Chemically, these products are based on synthetic polymers as well as derivatives of natural polymers, and they can be easily manufactured using conventional technology into films, sheets, and sponges. Film dressings are e.g. made from polyuretane, poly (hydroxyethylmethacrylate) (HEMA), silastic, copolymers of dimethylaminoethyl-methacrylate and acetonitrile, HEMA and polyethylene glycol, and many different biocompatible polymers, their copolymers or blends. Hydrocolloids and hydrogels are mostly based on natural polymers as gelable polysaccharides (agarose, starch, cellulose derivatives, etc.) or their combination with synthetic hydrophilic polymers.

3.1. HYDROCOLLOID DRESSINGS

Hydrocolloids are occlusive and adhesive wafer dressings, which combine absorbent colloidal materials with adhesive elastomers capable to absorb light to moderate amounts of wound exudate. Most hydrocolloids react with wound exudate to form a gel-like covering, which protects the wound bed and maintains a moist wound environment. Hydrocolloid powders and pastes are also available with increased absorptive capacity (Table 2).

TABLE 2. The examples of Hydrocolloids and Manufacturers Available

Product	Manufacturer
Tegasorb THIN	3M
DermAssist	AssisTec Medical
Cutinova hydro	Beiersdorf-Jobst
BGC Matrix	Brennen
CombiDERM ACD	ConvaTec
Hydrocol	Dow Hickam
Dermatell	Gentell
Hyperion HC Dressing	Hyperion Medical
NU-DERM	Johnson & Johnson
CURADERM	Kendall
ExuDERM	Medline
Ultec	Sherwood-Davis & Geck
RepliCare	Smith & Nephew, Inc.,
Comfeel	Coloplast

Best uses of hydrocolloid dressings are for granulating and epithelializing wounds that are draining low to moderate amounts of exudate. They are conformable for easy application and help reduce pain at the wound site. On the other hand, they have several disadvantages. The discharge of moderately to heavily exudating wounds may accumulate under the dressing, or break down of the product may produce a residue of varying colours and possible foul odour. This should not be confused with an infectious process. Usual dressing changes for hydrocolloids take place up to 3 times per week.

3.2. HYDROGEL DRESSINGS AND HYDROGEL SHEETS

Hydrogels are crosslinked water-swollen polymers whose water content ranges from 30 to 90%. Low molecular weight water-soluble active agents often diffuse through hydrogels at a too high a rate to be useful. Wound gels are excellent for helping to create or maintain a moist environment for wounds with minimal or no exudate. Hydrogels absorb more fluid than hydrocolloids.

Hydrogel sheets are cross-linked polymer gels in a sheet form. Some are available with an adhesive border. Some hydrogels and hydrogel sheets provide absorption, desloughing and debriding capacities to necrotic and fibrotic tissue. They help to provide and maintain a moist wound environment and by increasing moisture content, hydrogels have the ability to help cleaning and debriding of necrotic tissues. Several examples of hydrogels and hydrogel sheets commercially available are summarized in Table 3 and Table 4.

TABLE 3. The examples of Hydrogel and Manufacturers Available

Product	Manufacturer
Amerigel Topical Ointment	Amerx Health Care
DermAssist Glycerin	AssisTec Medical, Inc.
Bard Absorption Dressing	Bard Medical
CarraSorb	Carrington Laboratories
SAF-Gel	ConvaTec
Dermagran	Derma Sciences, Inc.
DermaMend	DERMARx Corp.
Gentell	Gentell, Inc.
CURASOL	HEALTHPOINT
Restore	Hollister Inc.
Hyperion Hydrophilic Gel	Hyperion Medical, Inc.
NU-GEL	Johnson & Johnson
CURAFIL	Kendall
SkinTegrity	Medline
MPM	MPM Medical, Inc.
Iamin	ProCyte
PanoPlex	Sage Laboratories
Hypergel	SCA MOLYLYCKE
IntraSite	Smith & Nephew, Inc
Elta Dermal	Swiss-American Products, Inc.

Hydrogels are non-adherent and can be removed without any pain and trauma to the wound bed. Their "soothing" effect promotes patient acceptance and comfort. Their disadvantages consist in a low absorption and therefore, hydrogels or hydrogel sheets may not be an appropriate choice for moderately to highly exudating wounds. Hydrogels require secondary dressing.

Smart hydrogel dressings represent a new saline-based approach to the intelligent treatment in wounds requiring autolytic debridement or in chronically recalcitrant wounds. They are osmotically balanced to promote rapid autolytic debridement in dry necrotic wound while providing optimal hydration to newly forming granulation tissue. In addition to sodium salts they are formulated with potassium, zinc and other inorganic salts. The osmotic action effectively cleanses the wound bed while allowing absorption of excess fluid during periods of exudate formation. The cross-linked hydrogel structure provides the ability to maintain constant viscosity over a wide range of temperatures or after the hydrogel has absorbed up to twice its weight in fluid.

TABLE 4. The examples of Hydrogel Sheets and Manufacturers Available

Product	Manufacturer
Tegagel	3M
AcryDerm	AcryMed
2nd Skin	AFASSCO, Inc.
THINSite	B. Braun
Transorbent	B. Braun
Vigilon	Bard
ClearSite	Conmed Corporation
AQUASORB	DeRoyal
FLEXDERM	Dow Hickam
NU-GEL	Johnson & Johnson
CURAGEL	Kendall
Derma-Gel	Medline Industries
Elasto-Gel	Southwest Technologies

The first prescription hydrogel therapy that utilizes genetically-engineered platelet-derived growth factor (becaplermin) is Regranex [12], which is indicated for the treatment of diabetic neuropathic foot ulcers that extend into the subcutaneous tissue or beyond and have an adequate blood supply. Regranex mimics a naturally occurring protein in the body by attracting wound-healing cells to the ulcer site. Recombinant human platelet-derived growth factor is not produced from blood, but by genetic engineering in yeast cells, formulated and preserved in a water-based gel.

4. Approaches to tissue engineering

The first products of the engineering of living tissues are skin substitutes. Considering tissue engineering and wound repair, several different approaches are available and products together with potential therapies fall into three categories: epidermal replacements, dermal replacements, and complex skin substitutes. Potential efficacy of such materials is dependent upon several factors summarised above in the characteristics of a previously described ideal skin substitute.

For skin, epidermal keratinocytes can represent a permanent wound closure. Fibroblasts have been incorporated into certain models to promote reestablishment of epithelial-mesenchymal interactions that are recognized [13] but remain poorly understood. Addition of epidermal melanocytes to cultured skin in vitro has also been reported [14]. In successful development of any tissue-engineered substitute is the selection of a cell source critical. Although the use of autologous cells circumvents any issue of immune rejection, it does not provide for the off-the-shelf availability, which is clinically important. On the other hand, the use of allogenic cells requires the engineering of immune acceptance as well. The well-established use of combined autologous and allogeneic grafts gives potential for preparing cultured skin also from chimeric populations [15]. As well, long term an important part of the future of tissue engineering will be the use of stem cells [16]. Gene therapy for treatment of local or systemic conditions is feasible with cultured skin substitutes [17].

Particular attention has been given to detection and quantification of growth factors: epidermal growth factor, transforming growth factors α and β, platelet-derived growth factor, insulin-like growth factor [18], and neurotrophic growth factors [19]. Growth factors are classified as proteins that act as intercellular signals to allow cells to communicate with one another. They are involved in all phases of wound healing and also have the ability to regulate many other functions within the cell, including protein synthesis. Growth factors are essential to wound healing. Specifically, they are capable to:

- Attract useful cells to the wound site, including immune cells to fight infection and other cells to form connective tissue.
- Stimulate and increase formation of connective tissue.
- Stimulate formation of new blood vessels to nourish the wound site.
- Promote remodeling.

Various authors have reported the use of allogenic keratinocytes as temporary dressings presuming that substances secreted by the allogenic cells, or released upon their death and dissolution will provide signals for enhancing healing [20].

4.1. EPIDERMAL SUBSTITUTES

Epidermal substitutes can be defined that they consist of keratinocytes grown either alone (on the surface of a tissue culture flask), or in close association with a carrier vehicle such as a polymeric film or bioresorbable matrix.

Of the epidermal replacement products the most advanced consist of cultured autologous epithelial cells grown to confluence in vitro. All methods for *in vitro* cultured epidermal grafts originate in the basic work of Rheinwald and Green, who offered a possibility for serial cultivation of human epidermal keratinocytes [21]. The cultured sheets of epithelial cells are usually attached to petrolatum gauze for easy handling. This technology has been widely applied to patients with very large burns, and products have been available by hospital tissue banks or commercial vendors, e.g. the product Epicel™ (Genzyme Tissue Repair, Cambridge, Massachusetts). Epicel™ is the patient's own epidermis cultured by commercial laboratories from a small patient's skin biopsy provided by the clinician. It is used to treat extensive deep wounds that require skin grafting (skin replacement), such as occurs with severe burns. The major drawback of cultured epidermis is, that it does not contain dermal support [22].

In excised full thickness wounds that are closed with epithelial cells alone, engraftment rates are suboptimal and long-term durability is less than ideal. Moreover, recognition of the fragility of thin epithelial cover has lead to several attempts to combine epithelial cells with a functioning dermal layer. Grafting of epihtelial cells on wounds previously covered with cell-free vascularized dermal allografts was successful [23].

Currently, the bioengineering approach in which autologous human keratinocytes grown on a hyaluronic acid membrane used as carrier, represents an efficient and useful semi-artificial substitute. The benzylester hyaluronic acid membrane (Laserskin, Fidia Advanced Biopolymers, Italy) is the carrier where the keratinocytes can easily adhere and proliferate. Hyaluronic acid as the natural compound of the extracellular matrix in the early phase of wound healing interacts with the fibrin clot where it plays a role in the migration of cells entering the tissue repairing processes [24].

The idea to combine epithelial cells with other dermal analogs was generally accepted and resulted in the development of generally acceptable synthetic dermal substitutes for constructing sophisticated composite total skin substitute. The previously mentioned Epicel™, is much more efficient in combination with something that restores also the dermal layer of skin.

4.2. DERMAL SUBSTITUTES AND ARTIFICIAL SKIN

In very severely burned patients who have little or no remaining intact skin, artificial skin is an extremely useful material not only to cover and thereby protect the wounded area, but to promote re-growth of a natural skin instead of scar tissue.

One of the ideas for solving the problem of closing large-surface and deep wounds was an original method of Burke and Yannas artificial skin [25,26], now called Integra™ Artificial Skin and commercialised by Integra LifeSciences Corporation (Plainsboro, New Jersey). Integra™ consists of two layers, just as the living skin. The bottom layer, designed to "regenerate" the lower, dermal layer of real skin, is composed of a matrix of interwoven bovine collagen (a dermal bovine collagen) and chondroitine-6-sulphate (glycosaminoglycan – GAG) that mimics the fibrous pattern of dermis. This matrix is then affixed to a temporary upper layer: a medical-grade, flexible silicon sheet that mimics the epidermal, or surface, layer of the skin. Key features of this material's design are the number and size of the pores in the collagen/glycosaminoglycan matrix as well as the rate at which the matrix disintegrates. The degradation rate of collagen-GAG sponge is controlled by glutaraldehyde-induced crosslinking. The collagen-GAG dermal layer functions as a biodegradable template that induces organised regeneration of dermal tissue (neodermis) by the body and the infiltration of fibroblasts, macrophages, lymphocytes, and endothelial cells that form a neovascular network [27]. As healing progresses, native collagen is deposited by the fibroblasts, and the collagen portion of the artificial skin is biodegraded over approximately 30 days. Following ingrowth of autologous tissue, which takes usually 3 to 4 weeks, the superficial silicon layer of Integra™ artificial skin is removed and replaced by thin dermo-epidermal autografts.

Another product that is similar to Integra™ is called AlloDerm™. This product, which is sold and manufactured by LifeCell Corporation of The Woodlands, Texas, is produced by removing all cell components that cause a burn patient's immune system to reject a graft from the dermal layer of the human cadaver skin. A key feature of the process is preserving to the greatest extent possible the "natural," three-dimensional structure of the dermis along with the basement membrane structures. First, the epidermis is enzymatically separated from dermis. Then, using hypertonic saline, the epithelial elements of the dermis are removed. The tissue is then treated in a detergent to inactivate any viruses and is freeze-dried. This process results in a theoretically non-antigenic complete dermal scaffold with basement membrane proteins, including intact laminin and type IV and VII collagen. It is designed to be combined with epithelial autograft at the time of wound closure [28].

The other commercially available skin equivalent Apligraft™ (Organogenesis, Sandoz/Ciba) is based on an original idea proposed by Bell et al [29]. This product, earlier also known as Graftskin, is made by separating out the cells (keratinocytes and fibroblasts) from normally–discarded infant human foreskin. The lower layer (dermis) consists of a collagen gel matrix formed by purified bovine type I collagen mixed with a suspension of dermal fibroblasts when the collagen matrix is believed to be condensed reducing of the volume. In the second step, human dermal keratinocytes are seeded on such a collagen matrix, forming an epidermis-like structure. The permanence of such grafts is scientifically questionable, as the long-term survival of allogenic keratinocytes and fibroblasts is usually not possible following grafting. Graftskin is highly effective in healing venous ulcers and other wounds, particularly those of long duration, that have proved hard to heal with conventional modalities probably due to temporary production of growth factors and cytokines by the living cells [30].

Particular interst can be focused on Dermagraft-TC (Advanced Tissue Sciences, La Jolla, CA). This is a product of tissue engineering combining a bioengineered humanised dermal layer and a synthetic epidermal layer. This bioengineered dermal layer is produced by culturing of human neonatal fibroblasts on polyglycolic acid and polyglactin-910 matrix bonded to a nylon mesh, which acts as a three-dimensional scaffold. When seeded with allogenic fibroblasts, they proliferate on the nylon mesh and secrete extracellular matrix proteins collagen type I – III – VI, elastin, fibronectin, decorin, and growth factors. After the fibroblasts cultivation period, Dermagraft-TC is sealed in transport packages, frozen to minus 70°C, and stored ready to be shipped for use. Designed to be combined with a thin autograft [31], it is a promising product in full thickness burns.

It is well accepted that the optimal skin substitute will provide for immediate replacement of both the lost dermis and epidermis. There have been a number investigations conducted in direction to develop an optimal dermal substitute and then successful composite substitute with understanding interactions between dermal and epidermal elements, in that one enhances the maturation of the other. Unfortunately, so far no such a substitute capable to replace simultaneously and permanently both layers of the human skin does exist.

5. Coladerm membrane

Although various biodegradable synthetic polymers show great promise, there is a reason to believe that naturally derived materials provide additional benefits that also warrant their investigation as biomaterials for tissue engineering applications. Natural materials on their own possess properties that are desirable for these applications. Natural biomacromolecules, e.g. of extracellular matrix glycoproteins, serve as intrinsic templates for cell attachment and growth. Generally, natural non-specific polymers can offer a range of advantages when compared with synthetic polymer based materials. Hence our membrane model Coladerm is also based on a complex of hyaluronan and enzymatically treated collagen. The complex is crosslinked with starch dialdehyde derivatives (SDD) [32]. This hybrid membrane is mechanically stable and suitable for suture, ready-to-use for covering and healing of chronic, acute, and surgical superficial, partial or full-thickeness wound in humans.

Atelocollagen (type I) prepared from bovine Achilles tendon is crystalline, native, lyophilised, with a content of non-collagenous peptides less 0.5 wt.%, and content of inorganic substances less 0.5 wt.%. The collagen does not contain bacteriostatic agents and is relatively resistant to microorganisms. Bacterial hyaluronan (HA) produced by *Streptococcus zooepidermicus* with MW 1.5 MDa was used in the form of a sodium salt. Protein content was less than 0.1 %. The content of HA in the complex with atelocollagen is 8 wt.%. The characteristic bubble macrostructure of the membrane is obtained by controlling gel rheology. The xenogenous product is free of bacterial and viral agents, especially BSE.

In order to consider the biologic behaviour of this complex membrane, a series of in vivo experiments were performed. Biocompatibility studies were performed on Wistar rats and IPR mice. The membrane samples were implanted symmetrically on dorsal sides of anesthesized animals subcutaneously. Samples for biopsies from the implantation sites were taken under general anaesthesia on different days during one month after implantation. The histological investigations were focused on biocompatibility, tolerability, local toxicity, vascularization, and remodelation rate of the membranes.

The new hybrid membrane, designed to serve as a synthetic dermal substitute, showed good biological response in tissue environment as the implant in this study. The preliminary histological studies proved low local irritability, good biocompatibility, ingrowth of autologous tissue starting on day 7 post implantation,and resorption within 4 weeks [33].

Then the study was focused on examination of biological response of tissue to the modified collagen/hyaluronan membranes explanted on Wistar rats [34]. The concept of the explantation studies used originated from the EU criteria for implantation [35]. However, the explantation needed a different approach in the technique of surgery and a different schedule for observation of the relation explant - tissue. In order to verify the adoption of the new material by the host after explantation, it was necessary to block the shrinkage of the skin defect caused by the existence of the subcutaneous muscle in rats. Therefore, the samples of explanted material were protected by implantation of the firm protective ring fixed to the defect edges to act against the pressure of surrounding tissues. On the other hand, the use of the tricky protective rings resulted in a slower interconnection of explants with surrounding tissues. The initial interconnections were observed after 7 days. After 14 days, these interconnections were secured with the formation of granulation and connective tissue.

In spite of this, we can conclude that the new material was accepted well by the host tissue in all animals, according to both macroscopic and microscopic observations. We found that giant cells which were present, may play a role of macrophages with ability to absorb also less soluble parts of the new material and then they disappear from the tissue after a few weeks. The most important finding was that no pathological changes or no more important destruction of the host tissues were detected.

In order to evaluate better the biologic behaviour of the hybrid membrane, we performed in vitro experiments [36] using human epidermal keratinocytes, human dermal fibroblasts and human cells from buccal mucosa before the product was accepted for the clinical trial. The character of cells growth on surface of the membrane and into the membrane was visualized by the immunohistochemical examination.

The hybrid membrane has several favourable properties: adheres easily to any wound bed; is hemostatic; does not show any inflammatory reactions or immunologic rejection; serves as a natural matrix for new tissue formation, coordinates fibroblast proliferation and vascularization, enhances the adhesion of allogenic keratinocytes as producer of cytokines and growth factors. The clinical trial has been recently extended by applications for guiding tissue regeneration in treatment of paradentosis. Currently the study is focused on further modification of the hybrid membrane for serving as a local donor for growth factors, antibiotics or other bioactive compounds.

6. References

1.   Nerem, R.M., Sabamis, A. (1995) Tissue engineering: From biology to biological substitutes, *Tissue Engineering* 1, 3-13.
2.   Boyce S.T. (2001) Design principles for composition and performance of cultured skin substitutes. Burns 27, 523-533
3.   Boyce, S.T. (1996) Cultured skin substitutes: A review, *Tissue Engineering* 2, 225-266.
4.   Tompkins, G.G., Burke, J.F. (1996) Alternative wound coverings, in Herndon, D.N.(ed.), *Total Burn Care*, W.B. Saunders, Philadelphia, pp. 164-172.

5.   Robson, M.C., Barnett, R.A., Leitch, I.O.W., and Hayward, P.G. (1992) Prevention and treatment of postburn scars and contracture, *World J. Surg.* **16**, 87-96.
6.   Tompkins, G.G., Burke, J.F. (1992) Burn wound closure using permanent skin replacement materials. (Review), *World Journal of Surgery* **16**, 47-52.
7.   Nimni, M.E., Cheung, D., Strates, B., Kodama, M., and Shikh, K. (1987) Chemically modified collagen: A natural biomaterial for tissue replacement, *J. Biomed. Mater. Res.* **21**, 741-771.
8.   Hansbrough, J.F., Cooper, M.L., Cohen, R., Spielvogel, R.L., Greenleaf, G., et al. (1992) Evaluation of a biodegradable matrix containing cultured human fibroblasts as a dermal replacement beneath meshed skin grafts on athymic mice, *Surgery* **111**, 438-446.
9.   No author listed (1995) *Proceedings of a conference on the indications for use of biobrane in wound management*, Houston, Texas, September 17, 1994., *J Burn Care Rehabil.* 16, 317-42.
10.   Williams, D.F. (1992) Biofunctionality and biocompatibility, in D.F. Williams (ed.), *Medical and Dental Materials*, VCH, Weinheim, pp. 2-27
11.   Silver, F., Doillon,Ch. (1989) *Biocompatibility – Interactions of biological and implantable materials, Vol.1: Polymers*, VCH Publishers, New York, pp.199-218.
12.   Miller, M.S. (1999) Use of topical recombinant human platelet-derived growth factor-BB (becaplermin) in healing of chronic mixed arteriovenous lower extremity diabetic ulcers, *J Foot Ankle Surg.* **38**, 227-31.
13.   Grant, D.S., Rose, R.W., Kinsella, J.K., and Kibbey, M.C. (1995) Angiogenesis as a component of epithelial-mesenchymal interactions (review), *EXS* **74**, 235-248
14.   Boyce, S.T., Medrano, E.E., Abdel-Malek, Z.A., Supp, A.P., Dodick, J.M., et al. (1993) Pigmentation and inhibition of wound contraction by cultured skin substitutes with adult melanocytes after transplantation to athymic mice, *J. Invest. Dermatol.* **100**, 360-365
15.   Rouabhia, M., Germain, L., Bergeron, J., and Auger, F.A. (1995) Allogeneic-syngeneic cultured epithelia. A successful therapeutic option for skin regeneration, *Transplantation* **59**, 1229-35.
16.   Solter, D., Gearhart, J. (1999) Putting stem cells to work, *Science* **283**, 1468-1470.
17.   Tompkins, R.G., and Burke, J.F. (1996) Alternative wound coverings, in D.N. Herndon (ed.), *Total Burn Care*, W.B. Saunders, Philadelphia, PA, pp. 164-172.
18.   Greenhalgh, D.G. (1996) The role of growth factors in wound healing, *J. Trauma* **41**,159-167.
19.   Aloe, L., Tirassa P., and Bracci-Laudiero, L. (2001) NGF in neurological and non-neurological diseases: Basic findings and emerging pharmacological prospective, *Current Pharmaceutical Desing* **7**, 113-123.
20.   Eming, S.A., Snow, R.G., Yarmush, M.L., Morgan, J.R. (1996) Targeted expression of insuline-like growth factor to human keratinocytes: Modification of the autocrine control of keratinocyte proliferation, *J. Invest, Dermatol.* **107**, 113-120
21.   Rheinwald, J.G., Green, H. (1975) Serial cultivation of strains of human epidermal keratinocytes: the formation of keratinizing colonies from single cells, *Cell* **6**, 331-344.
22.   Carsin, H., Ainaud, P., Le Bever, H., Rives, J., Lakhel, A., Stephanazzi, J., Lambert, F., and Perrot, J. (2000) Cultured epithelial autografts in extensive burn coverage of severely traumatized patients: a five year single-center experience with 30 patients, *Burns* **26**, 379-87.
23.   Cuono, C.B., Langdon, R., Birchall, N., Barttelbort, S., and McGuire, J. (1987) Composite autologous-allogeneic skin replacement: Development and clinical aplication, *Plast. Reconstr. Surg.* **80**, 626-637.
24.   Andreassi L, Pianigiani E, Andreassi A, Taddeucci P, Biagioli M. (1998) A new model of epidermal culture for the surgical treatment of vitiligo, *Int J Dermatol.* **37**, 595-598.
25.   Burke, J.F., Yannas, I.V., Quinby, W.C.Jr., Bondoc, C.C., and Jung, W.K.(1981) Successful Use of a Physiologically Acceptable Artificial Skin in the Treatment of Extensive Burn Injury, *Ann. Surg.* **194**, 413-428.
26.   Yannas, I.V., Burke, J.F., Orgill, D.P., and Skrabut, E.M. (1982) Wound Tissue Can Utilize a Polymeric Template to Synthesize a Functional Extension of Skin, *Science* **215**,174-176.
27.   Boyce ST, Goretsky MJ, Greenhalgh DG, Kagan RJ, Rieman MT, Warden GT. Comparative Assessment of Cultured Skin Substitutes and Native Skin Autograft for Treatment of Full-Thickness Burns. *Ann Surg* 1995;222:743-52.
28.   Wainwright, D., Madden, M., Luterman, A., et al. (1996) Clinical evaluation of an cellular allograft dermal matrix in full-thickeness burns, *J. Burn Care Rehabil.* **17**, 124-136.

29.   Bell, E., Erlich, H.P., Buttle, D., and Nakatsuji, T. (1981) Living tissue formed in vitro and accepted as skin-equivalent tissue of full thickness, *Science* **211**,1052-1054.
30.   Falanga, V.J. (2000) Tissue engineering in wound repair, *Adv Skin Wound Care* **13**, 15-9
31.   Stoppie, P., Borghgraef, P., De Wever, B., Geysen, J., and Borgers, M. (1993) The epidermal architecture of an in vitro reconstructed human skin equivalent, *Eur J Morphol*. **31**, 26-29.
32.   Reháková, M., Bakoš, D., Vizárová, K., Soldán, M., and Juríčková, M. (1996) The study of properties of collagen and hyaluronic acid composite materials. The modification by chemical cross-linking, *J. Biomed. Mat. Res.* **30**, 369-372.
33.   Koller, J., Bakoš, D. and Sadloňová, I. (2000) Biocompatibility studies of a new biosynthetic dermal substitute based on collagen/hyaluronan conjugate, *J. Cell. Tissue Baking* **1**, 75-80
34.   Koller J., Bakoš D., Sadloňová I.(2001) Biocompatibility studies of modified collagen/hyaluronan membranes after explantation. *J. Cell. Tissue Baking*, in press.
35.   ISO 10993 – 6 (1994) Tests for local effects after implantation, *Biological evaluation of medical devices – Part 6.*
36.   Vizárová, K, Bakoš, D., Reháková, M., Petríková, M., Panáková, E., and Koller, J. (1995) The Modification of Layered Atelocollagen: Enzymatic Degradation and cytotoxicity Evaluation, *Biomaterials* **16**, 1217- 1221.

KEY-PROPERTIES AND RECENT ADVANCES IN BONE CEMENTS TECHNOLOGY

B. VÁZQUEZ, G.A. ABRAHAM, C. ELVIRA, A. GALLARDO
AND J. SAN ROMÁN

Instituto de Ciencia y Tecnología de Polímeros, CSIC
Juan de la Cierva 3, 28006 Madrid, SPAIN

1. Introduction

At the beginning of the 60's decade Sir John Charnley presented the preliminary results of a new method for bone prosthesis fixation [1, 2, 3] based on the load distribution between the bone and the implant by a filling compound called bone cement which is basically autopolymerizable PMMA. This technique has been accepted worldwide in the fixation of knee and hip prosthesis giving excellent results, but in some cases, revisions and prosthesis replacement are required with the corresponding social and economical costs. The main advantages of bone cemented prosthesis rely on the excellent primary fixation between bone and implant and, consequently, in a faster patient recuperation. It is also a low damaging and facile technique to be applied as the bone cements are easily moulded and well adapted to bone complex cavities. However, the technique presents disadvantages such as long-term prosthesis loosening as a consequence of the absence of secondary fixation of the cement. Moreover, the polymerization reaction of methyl methacrylate, MMA, is highly exothermic, and can provoke cellular necrosis in the surrounding tissues. It has also to be considered the toxicity of the aromatic tertiary amine used in these formulations to activate the initiation of the polymerization process.

69

R.L. Reis and D. Cohn (eds.),
Polymer Based Systems on Tissue Engineering, Replacement and Regeneration, 69–92.
© 2002 *Kluwer Academic Publishers. Printed in the Netherlands.*

Taking into consideration theses aspects, it has to be pointed out that the cementing technique balance is highly positive. Its effectiveness is total after approximately 7 years of implantation and it is reduced to 80% after 15 years, in comparison to the non-cemented prosthesis. These statistics clinical experiences are based on revisions carried out in hospitals worldwide in which new surgical intervention has been performed in periods of 20 years time. The results concluded that non-cemented prosthesis (Al_2O_3, metal-metal, or hydroxyapatite plasma spray prosthesis) present effectiveness much lower than cemented ones, over short and long periods of time.

Several research groups are involved in projects to avoid the bone cements disadvantages and to develop new improvements, which are presented in this chapter.

2. General Concepts

2.1. CHEMICAL COMPOSITION

Acrylic bone cements are basically poly(methyl methacrylate), PMMA, a selected material due to its biostability as well as the adaptation possibilities that pre-polymerized formulations can offer. Commercially available formulations are presented in two separated phases: a transparent liquid in a dark ampoule and a white powder in a sterilized bag [4].

2.1.1. *Liquid phase*

The liquid phase contains methyl methacrylate 97% vol-%, N,N-dimethyl-*p*-toluidine (DMT) 1.2-2.7 vol-% and 75 ppm of hydroquinone. The DMT acts as an activator of the initiation step of the polymerization reaction which is produced in periods of time in accordance to the clinical requirements. Hydroquinone acts as inhibitor of the MMA polymerization during its storage.

2.1.2. *Solid phase*

The basic components of the solid phase are PMMA in 89-90 wt-% as beads or microspheres of defined size, in some cases with addition of other polymers or copolymers. Benzoyl peroxide (BPO), is also in this phase in 0.75-2wt-% which acts as initiator of the MMA bulk polymerization. The 10 wt-% is constituted by barium sulphate or zirconium oxide, which permit periodical revisions due to its radiopaque character.

2.2. POLYMERIZATION: CURING REACTION

Physical and chemical processes constitute the curing reaction of acrylic bone cements when the solid and liquid phases are mixed. Initially, the cement is a viscous mass that in a few minutes looses fluidity, changing its consistency over the 15-20 minutes of the curing process, to attain a rigid state.

The physical processes involved in the curing reaction can be classified in four steps that are called: sandy, fibrous, paste and cured, in which the following phenomena take place: solvation of the PMMA beads and BPO by MMA, diffusion of MMA through PMMA beads, polymer-polymer diffusion, and monomer evaporation. One of the most important aspects to be considered is the monomer diffusion through the polymer beads that is influenced by the free volume of the polymer network determined by its glass transition temperature and its microstructure. On the other hand, PMMA solvation depends on the molecular weight, and the lower the molecular weight, the faster the dissolution of the PMMA beads becomes [5].

The chemical aspects of the curing process are predominantly based on the MMA bulk polymerization, which leads to a biphasic compound of PMMA particles dispersed into a polymerised MMA matrix. The MMA polymerization is a typical radical mechanism in which the generation of radicals is produced by the redox reaction between BPO and DMT that initiate the reaction, followed by the propagation and termination steps [6,7]. The initiation reaction mechanism in the presence of aromatic tertiary amines is shown in Figure 1. Experimental results have demonstrated that tertiary amines participate in the generation of free radicals by forming cyclic complexes with BPO, and also can form free radicals through their N-methyl groups as shown in Figure 1. These aminomethyl radicals can participate in the polymerization reaction and the amine can be incorporated in the chain end of the macromolecules formed during the propagation and termination steps.

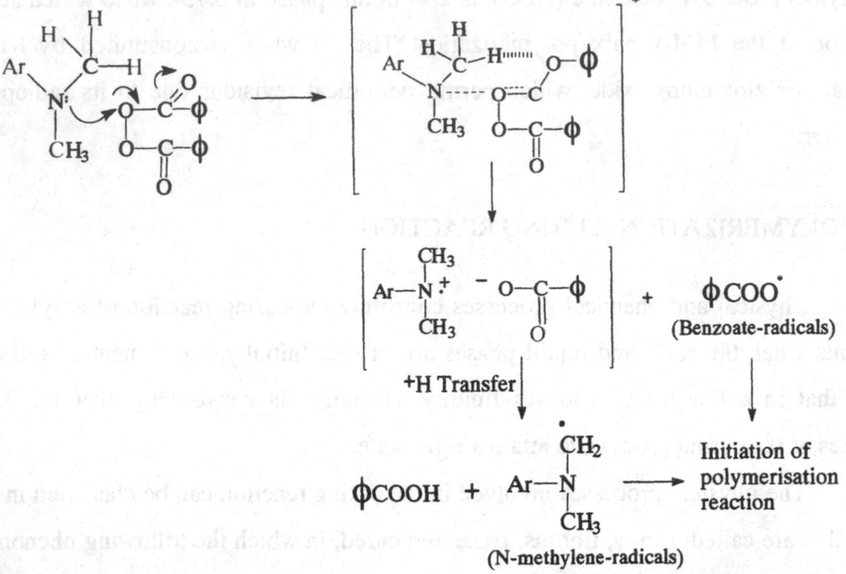

Figure 1. DMT activation mechanism in the initiation sep of the acrylic bone cements polymerization.

The MMA bulk polymerization is a highly exothermic reaction (130 cal/g by monomer unit), the temperature of the reacting mass being increased as a consequence of the gel or Trommsdorff [8] effect. This effect is characteristic of the radical polymerization of acrylates when the termination rate is dramatically decreased due to the high viscosity of the medium. Termination step can be produced by combination or disproportion by a typical hydrogen transfer in β-position.

2.3. ASPECTS INFLUENCING THE CURING PROCESS OF ACRYLIC BONE CEMENTS

The curing process of acrylic bone cements is mainly determined by the chemical composition of the cements. However, factors such as room temperature, relative humidity, solid-liquid ratio and mixing frequency, insertion time in the bone cavity and thickness of the cement, also influence the curing process. When room temperature is increased, the dough state and the setting time are decreased, increasing the maximum temperature of the reaction [9]. The most optimal solid:liquid ratio is 2:1 taking into consideration that an increase in amount of monomer leads to higher temperatures and setting time [10, 11]. A factor such as delay in introducing the cement thickness of 2-3 mm in order to dissipate the heat of the reaction, avoiding the damage of the surrounding tissues [13]. Time and phases mixing rate, as well as insertion force into the bone cavity have also influence the curing process. The average size of the PMMA beads have been found to influence the curing parameters. Decreasing curing maximum temperatures and increasing dough and setting times with respect to commercial formulations were observed when particles with relatively high average diameters (50-60 μm) were used.

3. Characterization of Acrylic Bone Cements

3.1. CURING PARAMETERS

The determination of the curing parameters of the acrylic bone cements polymerization is detailed in the ASTM F451-86 specification for medical and surgical devices and materials [14], in which the range and values of dough time, setting time and peak temperature are established. The dough time (t_{dough}) is defined as the time between the solid-liquid mixing moment and the non-adherence to a surgical glove. This time is limited by the specification to a maximum of 5 minutes. From that moment on, the cement is introduced into the bone cavity or a Teflon cavity in order to register the temperature changes. The maximum or peak temperature (T_{max}) is that registered in the exothermic maximum, the value being limited to 90°C. The setting time (t_{set}) is the time at which the temperature of the mass is the addition of the room temperature (23 ± 1°C) and the maximum temperature, divided by two. The ASTM specification establishes that the range time of the setting must be 5-15 minutes. It is also detailed the compression mechanical requirements to be 70 MPa as the minimum compressive strength in a testing machine operating at a crosshead speed of 22 mm/min.

3.2. RESIDUAL MONOMER

During the curing of the cement in the bone cavity, and as a consequence of the high viscosity of the reacting mass, the polymerization process evolves with difficulties stopping after certain time without consuming all the present MMA monomer, due to the macromolecular rigidity of the vitreous state attained during the process. In the curing process the average temperature, with exception of the peak temperature, is about 40° C which is very low in comparison to the PMMA glass transition temperature (T_g = 100-120° C), avoiding the total conversion of monomer into polymer. Therefore, a small quantity of non-reacted monomer remains trapped in the cement mass due to the rate decrease of free radicals diffusion as a consequence of the increase in viscosity of the reacting mass [8]. The determination of the residual monomer content, in the range of 1.5-5 wt-% [15] (wt), can be performed by different techniques as Nuclear Magnetic Resonance (NMR) [16, 17], Gas Chromatography (GC) [18] or High Performance Liquid Chromatography (HPLC) [19] The trapped monomer stays for long periods of time in the cement surrounded by the physiological fluid, and can be released by diffusion processes.

3.3. MECHANICAL PROPERTIES

The main function of the bone cements is to transfer stress from the prosthesis to bone by increasing the load distribution in the prosthesis. If the transferred stress if higher than the capacity of load distribution, the cement can be fractured, and the prosthesis [20, 21] can fail. The bone cement mechanical properties are very important in terms of clinical success as during the *in vivo* applications the cement has to stand compressive and shear stresses. Taking into consideration the polymeric nature of the acrylic bone cements, exhibiting viscoelastic behaviour, their mechanical properties such as tensile, compression, flexure and shear, will depend on the temperature and strain speed. When strain speed increases, both elastic modulus and strength also increase, but when the temperature raises both parameters decrease [22, 23]. In this sense, dynamic-mechanical thermal analysis (DMTA) is an excellent technique to characterize the viscoelastic behaviour of these materials.

Static mechanical properties such as elastic limit in compression, tensile, flexural and shear strengths, as well as their corresponding elastic moduli are the most important parameters to be evaluated in terms of the biomedical application. Surgical PMMA is relatively weak in terms of tensile strength but resistant in terms of compression. Compressive strength values are in the range of 64-103 MPa which are related to 50-70% strength of the cortical bone, and elastic modulus values are in the

PMMA is relatively weak in terms of tensile strength but resistant in terms of compression. Compressive strength values are in the range of 64-103 MPa which are related to 50-70% strength of the cortical bone, and elastic modulus values are in the range of 2.1-3.4 GPa, same order of magnitude of the cortical bone values (Table 1). In terms of tensile strength, which depends on the testing conditions and type of specimens, the values of tensile strength and elastic modulus are in the ranges of 22-48 MPa and 1.7-3.2 GPa, respectively [24]. On the other hand, the determination of mechanical properties such as flexure and shear, as well as the analysis of toughness to fracture and dynamic properties as fatigue and impact, gives good information about the mechanical behaviour of these materials. The most important regions in the prosthesis loosening are the bone/cement and cement/prosthesis interfaces, which are intimately related to the cemented prosthesis durability. In that sense, an initially connected interface can be loosen in a localized point due to a failure in tensile, compression or loosening of the interface components leading to non load transmission in the affected area [13].

TABLE 1. Mechanical properties of cortical bone and biomaterials applied as orthopedic implants.

Biomaterial	E (GPa)	σ_{max} (MPa)	K_{IC} (MN/m^2)	G_{IC} (J/m^2)
Cortical bone	7-30	50-150	2-12	600-500
Alumina	365	6-55	~3	~40
Al-Co alloy	230	430-1028	~100	~50000
Austenitic Steel	200	230-1160	~100	~50000
Ti-6% Al-4% Ti	106	78-1050	~80	~10000
Bone cement	2.1-3.4	40-45	1.5	~400
Polyethylene	1	30	-	8000

E = Elastic modulus, σ_{max}= Tensile strength, K_{IC}= Critical factor of stress intensity, G_{IC}= Speed of strain energy loosening

Factors affecting mechanical properties of acrylic bone cements, apart from its biphasic nature, can be found in the technique for mixing the components, the presence of corporal fluids, the thickness of the cement, the type of created interfaces, or the water absorption [13, 25]. Factors such as the pressure applied when introducing the cement into the bone cavity, increasing the cement penetration in the tissue and the mechanical resistance but decreasing the cement thickness, affect directly heat dissipation [26, 27]. The cement thickness has been found to be related to the volume shrinkage associated to the polymerization process, which can be avoided by increasing the moulding pressure and by vacuum mixing [28].

One of the most important factors that affect acrylic bone cements is the porosity of the samples. Pores can act as weak points concentrating tensions and initiating a fracture [29]. These pores are produced by the absence of air displacement in the initial powder and during the manual mixing, providing formulations with a 2-10% pores volume fraction [24]. To reduce its formation treatments as ultrasound or vacuum mixing and ultracentrifugation have been proposed [30, 31].

Finally, fluids absorption as well as aging also affect mechanical properties. Tests performed on samples extracted from patients showed an initial increase in the mechanical properties that lower slightly over the pass of years [32]. The introduction of water into the cement matrix acts as a plasticizer, increasing the mobility and allowing higher deformations, inhibiting the cracking in defective points of the material and dissipating stresses along the cement [33, 34]. The addition of radiopaque agents produces in general a deterioration of the mechanical properties.

4. Particulate Reinforcements in Acrylic Bone Cements.

4.1. RUBBER-TOUGHENED PMMA

Particulate fillers have been incorporated in the PMMA matrix with the aim of improving mechanical properties. One iniciative has been the rubber modification of bone cements as a means of increasing fracture toughness. The modification of standard bone cement by incorporating rubber-toughened PMMA powder into PMMA beads of commercial formulations was studied by Murakami et al. [35] The resulting cement presented increased elongation and fracture toughness with a reduced setting time. Vila et al. [36] incorporated tough particles of ABS (acrylonitrile-butadiene-styrene) copolymer in acrylic bone cements. A volume fraction of ABS particles of 20% provided an increase of fracture toughness from 1.39 up to 2.24 MPa m$^{1/2}$ and for 10% ABS roughened bone cement the fatigue crack propagation rate was 100 times slower than for vacuum-mixed conventional bone cement. PMMA bone cement has also been reinforced with polyisobutylene covalently linked to the PMMA matrix to improve fracture toughness [37].

4.2. HYDROXYAPATITE

Incorporation of hydroxyapatite (HA) to acrylic bone cements has been attempted to improve mechanical properties but also to enhance the biocompatibility and interlocking with bone. HA is the main mineral component of calcified tissues, it is totally biocompatible and it forms a direct bond with the surrounding bone. Olmi et al. [38] incorporated different quantities of HA in the formulation of commercial Simplex and have found that the reinforced cement presented a higher velocity of stabilization in creep experiments. They also found that low proportions of HA ($\approx$ 3%) provided a significant increase in flexural strength. Castaldini et al. [39] studied the influence of HA on the elastic modulus through dynamic and static tests finding significant differences between the pure polymer and the composite. Setting parameters were changed with the addition of HA and a significant relative minimum of the exothermic peak value at 5 wt-% HA was observed and attributed to the HA "crumbling" action on the microporosity features of the composite [40]. Other authors reported that fracture toughness increased from 1.23 to 1.55 MPa m$^{1/2}$ with increasing content of HA volume fraction from 3.6 to 17.8% in comparison to the value reported by the control (Zimmer LVC) of 1.07 MPa m$^{1/2}$ [41]. Fracture and mechanical characterization of PMMA bone cements modified with different contents of HA has been carried out more recently by Montemartini et al. [42]. The results showed that there is a limit in the HA content to enhance fracture resistence, flexural modulus and yield stress, and beyond that limit properties started to decrease because the addition of HA also affected the cement porosity. The effect of storage conditions revelead that water uptake acted as a plasticizer leading to a decrease in mechanical properties. The influence of adding sintered or not sintered HA to a bone cement formulated with methylmethacrylate-styrene copolymer beads has been reported [43]. It was found that composites prepared with sintered HA cured in shorter times and presented improved compressive strength than the corresponding cements modified with not sintered HA, but tensile strength showed a contrary behaviour.

5. Radiopaque Acrylic Bone Cements

Acrylic bone cements usually become radiopaque by the addition of an inorganic compound, the most common being barium sulfate or zirconium dioxide [44]. However, small quantities of finely divided inorganic salts are not compatible with the organic matrix, PMMA. Studies on the effect of addition of this component have revealed a substantial loss in mechanical properties. Barium sulfate reduced considerably tensile strength [45, 46] and fracture toughness [47]. Zirconium dioxide does not produce a drastic reduction in tensile strength as barium sulfate does in the same concentration, and this fact has been explained by the agglomeration of ZrO_2 particles in cauliflower-like inclusions in the bone cement whereas $BaSO_4$ forms a disseminated network. It has been demonstrated by scanning electron microscopy that there is no adhesion between the polymerized PMMA and the filler particles [48].

Alternatives to the traditional radiopaque agent is the use X-ray-opaque methacrylate, one choice being the use of iodine containing methacrylates [49, 50]. Davy *et al.* [51] have prepared polymer beads based on copolymers of methacrylic monomers containing the group triiodobenzoate and they have used them to prepare a cold-cure system with good mechanical properties. Koole *et al.* [52] developed a radiopaque cement by using polymer beads prepared from a copolymer of methyl methacrylate and 2-[4'-iodobenzoyl]-oxo-ethyl-methacrylate in a 1:1 wt/wt ratio. This novel experimental cement presented a 6-7 fold extended fatigue life and significantly better fracture toughness compared to Simplex® P due to the integration of the contrast particles in the cement matrix. The biological evaluation [53] of this novel radiopaque cement *in vitro* and *in vivo* showed that the iodine-containing cement did not differ in its biocompatibility from translucent PMMA or $BaSO_4$-containing cement.

Other authors have reported the incorporation of iodine containing methacrylates in the liquid phase of the acrylic cement. 2,5-diiodo-8-quinolyl methacrylate (IHQM) has been synthesized and used in the preparation of acrylic bone cements [54]. Peak temperature decreased and setting time increased with the content of IHQM in comparison to PMMA cements. The addition of 5 wt-% IHQM provided a statistically significant increase in the tensile strength, fracture toughness and ductility with respect to the $BaSO_4$ containing cement [55]. This effect can be attributed to the fact that the use of a radiopaque monomer eliminates the porosity associated with the barium sulfate particles which show no adhesion to the matrix. In addition, the monomer IHQM must present some reinforcing effect since the tensile and fracture toughness values reached are even higher than those shown by the radiolucent cement. Artola *et al.* have synthesized 4-iodophenyl methacrylate (IPMA) [56], 2-[2',3',5'-triiodobenzoyl] ethyl methacrylate (TIBMA) [57] and 3,5-diiodosalicylic metacrylate (DISMA) [58] and used them in the formulation of acrylic bone cements by incorpating different proportions of the corresponding radiopaque monomer in the liquid phase (Figure 2). The mechanical evaluation of the resulting cements showed that in all cases the introduction of the iodine-containing monomer produces an increase in the compressive strength and elastic modulus [59] in comparison with BaSO$_4$ containing cements. The cement prepared with 5% DISMA presented the highest value of compressive strength in dry specimens although for wet samples the opposite behaviour was observed probably due to the higher water uptake of these cements bearing carboxylic groups.

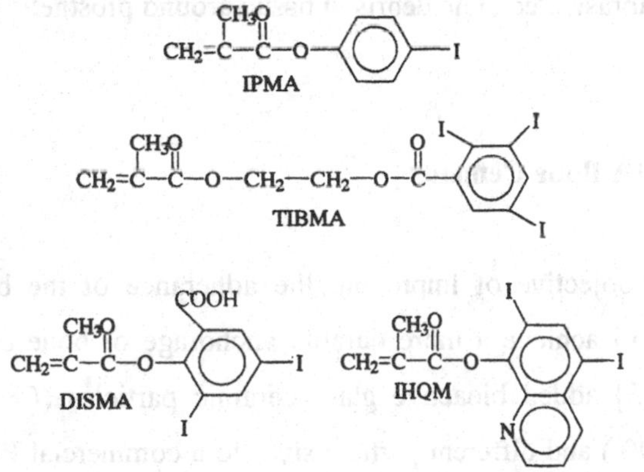

Figure 2. Radiopaque monomers used in the formulation of acrylic bone cements.

On the other hand, triphenyl bismuth (TPB) is known to render radiopacity to poly(methyl methacrylate) polymers [60] and it is noticeably insensitive to moisture which avoids the compound being leached into the aqueous environment. Triphenyl bismuth has been investigated as a potential radiopaque agent for acrylic bone cements [61]. Two different formulations of bone cements were prepared. One of them was prepared by blending TPB powder with the powder component of a commercial bone cement (CMW1) and the other one by dissolving different amounts of TPB in methyl methacrylate prior to cement formation. The latter formulation provided cements with improved mechanical properties. Ultimate tensile strength and elastic modulus increased with the content of TPB in the liquid phase. This can be due to the solubilization of TPB into the polymer matrix during polymerization of the corresponding monomer in which the compound is up to 70% soluble and therefore, the homogeneity of the matrix can be responsible for the improvement in mechanical properties.

Another route to prepare radiopaque bone cements has been reported by Abboud et al. [62]. They formulated acrylic bone cements from alumina particles previously treated by 3-(trimethoxysilyl)propylmethacrylate and embedded in poly(methylmethacrylate-co-ethylacrylate) beads with about 7 mol-% of ethyl acrylate repeating units. Compressive strength of cured cement decreased with alumina content whereas compressive modulus remained roughly constant. However, covalent bonding established between the acrylic matrix and the alumina fillers could act as a brake upon the production of abrasive ceramic debris in tissue around prosthetic joints.

6. Bioactive PMMA Bone Cements

With the objective of improving the adherance of the bone cement at the interface in order to achieve a more durable anchorage of bone cement in the tissue Henning et al. [63] added bioactive glass ceramic particles (Ceravital) in different proportions (50-70%) and different particle sizes to a commercial PMMA bone cement and tested the cements in vivo. The experiments with an implantation period of six months demonstrated a tight bonding between the newly formed osseous tissue and the glass ceramic particles at the interface. The inflammatory reaction in the vicinity of the implant was small. Heikkila et al. [64] studied the effects of PMMA on the bone formation in composites of PMMA with hydroxyapatite and bioactive glass in the rabbit subchondral femur. They found that the osteoconductive bone formation at the surface of hydroxyapatite was disturbed by PMMA and fibrous tissue was always found at the PMMA-tissue interface. Bone was formed at the interface between the implant and bone only when HA or bioglass was present. The amount of bone formed at the interface with HA was significantly lower than with bioglass particles. Recently, Shinzato *et al.* [65] have reported the preparation of new bioactive PMMA-based bone cement by incorporating glass beads, apatite- and wollastonite-containing glass-ceramic and hydroxyapatite fillers and studied their mechanical properties *in vitro* and their osteoconductivity *in vivo*. Each filler added to the cement amounted to 70 wt%. The results revealed that the higher osteoconductivity of the cements was due to the higher bioactivity of the bioactive glass beads at the cement surface and the lower solubility of the new PMMA powder to MMA monomer. In addition, it was found that the smaller spherical shape and glassy phase of the glass beads provided cements with strong enough mechanical properties to be used under load-bearing conditions.

Our group of investigation is developing bioactive PMMA bone cement by addition of phosphate glasses (PG) in the system P_2O_5-CaO-Na_2O. Addition of 20-47 wt-% PG with respect to total mass provided cements with good mechanical properties. Compressive strength and elastic modulus of any cement were superior to those of PMMA. These cements were evaluated as carriers of antibiotic controlled delivery and vancomycin was selected as the antibiotic. Release profiles were found to be influenced by the content of phosphate glasses present in the cement. The cement containing 47%PG provided the higher drug release. After an initial release of 30% of the total amount in the first half an hour, the composite continued releasing the drug at a uniform rate until about 94% release was over in a period of 50 days. The high total drug released can be attributed to the dissolution of the phosphate glasses in the simulated body fluid [66].

7. Chemically Modified Acrylic Bone Cements

New acrylic bone cements formulations have been developed with modifications in the liquid phase, solid phase or both phases by using new activators, incorporation of crosslinking agents and other monomers, and different polymers beads in the solid phase.

7.1. ACTIVATORS OF REDUCED TOXICITY AND FUNCTIONALISED ACTIVATORS

The substitution of the traditional DMT by other more biocompatible tertiary aromatic amines has been a subject of concern due to the recognised toxicity of DMT. DMT belongs to a chemical class structurally alert to DNA reactivity. It is a chromosome-damaging agent and shows a significant clastogenic effect [67, 68]. Stea *et al.* [69] verified that DMT toxic effect on cell cultures is dose related and a delay in the cell replication cycle is induced *in vitro*. A way to reduce the adverse biological effects due to the leaching of the amine is to use other amines with high molecular weight [70]. Brauer *et al.* reported the use of amine accelerators based on 4-N,N-dialkylaminophenalkanoic acids and their corresponding methyl esters in the curing of commercial bone cements [71]. More recently, the amine 4-N,N-dimethylaminophenethanol has been incorporated to a commercial low-viscosity bone cement Sulfix 60 [72]. Polymerizable tertiary amines bearing an acrylic residue susceptible of reacting with the growing macroradicals have also been investigated [73]. Tanzi *et al.* [74] have synthesized unsaturated tertiary-aryl-amine accelerators acryloyl- and methacryloyl-N-phenylpiperazine which can be chemically incorporated in the polymerizing resin, and they studied their efficiency as activators with the commercial cement CMW. The compressive strength of the resulting cements were very similar to those of DMT-cured PMMA. Trap *et al.* [75] have developed a new bone cement (Boneloc) which contains dihydroxypropyl-4-toluidine as the activator. The residual content of aromatic amines was considerably reduced with the new cement formulation and the cement cured faster and more completely than that of based on DMT/PMMA.

The substitution of DMT by 4-N,N-dimethylaminobenzyl alcohol (DMOH) and the polymerizable amine 4-N,N-dimethylaminobenzyl methacrylate (DMMO) has been attempted by Vázquez *et al.* [76] Acute toxicity of these amines was determined by intravenous injection of the corresponding saline solution of the chlorhydrates in mice [77]. LD_{50} values of DMOH and DMMO were 3.5 and 2.7 times higher than the traditional DMT under the same experimental conditions. The use of DMOH and DMMO in the self-curing of acrylic formulations provided a decrease in the peak temperature, with peak temperature values around 50°C for DMMO. Mechanical properties, molecular weight distributions and residual monomer content were comparable to commercial formulations [76].

Activators based on a long-chain naturally occuring fatty acid have been synthesized [78,79] by our group. N,N-dimethylamino-4-benzyl lautare (DML) and N,N-dimethyl-4-benzyl oleate (DMAO) have been successfully used in the curing of PMMA and the standard commerical Palacos®R cements. These tertiary amines are hydrophobic in nature and therefore less prone to leaching out from the materials. The application of DMAO is beneficial over conventional DML due to its greater molecular size and the presence of the unsaturated group which would allow it to be incorporated chemically into the final matrix. The use of fatty acid derivatives activators with the commercial cement provided a decrease in peak temperature of approximately 20°C and an increase in setting time of 7 and 14 min for DMAO and DML respectively. The curing of PMMA cements with the novel activators followed the same trend but with improved setting time values. Exotherms of polymerization are plotted in Figure 3.

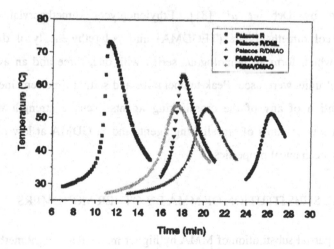

Figure 3. Time-temperature profiles of cements formulated with activators conatining fatty acid residues.

Mechanical properties in tensile test of cements prepared with DML did not appreciably change with respect to the commercial Palacos®R, however the use of DMAO provided cements with improved strain to failure and lower elastic modulus which can be attributed to the presence of a longer aliphatic chain of the fatty acid residue. Ultimate tensile strength did not significantly change for these cements.

Cytotoxic and biological evaluation of bone cements formulated with DML showed no adverse effects on the novel activator. Furthermore, osteoblast-like cells were seen to proliferate and differentiate more readily on DML containing cements [80].

7.2. CROSSLINKING AGENTS

Incorporation of crosslinking agents causes an insoluble network presenting improved properties such elastic modulus, heat distortion, shrinkage and glass transition temperature. Different crosslinking agents may have different effects on the properties depending on the inherent structure of the crosslinking molecules. One of the modifications that affect the liquid phase of acrylic cements has been the incorporation of low concentrations of a crosslinking agent. The addition of ethylene dimethacrylate or hexamethylene dimethacrylate did not appreciably change the curing parameters although compressive strength decreased slightly [71]. The influence of the chain length between the functional ends of three different dimethacrylate crosslinking agents has been studied by Deb et al. [81]. Ethyleneglycol dimethacrylate (EGDMA), triethyleneglycol dimethacrylate (TEGDMA) and poly(ethyleneglycol dimethacrylate (PEGDMA) which form an homologous series with one, three and an average of ten ethyleneglycol units were used. Peak temperature and setting time remained unchanged with the addition of any of the crosslinking agents. Tensile strength was found to increase with low amounts of crosslinking agent, and PEGDMA at low mole fractions enhanced the mechanical properties.

7.3. PARTIAL SUBSTITUTION OF MMA BY OTHER MONOMERS

The partial substitution of MMA by higher molecular weight methacrylates has been known to produce a beneficial effect on the exotherm decreasing the maximum temperature. The incorporation of dicyclopentenyloxyethyl methacrylate provided an increase in the setting time up to 9 min and a decrease of peak temperature to 62°C and the cement presented a compressive strength of 78 MPa [46]. Pascual et al. [82] have prepared new acrylic cements by the substitution of high proportions of MMA (up to 60 vol-%) by a more hydrophilic monomer, ethoxytriethyleneglycol monomethacrylate (TEGMA). The essential advantage of these formulations was the decrease of maximum temperature and residual monomer content of the cured materials compared to conventional PMMA formulations. In addition, these modified bone cements had reduced polymerization shrinkage and similar levels of porosity. Tensile test and SEM characterization revealed a noticeable increase in fracture strain.

Elvira *et al.* [83] have incorporated a methacrylic monomer derived from salicylic acid, 5-hydroxy-2-methacrylamido benzoic acid (5-HMA) with 2-hydroxyethyl methacrylate (HEMA) to the classical bone cement formulations. The incorporation of the acrylic derivative of salicylic acid provides the possibility of formation of intermolecular complexes with salts containing calcium ions as well as the potential pharmacological effect (antiinflammatory and analgesic) associated to the salicylic residue. Lower peak temperature values in the formulations containing 5-HMA were obtained. The formulation with 10% 5-HMA presented the highest value of tensile strength (62 MPa) and elastic modulus (2.92 GPa) whereas strain to failure decreased. Contact angle measurements revealed a higher hydrophilicity of the modified cements.

Another modification of the liquid phase has been the use of either methacrylic acid or diethylaminoethyl methacrylate as comonomers of MMA [84]. Residual monomer content of the final materials was not affected by the presence of either acid or alkaline monomer. However, molecular weight, curing time and glass transition temperature were composition dependent. A fast curing time, high molecular weight and high glass transition temperature were observed for formulations prepared with the acid comonomer. Selected formulations containing these functionalised methacrylates were filled with hydroxyapatite. These cements fullfilled the minimum compressive strenght (>70 MPa) required for bone cement use. However, the minimum tensile strength was only fullfilled by the cements containing methacrylic acid.

Acrylic bone cement formulations with anti-oxidant character were prepared by incorporation of a methacrylic monomer derived from vitamin E (MVE) in concentrations ranging between 10-25 wt-% [85]. Peak temperature values of the formulations ranged between 62-36°C and setting time values between 17-25 min with increasing MVE content. The strong reduction of the peak temperature is expected to contribute to the biological stabilization of the prosthesis. Compressive strength of the modified cements were superior to 70 MPa in all cases and Young's modulus showed a significant increase for the cements prepared with 15-25 wt-% MVE. The biocompatibility of these cements was studied in vitro. Cements containing 15-25 wt-% MVE provided the best results. Cell viability, proliferation and differentiation indicated that the presence of the vitamin E containing monomer contributed to cytocompatibility.

7.4. ACRYLIC BONE CEMENTS MODIFIED IN THE SOLID PHASE

Modifications in the composition of the solid phase has been carried out by Lytsky *et al.* [86]. They formulated a cement based on poly(butylmethacrylate) as the solid phase in a methyl methacrylate matrix. The resulting materials presented a glass transition temperature of 27°C. The exotherm was lower and the tensile elastic modulus was lower by a factor of eight in comparison to PMMA cements. The use of PBMA beads provides low viscosity cements and BMA/MMA copolymers are known to be used in some commercial cements such as Sulfix60.

Partially-biodegradable acrylic cements have been formulated with poly(methylmethacrylate)-poly(ε-caprolactone) (PMMA/PCL) beads which were obtained by suspension polymerization of MMA in presence of PCL [87]. Beads of PMMA/PCL of 89/11-77/23 weight ratio were mixed with MMA in a solid:liquid ratio of 1.5:1 to prepare the cements [88] (Figure 4).

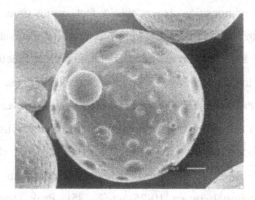

Figure 4. SEM micrograph of PMMA/PCL beads (89/11) obtained by suspension polymerization and used in the formulation of acrylic bone cements.

Approximatly 2 wt-% weight loss was observed after immersion of samples in simulated body fluid (SBF) for a period of 8 weeks. In all cases the presence of PCL provided a significant decrease in both compressive strength and elastic modulus compared to PMMA. These cements were loaded with 3 wt% vancomycin and evaluated as carriers for local release of antibiotics.The composite prepared with beads of PMMA/PCL ratio 86/14 provided the best results, allowing to release nearly the total amount of the initial drug (90%) in approximately 2 months.

7.5. ACRYLIC BONE CEMENTS MODIFIED IN BOTH PHASES

A new cement in which both liquid and solid phase have been changed was developed by Braden *et al.* [89, 90]. The new cement consisted of poly(ethylmethacrylate) (PEMA) as the polymer phase and n-butylmethacrylate as the monomer component. The exotherms of these cements decreased considerably and maximum temperatures ranged between 50-55°C. Tensile strength ranged from 20-25 MPa and elastic modulus between 0.7 to 1.0 GPa. PEMA/BMA cement also had a superior fatigue life and fracture toughness in comparison to PMMA. The biocompatibility of this cement has been studied by intramuscular implantation in rats and intraosseous implantation in dogs. A better biological response of the new cement was observed compared to PMMA [91].

New partially degradable and bioactive acrylic bone cements have been developed by Espigares *et al.* [92]. The new cements were prepared by free radical polymerization of methyl methacrylate (MMA) and acrylic acid (AA) in presence of corn starch/cellulose acetate blends (SCA). HA (sintered and non-sintered) was incorporated to confer a bone-bonding character to the cements. The best results were obtained with a solid:liquid ratio of 55:45 and a content of 20 wt-% sintered HA, giving rise to a material with a compressive strength of 98 MPa. The optimum balance of hydrophobicity was attained for a composition of 38 %MMA and 7 %AA. The heterogeneous morphology of the cured cements can be applied to the formation of a relatively porous material with a compensation of the contraction of volume after the polymerization and in addition, the induction of bioactivity by the presence of HA (Figure 5).

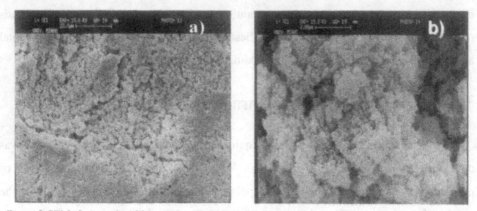

Figure 5. SEM photographs of formulation 55/45 30N (S/L 30 % non-sintered HA) bone cement formulation after 7 days of immersion in SBF solution; original magnifications: **a)** x 1000, **b)** x 10000.

8. References

1. Charnley, J. (1960) Anchorage of the femoral head prosthesis to the shaft of the femur, *J. Bone Joint Surg.* **42B**, 28-35.
2. Charnley, J., Follacci, F.M. and Hammond B.T. (1968) The long-term reaction of bone to self-curing acrylic cement, *J. Bone Joint Surg.* **50B**, 822-829.
3. Charnley, J. (1970) *Acrylic Cement in Orthopaedic Surgery*, E. & S, Livingstone, London.
4. Lautenschlager, E. P., Stupp, S. I. and Keller, J. C. (1984) Structure and properties of acrylic bone cement in P. Ducheyene and G.W. Hastings (eds.) *Functional Behavior of Orthopaedic Biomaterials. Vol II. Applications*, CRC Press, Boca Raton, Florida, pp. 87-119.
5. Elliot, J. F. (1968) *Polymer Fractionation*, Academic Press, New York.
6. Brauer, G. M., Davenport, R. M. and Hansen, W. C. (1956) Accelerating effect of amines on polymerization of methyl methacrylate, *Mod. Plastics* **34**, 153-256.
7. Pryor, W. A. (1966) *Free Radicals*, McGraw-Hill, New York.
8. Odian, G. (1991) *Principles of Polymerization*, Wiley Interscience, New York.
9. Meyer, P. R., Lautenschlager, E. P. and Moore, B. K. (1973) On the setting properties of acrylic bone cements, *J. Bone Joint Surg.* **55A**, 149-156.
10. Schoenfeld, C. M., Conard, G. J.and Lautenschlager, E. P. (1979) Monomer release from methacrylate bone cements during simulated in vivo polymerization, *J. Biomed. Mater. Res.* **13**, 135-147.
11. Turner, R. C., Atkisn P. E. Ackley, M. A. and Park, J. B. (1981) Molecular and macroscopic properties of PMMA bone cement: Free radical generation and temperature change versus mixing ratio, *J. Biomed. Mater. Res.* **15**, 425-432.
12. Bayne, S. C., Lautenschlager, E. P., Compere, C. L. and Wildes, R. (1975) Degree of polymerization of acrylic bone cement, *J. Biomed. Mater. Res.* **9**, 27-34.
13. Huikes, R. (1985) *The Bone-Implant Interface*, American Academy of Orthopaedic Surgeons.
14. ASTM Legislation, F451-86, *Medical and Surgical Materials and Devices*.
15. Willert, H. G., Frech, H. A. and Bechter A. (1975) Polymer Science and Technology, Vol. 7, Plenum Press, New York.
16. Brauer, G. M. Termini, D. J. and Dickson, G. (1977) Analysis of the ingredients and determination of the residual components of acrylic bone cements, *J. Biomed. Mater. Res.* **11**, 577-607.
17. Trap, B., Wolff, D. and Jensen, J.S. (1992) Acrylic bone cements: Residuals and extractability of methacrylate monomers and aromatic amines, *J. Appl. Biomat.* **3**, 51-57.
18. Vallo, C. I., Montemartini, P. E. and Cuadrado T.R. (1998) Effect of residual monomer content on some properties of poly(methyl methacrylate)-based bone cement, *J. Appl. Polym. Sci.* **69**, 1367-1383.
19. Davy, K. W. M. and Braden, M. (1991) Residual monomer in acrylic polymers, *Biomaterials* **12**, 540-544.
20. Noble, P. C. (1983) Selection of acrylic bone cements for use in joint replacement, *Biomaterials* **4**, 94-100.
21. Wang, C. T. and Pilliar, R. M. (1989) Fracture toughness of acrylic bone cements *J. Mater. Sci.* **24**, 3725-3738.
22. Saha, S. and Pal, S. (1984) Mechanical properties of bone cement: A review, *J. Biomed. Mater. Res.* **18**, 435-462.
23. Oysaed, H. (1990) Dynamic mechanical properties of multiphase acrylic systems, *J. Biomed. Mater. Res.* **24**, 1037-1048.
24. Kusy, R. P. (1978) Characterization of self-curing acrylic bone cements, *J. Biomed. Mater. Res.* **12**, 271-305.
25. Lewis, G. (1997) Properties of acrylic bone cements: State of the art review, *J. Biomed. Mater. Res. (Appl. Biomater.)* **38**, 155-182.
26. Brown, S. A. and Bargar, W. L. (1984) The influence of temperature and specimen size on the flexural properties of PMMA bone cement, *J. Biomed. Mater. Res.* **18**, 523-536.
27. Huiskes, R. (1980) Some fundamental aspects of human replacement. Analysis of stresses and heat conduction in bone-prosthesis structures, *Acta Orthop. Scand. Suppl.* **185**, 1-208.
28. Lidgren, L., Bodelind, B. and Moller, J. (1987) Bone cement improved by vacuum mixing and chilling, *Acta Orthop. Scand.* **58**, 27-32.
29. James, S.P., Jasty, M., Davies, J., Piehler, H. and Harris, W.H. (1992) A fractographic investigation of PMMA cement focusing on the relationship between porosity reduction and increased fatigue life, *J. Biomed. Mater. Res.* **26**, 651-662.
30. Linden, U. (1991) Mechanical properties of bone cement. Importance of the mixing technique, *Clin. Orthop.* **272**, 274-278.

90

31. Hansen, D. and Jensen, J.S. (1992) Mixing does not improve mechanical properties of all bone cements. Manual and centrifugation-vacuum mixing compared for 10 cements brands, *Acta Orthop. Scand.* **63**, 13-18.
32. Looney, M. A. and Park, J. B. (1986) Molecular and mechanical property changes during aging of bone cement in vitro and in vivo, *J. Biomed. Mater. Res.* **20**, 555-563.
33. Treharne, R. W. and Brown, N. (1975) Factors influencing the creep behavior of poly(methyl methcrylate) cements, *J. Biomed. Mater. Res. Symposium* **6**, 81-88.
34. Hailey, J.L., Turner, I.G. and Miles, A.W. (1994) An in vitro study of the effect of environment and storage time on the fracture properties of bone cement, *Clin. Mater.* **16**, 211-216.
35. Murakami, A., Behiri, J.C. and Bonfield, W. (1988) Rubber-modified bone cement, *J. Mater. Sci.* **23**, 2029-2036.
36. Vila, M.M., Behiri, J.C. and Planell, J.A. (1992) Fatigue crack propagation in acrylic bone cements, in P.J. Doherty et al. (eds), *Biomaterials-Tissue Interfaces. Advances in Biomaterials*, **10**, Elsevier, p.187.
37. Kennedy, J.P., Askew, M.J. and Richard, J.C. (1993) Polyisobutylene-toughned poly(methyl methacrylate). 3. PMMA-L-PIB networks as bone cements, *J. Biomater. Sci. Polymer Edition* **4**, 445-449.
38. Olmi, R., Moroni, A., Castaldini, A., Cavallini, A. and Romagnoli, R. (1983) Hydroxyapatites alloyed with bone cement: Physical and biological characterization, in P. Vicenzini (ed.) *Ceramics in Surgery*, Elsevier Scientific Publishing, Amsterdam.
39. Castaldini, A., Cavallini, A., Moroni, A. and Olmi, R. (1984) Young's modulus of hydroxy-apatite mixed bone cement, in P. Ducheyne, G Van der Perre and A.E. Aubert (eds) *Biomaterials and Biomechanics*, Elsevier Science Publishers, Amsterdam.
40. Castaldini, A., Cavallini, A. (1985) Setting properties of bone cement with added synthetic hydroxyapatite, *Biomaterials* **6** 55-60.
41. Perek, J. and Pilliar, R.M. (1992) Fracture toughness of composite acrylic bone cements, *J. Mater. Sci: Materials in Medicine* **3**, 333-344.
42. Montemartini, P., Cuadrado, T. and Frontini, P. (1999) Fracture evaluation of acrylic bone cement modified with hydroxyapatite: Influence of the storage conditions, *J. Mater. Sci: Materials in Medicine* **10**, 309-317.
43. Alonso, L.M., Mijares, E.M., Davidenko, N., Arevalo, J.E.P., Gastinel, C.F.J., García-Menocal, J.A.D. and Mendel, L.M. (1999) Effect of molecular weight and amount and type of hydroxyapatite in acrylic bone cement formulations, *Adv. Sci. Technol. (Materials in Clinical Applications)* **28**, 267-273,
44. Rudigier, J., Kirschner, P., Richter, I.E. and Schweikert, C.H. (1980) Influence of different x-ray contrast materials on structure and strength of bone cements, in G.W. Hastings and D.F. Williams (eds) *Mechanical Properties of Biomaterials*, Wiley & Sons, p. 289
45. Beaumont, P.W.R. (1977) The strength of acrylic bone cements and acrylic cement-stainless steel interfaces. Part 1. The strength of acrylic bone cement containing second phase dispersions, *J. Mater. Sci.* **12**, 1845-52.
46. Vázquez, B., Deb, S. and Bonfield, W. (1997) Optimisation of benzoyl peroxide concentration in an experimental bone cement based on poly(methyl methacrylate), *J. Mater. Sci. Mater. Med.* **8**, 455-460.
47. Owen, A.B. and Beaumont, P.W.R. (1980) Fracture characteristics of surgical acrylic bone cements. in G.W. Hastings and D.F. William (eds), *Mechanical properties of biomaterials*, John Wiley, New York, pp. 270-87.
48. Vila, M.M. (1992) Ph.D Thesis, Universidad Politécnica de Cataluña.
49. Jayakrishnan, A. and Chithambara Thanoo B. (1992) Synthesis and polymerization of some iodine-containing monomers for biomedical applications, *J. Appl. Polym. Sci.* **44**, 743-8.
50. Kruft, M.A.B., Benzina, A., Bar, F., van der Veen, F.H., Bastiaansen, C.W.M., Blezer, R., Lindhout, T. and Koole, L.H. (1994) Studies on two new radiopaque polymeric biomaterials, *J. Biomed. Mater. Res.* **28**, 1259-66.
51. Davy, K.W.M., Anseau, M.R., Odlyha, M. and Foster, C.M. (1997) X-ray opaque methacrylate polymers for biomedical applications, *Polym. Int.* **43**, 143-54.
52. Koole, L.H., Kruft, M.A.B., Colnot, J.M., Kuijer, R. and Bulstra, S.K. (1999) Studies on a new, all-polymeric radiopaque orthopaedic bone cement, *Society for Biomaterials, 25th Annual Meeting Transactions*, p. 316.
53. Kuijer, R., Bulstra, S.K., Kruft, M.A.B., Vermeulen, A. and Koole, L.H. (1999) Biocompatibility of iodine-containing acrylic bone cement in vitro and in vivo, *Society for Biomaterials, 25th Annual Meeting Transactions*, p. 95.
54. Vázquez, B., Ginebra, M.P., Gil, F.J., Planell, J.A., Lopez-Bravo, A. and San Román, J. (1999) Radiopaque acrylic cements prepared with a new acrylic derivative of yodo-quinoline, *Biomaterials* **20**, 2047-2053.

91

55. Ginebra, M.P., Aparicio, C., Albuixech, E., Fernandez-Barragan, E., Gil, F.J., Planell, J.A., Morejon, L., Vázquez, B. and San Román, J. (1999) Improvement of the mechanical properties of acrylic bone cements by substitution of the radio-opaque agent, *J. Mater. Sci. Mater. Med.* **10**, 733-737.
56. Artola A, Gurruchaga M, Goñi I, Studies on a new radiopaque methacrylic bone cement, 15th European Conference on Biomaterials, 1999.
57. Artola, A., Gurruchaga, M., Ginebra, P., Gil, X. and Planell, J.A. (2000) Studies on two new radiopaque methacrylic bone cements, *Sixth World Biomaterials Congress.*
58. Artola, A., Goñi, I., Vázquez, B., San Román, J. and Gurruchaga, M.D. (2000) Radiopacity and antiinflammatory activity in a new acrylic bone cement, *I Iberian Congress on Biomaterials and Biosensors. BioAvila 2000.*
59. Artola, A., Gurruchaga, M., Gil, J., Ginebra, P., Manero, J.M. and Goñi, I. (2001) Obtención de cementos óseos a partir de matrices poliméricas radiopacas, *VII Reunión del GEP.*
60. Rawls, H.R., Granier, R.J., Smid, J. and Cabasso, I (1996) Thermomechanical investigation of poly(methylmethacrylate) containing an organobismuth radiopacifying additive, *J. Biomed. Mater. Res.* **31**, 339-343.
61. Deb, S., Abdulghani, S. and Behiri, J.C. (2001) Radiopacity in bone cements using an organo-bismuth compound, *European Society for Biomaterials, 2001 Conference.*
62. Abboud, M., Casaubieilh, L, Morvan, F., Fontanille, M. and Duguet, E. (2000) PMMA-based composite materials with reactive ceramic fillers: IV. Radiopacifying particles embedded in PMMA beads for acrylic bone cements, *J. Biomed. Mater. Res. (Appl. Biomater)* **53**, 728-736.
63. Henning, W., Blencke, B.A., Bromer, H., Deutscher, K.K., Gross, A. and Ege, W. (1979) Investigations with bioactivated polymethylmethacrylates, *J. Biomed. Mater. Res.* **13**, 89-99.
64. Heikkila, J.T., Aho, A.J., Kangasniemi, I. and Yli-Urpo, A. (1996) Polymethyl methacrylate composites disturbed bone formation at the surface of bioactive glass and hydroxyapatite, *Biomaterials* **17**, 1755-1760.
65. Shinzato, S., Koyashi, M., Mousa, W.F., Kamimura, M., Neo, M., Kitamura, Y., Kokubo, T. and Nakamura, T. (2000) Bioactive polymethyl methacrylate-based bone cement: Comparison of glass beads, apatite- and wollastonite-containing glass-ceramic and hydroxyapatite fillers on mechanical and biological properties, *J. Biomed. Mater. Res.* **51**, 258-272.
66. Clement, J., Bjelkemyr, A., Martínez, S., Fernández, E., Ginebra, M.P. and Planell, J.A. (1999) Analysis of the kinetics of dissolution and the evolution of the mechanical properties of a phosphate glass stored in simulated body fluid, *Bioceramics* **12**, 375-378.
67. Taningher, M., Pasqini, R. and Bonatti, S. (1993) Genotoxicity analysis of N,N-dimethyl-p-toluidine, *Environ. Mol. Mutagen.* **21**, 349-356.
68. Bigatti, M.P., Lamberti, L., Rizzi, F.P., Cannas, M. and Allasia, G. (1994) In vitro micronucleus induction by polymethyl methacrylate bone cement in cultured human lymphocytes, *Mutat. Res.* **321**, 133-137.
69. Stea, S., Granchi, D., Zolezzi, C., Ciapetti, G., Visentin, M., Cavedagna, D. and Pizzoferrato, A. (1997) High-performance liquid chromatography assay of N,N-dimethyl-p-toluidine released from bone cements evidence for toxicity, *Biomaterials* **18**, 243-246.
70. Dulik, D.M. (1979) Evaluation of commercial and newly-synthesized amine accelerators for dental composites, *J. Dent. Res.* **58**, 1308-1316.
71. Brauer, G.M., Steinberger, D.R. and Stansbury, J.W. (1986) Dependence of curing time, peak temperature and mechanical properties on the composition of bone cement, *J. Biomed. Mater. Res.* **20**, 839-852.
72. Fritsch, E.W. (1996) Static and fatigue properties of two new low-viscosity PMMA bone cements improved by vacuum mixing, *J. Biomed. Mater. Res.* **31**, 451-456.
73. Dnebosky, J., Hynkova, V. and Hrabak, F. (1984) Polymerizable amines as promoters of cold curing resins and composites, *J. Dent. Res.* **54**, 773-776.
74. Tanzi, M.C., Sket, I., Gatti, A.M. and Monari, E. (1991) Physical characterization of acrylic bone cement cured with new accelerator systems, *Clinical Materials* **8**, 131-136.
75. Trap, B., Wolff, P. and Jensen, J.S. (1992) Acrylic bone cements: Residuals and extractability of methacrylate monomers and aromatic amines, *J. Appl. Biomater.* **3**, 51-57.
76. Vázquez, B., Elvira, C., Levenfeld, B., Pascual, B., Goñi, I., Gurruchaga, M., Ginebra, M.P., Gil, F.X., Planell, J.A., Liso, P.A., Rebuelta, M. and San Román, J. (1997) Application of tertiary amines with reduced toxicity to the curing process of acrylic bone cements, *J. Biomed. Mater. Res.* **34**, 129-136.
77. Liso, P.A., Vázquez, B., Rebuelta, M., Hernáez, M.L., Rotger, R. and San Román, J. (1997) Analysis of the leaching and toxicity of new amine activators for the curing of acrylic bone cements and composites, *Biomaterials* **18**, 15-20.
78. Vázquez, B., San Román, J., Deb, S. and Bonfield, W. (1998) Application of long chain amine activators in conventional acrylic bone cement, *J. Biomed. Mater. Res. (Appl. Biomater.)* **43**, 131-139.

92

79. Vázquez, B., Deb, S., Bonfield, W. and San Román J. (in press) Characterization of new acrylic bone cements prepared with oleic acid derivatives, *J. Biomed. Mater. Res. (Appl. Biomater.)*.

80. Deb, S., Di Silvio, L., Vázquez, B. and San Román, J. (1999) Water absorption characteristics and cytotoxic and biological evaluation of bone cements formulated with a novel activator, *J. Biomed. Mater. Res. (Appl. Biomater.)* **48**, 719-725.

81. Deb, S., Vázquez, B. and Bonfield, W. (1997) Effect of crosslinking agents on acrylic bone cements based on poly(methyl methacrylate), *J. Biomed. Mater. Res.* **37**, 465-473.

82. Pascual, B., Gurruchaga, M., Ginebra, M.P., Gil, F.J., Planell, J.A., Vázquez, B., San Román, J. and Goñi, I. (1999) Modified acrylic bone cement with high amounts of ethoxytriethyleneglycol methacrylate, *Biomaterials* **20**, 453-463.

83. Elvira, C., Vázquez, B., San Román, J., Levenfeld, B., Ginebra, M.P., Gil, X. and Planell, J.A. (1998) Acrylic bone cements incorporating polymeric active components derived from salicylic acid: curing parameters and properties, *J. Mater. Sci. Mater. Med.* **9**, 679-685.

84. Islas-Blancas, M.E., Vargas-Coronado, R. and Cauich-Rodriguez, J.V. (2001) Mechanical characterization of bone cements prepared with HA or α-TCP and functionalized methacrylates, *European Society for Biomaterials, 2001 Conference*.

85. Mendez, J.A., Aguilar, M.R., Abraham, G.A., Vázquez, B., Dalby, M., Di Silvio, L. and San Román, J. (submitted) New acrylic bone cements conjugated to vitamin E: Curing parameters, properties and biocompatibility, *J. Biomed. Mater. Res.*

86. Litsky, A.S., Rose, R.M., Rubin, C.T. and Thrasher, E.L. (1990) A reduced modulus acrylic bone cement: preliminary results, *J. Orthop. Res.* **8**, 623-626.

87. Abraham, G.A., Gallardo, A., Motta, A., Migliaresi, C. and San Román, J.(2000) Microheterogeneous polymer systems prepared by suspension polymerization of methyl methacrylate in the presence of poly(ε-caprolactone), *Macromol. Mater. Eng.* **282**, 44-50.

88. Méndez, J.A., Abraham, G.A., Fernández, M.M., Vázquez, B. and San Román, J. (in press) Self-curing formulations containing PMMA/PCL composites: Properties and antibiotic release behaviour, *J. Biomed. Mater. Res.*

89. Braden, M. (1988) Composites for use in making bone cements, *US Patent*, 4, 791, 150.

90. Weightman, B., Freeman, M.A.R., Revell, P.A., Braden, M., Albrektsson, B.E.J. and Carlson, L.V. (1987) The mechanical properties of cement and loosening of the femoral component of hip replacements, *J. Bone & Joint Surg.* **69B**, 558-564.

91. Revell, P.A., Braden, M. and Freeman, M.A.R. (1998) Review of the biological response to a novel bone cement containing poly(ethyl methacrylate) and n-butyl methacrylate, *Biomaterials* **19**, 1579-1586.

92. Espigares, I., Elvira, C., Mano, J.F., Vázquez, B., San Román, J. and Reis, R.L. (in press) New partially degradable and bioactive acrylic bone cements on starch blends and ceramic fillers, *Biomaterials*.

SOY PROTEIN-BASED SYSTEMS FOR DIFFERENT TISSUE REGENERATION APPLICATIONS

C.M. VAZ[1], L.A. de GRAAF[2], R.L. REIS[1,3],A.M. CUNHA[1]
[1] Department of Polymer Engineering, University of Minho, Campus de Azurém, 4800-058 Guimarães, Portugal
[2] ATO B.V., P.O. Box 17, 6700-AA Wageningen, the Netherlands
[3] 3B's Research Group – Biomaterials, Biodegradables and Biomimetics, University of Minho, Campus de Gualtar, 4710-057 Braga, Portugal

Abstract
With the main aim of developing soy protein drug delivery systems for different tissue regeneration applications, various processing techniques are being studied in our research group. A wide range of shapes and processing methods could already be developed. This includes membrane, microparticles and thermoplastic systems. The resulting soy membranes, microparticles, and thermoplastics are intensively characterised and their potential use as drug delivery systems, especially as pH-sensitive systems, is discussed. The use of protein modifications, namely crosslinking, for the improvement of functional properties relevant for each specific application is also addressed in this chapter.

1. Introduction

Proteins and protein-based materials find increasingly application in the emerging field of tissue regeneration, aiding to the recovery and repair of tissues and organs in the human body [1]. Tissues consist of cells embedded in an extracellular matrix composed of a network of macromolecules. Tissues like bone, cartilage or skin are mostly composed of extracellular materials, whereas cells dominate in the very soft ones, as liver or brain. The tissue regeneration process is based on the design and processing of a biodegradable scaffold or matrix, made of a synthetic or natural polymer or of a natural material found in the human body, such as collagen. Various materials can be used as matrices: poly(lactic acid) [2], poly(glycolic acid) [3], chitosan [4], starch [5], calcium alginate [3], poly(hyaluronic acid) [6] and collagen [7].

Rejection of the repair tissue by the patient's immune system is an important aspect that should be taken into consideration when choosing the right material for a particular application. A list summarising the principal demands for scaffold materials is presented in Table 1.

R.L. Reis and D. Cohn (eds.),
Polymer Based Systems on Tissue Engineering, Replacement and Regeneration, 93–110.
© 2002 Kluwer Academic Publishers. Printed in the Netherlands.

TABLE 1. Principal requirements of matrix materials (adapted from ref. [8])

High biocompatibility	Adequate tensile strength
High bioactivity	High chemical versatility
Little foreign body response	High purity
Adequate biodegradability	High availability and repeatability
Structural stability	No heath risks

1.1. NATURAL PROTEINS VERSUS SYNTHETIC POLYMERS

1.1.1. *Synthetic Biodegradable Polymers*
Both synthetic and natural polymers can be used as scaffold materials.
The selection criteria should be based on: i) versatility of the material to
be tailored in terms of composition and properties; ii) easy processability
into three-dimensional forms; and iii) nature, cytotoxicity and pH of the
degradation products.

1.1.2. *Natural Proteins*

Protein Structure
A variety of modifications can be carried out on proteins, as there are
twenty common amino acids with side chains of different sizes, shapes,
charges, and chemical reactivity (some examples in Table 2) [9]. Their
degree of hydrophobicity and hydrophilicity is one of the major
determinants of the three-dimensional structure of proteins[9,10].
Glycine, alanine, valine, leucine, isoleucine, methionine, and praline
have non-polar aliphatic side chains while phenylalanine and tryptophan
have non-polar aromatic side groups [9,10].

TABLE 2. Reactive amino acid groups in proteins (adapted from ref. [10])

$$\text{Basic Protein Structure } (- NH - CH - C = O)_n \quad \overset{R}{\underset{|}{|}}$$

R	Structure	Amount (%)
Acidic		2 – 10%
	- CO = O	
Neutral		6 – 10%
	- CH$_2$ – OH	
	- CH(CH$_3$) – OH	
	- Ph – OH	
Basic		13 – 20%
	- NH$_2$	
	- NH – C(NH-) = NH$_2^+$	
Sulphur-containing		0 – 3%
	- CH$_2$ – SH	
Amide		15 – 40%
	- C(NH$_2$) = O	

These hydrophobic amino acids are generally found in the interior of
proteins characteristic coil, forming the so-called hydrophobic core
[9,10]. Other amino acids such as, arginine, aspartic acid, glutamic acid,
cystein, histidine, lysine, and tyrosine have ionisable side chains [9,10].

Together with asparagine, glutamine, serine, and threonine, which contain non-ionic polar groups, they are frequently located on the exterior of the protein coil where they can interact strongly with the aqueous environment [9,10]. The reactivity of the protein, in terms of its ability to be chemically modified, will be mainly determined by its amino-acid composition (Table 2) and the sequential location of the individual amino acids in the three-dimensional structure of the molecule.

Proteins Modification

The high amount of reactive groups makes proteins very suitable materials for modification. Modification can be chemical (Figure 1), physical or enzymatic.

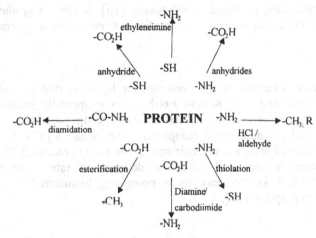

Figure 1. Some commonly used protein modifications.

For instance, hydrophobic alkyl groups can be introduced by reaction of the carboxylic groups, present in the protein, with alcohols, or by reaction of aldehydes with amine terminated residues [10,11]. Conversely, additional reactive groups can be incorporated by hydrophilization of the protein [10,11]. For example, introduction of carboxylic groups can be achieved by deamidation of the amide side group of the protein [10,11]. The amine group content can be increased by reaction of basic residues with imines [10,11]. The specificity of the reactions depends among others on the reactants used, temperature and the pH of the reaction medium. As consequence, there are several versatile routes to tailor protein properties towards the diverse requirements of specific biomedical applications.

Proteins Processing Techniques

Proteins can be processed in the presence of a high amount of water (e. g. coatings/films, adhesives, surfactants), or under low-moisture conditions (extrusion) [12]. Generally, protein based coatings/films and adhesives are produced by dissolving the protein in water at high or low pH and/or using denaturants such as urea [12,13]. Often [12] the temperature is increased to promote dissolution. In most cases the dry solids matter does not exceed 20% w/w. These solutions can be cast into coatings/films, or formulated into adhesives by means of urea and borax [14]. Protein surfactants are usually relatively small molecules (Mw << 40 kD) which tend to accumulate at interfaces, stabilising them [12]. Thermoplastic materials are obtained by means of extrusion. Dry powder is fed into an extruder with a sufficiently high amount of water (usually >> 20%) and other plasticizers such as glycerol [15]. The material is intensively mixed at temperatures around 100 °C, and molten into a dough-like texture-the thermoplastic protein [15].

1.1.3. Natural Proteins as Biomaterials

Elastin
Elastin is a rubber-like, hydrophobic protein that provides elasticity to, for example, lung or blood vessel tissues [16]. It can be synthetically produced and it has interesting properties for use in tissue regeneration [16].

Keratin
Keratin is an example of a fibrous structural protein that is present in wool, feathers and hair. Keratin has been experimentally studied as a scaffold material and it was shown to have good properties for cell adhesion and proliferation in comparison with collagen [17]. It can be readily extracted from wool or hair and can be easily processed. The high tensile strength combined with low degradation rates lead to the conclusion that keratin may be a promising biomaterial for tissue engineering purposes [17].

Gelatin
Gelatin is another example of protein used in tissue regeneration [18]. Gelatin membranes may form complexes with growth factors and thus effectively regulate their release in applications for *in vivo* tissue regeneration [18].

Collagen
In the search for biomaterials that are both versatile and biocompatible with human-tissues, considerable interest has also been paid [19,20] to collagen-based materials for the repair and replacement of soft body tissues such as tendons, skin, vascular grafts and heart valves. Collagen presents very interesting properties which make it atractive to be used as a biomaterial, including the high strength of the fibres, low extensibility, minimal antigenicity, its suitability as a substrate for cell growth, and its controllable stability by chemical or physical crosslinking [20].

An important disadvantage of collagen, gelatin, keratin and elastin is their animal origin. The recent BSE (Bovine Spongiform Encephalopathy) crises lead to an increasing concern about the use of animal proteins in biomedical devices [21]. In fact, an intense research effort is presently focused on alternative non-animal sources of proteins.

1.1.4. *Soy-An Alternative Plant-Protein*

Soy protein, the major component of the soybean (30-45%) [9] is a readily available biopolymer and an economically competitive renewable source for biodegradable polymers. Soy protein is based on amino acids of aspartic acid (aspargine) and glutamic acid (glutamine), nonpolar amino acids (glycine, alanine, valine and leucine), basic amino acids (lysine and arginine), and less than 1% of cysteine [9]. About 90-95% of the soy is storage protein, with two subunits, namely 35% conglycinin (7S) and 52% glycinin (11S), with molecular weights of ca. 20,000 and 35,000 daltons, respectively [9]. At pH = 4.5, soy protein presents a minimal charge density (isoelectric point, pI) and is highly water-resistant [9]. Among the different types of biodegradable polymers and natural proteins, soy has the following advantages: i) its non-animal origin; ii) being highly economically competitive [22]; iii) its good water resistance [22]; and iv) its good storage stability [22]. The combination of these properties with a similarity to tissue constituents [9] and a reduced susceptibility to thermal degradation (allowing for its easy processing into 3-D shapes by conventional melt based technologies) [23] makes soy an ideal template to be used as a biomaterial. Nevertheless, this potential for biomedical applications has not been really explored in the past.

This paper reports several potential applications of soy materials for tissue regeneration, namely: i) membranes; ii) microparticles; and iii) thermoplastics. It describes several research approaches that are being studied in our research group with the aim of developing adequate soy based biomaterials.

2. Materials & Methods

2.1. MATERIALS

A soy protein isolate was provided by Loders Crocklaan BV (Wormerveer, The Netherlands), with a protein content of 90-91% (dry basis) and an isoelectric point of 4.2-4.5. Glycerol (a food-grade plasticizer) and glyoxal (40% v/v) were obtained from Sigma-Aldrich Chemie BV (Zwijndrecht, The Netherlands). Two different bioactive compounds, purchased from Sigma-Aldrich Chemie BV (Zwijndrecht, The Netherlands) were selected as model drugs to be used in these studies:

i) meclofenamic acid ((2-[2,6-dichloro-3-methyl-phenyl] amino) benzoic acid) sodium salt), $C_{14}H_{10}Cl_2NO_2Na$), a non-steroid anti-inflammatory drug (Mw 318.1, melting point (T_m) 289-291°C and solubility 15mg/ml (H_2O, RT));

ii) theophylline, $C_7H_8N_4O_2$, a xanthine derivative with diuretic, cardiac stimulant and smooth muscle relaxant activities (Mw 180.2, melting point (T_m) 270-274 °C and solubility 1-5 mg/ml (H_2O, RT)).

The o-phthaldialdehyde (OPA), NaCl, NaOH and HCl were all of analytical grades.

2.2. SOY MEMBRANES

2.2.1. *Membranes Preparation*

Film-forming solutions were prepared at room temperature by slowly suspending the soy protein isolate powders (10%, w/w), under constant stirring, in distilled water with glycerol (20% w/w relative to the protein content). After adjusting the pH to 8.0 ± 0.1 with 1M sodium hydroxide, glyoxal solution was added to the referred to solutions at a level of 0 and 0.9% (w/w relative to the protein content). The membranes were cast on square Petri dishes followed by air-drying for about 24 hours at room temperature (RT) and humidity. After drying, the obtained non-crosslinked (SI_m) and crosslinked ($0.9XSI_m$) membranes, of about 100 μm thickness, were peeled off from the dishes and cut into appropriate shapes to be evaluated using different characterization techniques. All specimens were stored in an exsiccator (58% RH and RT) until further use.

2.2.2. *Membranes Characterization*

Scanning electron microscopy

The surface morphology of the air-dried membranes was analysed by scanning electron microscopy-SEM (Leica Cambridge S360).

Moisture Content

Membranes were stored for 1 week in an exsiccator maintained at 20°C and 58% RH (relative humidity). Subsequently, they were weighed in aluminium dishes and dried for 24 h in a vacuum oven at 40°C. Moisture content (MC) was as percentage of initial weight (W_0) lost during drying (W_{0d}).

$$MC = [(W_0 - W_{0d})/W_0] \times 100 \tag{1}$$

Tensile testing

The mechanical performance of the membranes was assessed by means of tensile tests. The specimens were also pre-conditioned for its moisture content for at least 1 week at 58% RH and 20°C, before testing. The experiments were performed in a universal testing machine, in a controlled chamber (20°C and 55% RH). A 200N load cell was used; a pre-load of 0.03N was applied and the samples were tested at a constant loading speed of 1 mm/min. E-modulus ($E_{0.05-0.25\%}$), maximum tensile strength (σ_{max}) and strain at break (ε_b) were computed.

Water-vapor transmission rate

The moisture permeability of the membranes was determined by measuring the water-vapor transmission rate (WVTR) across the material as stipulated by a modified ASTM standard method E96-80 [24], performed at 37°C and 58% RH.

Swelling studies

To determine the pH-dependence of the water sorption capacities of the air-dried membranes, swelling tests were performed. The specimens were immersed in an isotonic saline solution (ISS – NaCl 9g/l) at pH values of 4.2, 7.4 and 10, at 37°C. Pre-weighed dry specimens were placed in the media for 24h (equilibrium time). The following equation was used to determine the swelling degree (Q_s):

$$Q_s = W_s/W_0 \qquad (2)$$

In-vitro degradation

Soy membranes were submitted to in-vitro degradation tests. Pre-weighed dry specimens were immersed for 24 h in an isotonic saline solution (ISS - NaCl 9g/l, pH = 7.40 ± 0.02) at 37°C. After the aging period, the membranes were removed and dried in a vacuum oven (40°C/24 h). The percentage weight loss of the soy membranes was then calculated from the equation (3)

$$WL_t = [(W_0 - W_t)/W_0] \times 100 \qquad (3)$$

Anti-inflammatory release studies

Meclofenamic acid sodium salt (Mc) (20% w/w) was added to 5 g of soy film-forming solution with continuous stirring for 15 min. After adjusting the pH, the film-forming solution was crosslinked as described in section 2.2.1. The membranes containing anti-inflammatory drug (0.9XSI$_m$+20Mc) were then air-dried, as mentioned above. To study the release pattern of this anti-inflammatory, membrane specimens were immersed, at 37°C, in ISS – NaCl 9g/l with pH values of 4.2, 7.4 and 10. At predetermined time intervals, the released meclofenamic acid sodium salt was quantified by UV/Vis spectrophotometry.

2.3. SOY MICROPARTICLES

2.3.1. *Microparticles Preparation*
Spray-drying solutions were prepared as described in 2.2.1. Once the suspension was prepared, it was fed as droplets into the heated chamber of a spray-dryer. When sprayed, the droplets were rapidly dehydrated, thereby producing dry capsules (SI$_{mp}$). The capsules were then collected in the product vessel at the end of the spray dryer.

2.3.2. *Microparticles Characterization*
The surface morphology of the spray-dried microparticles was also analysed by SEM.

Anti-inflammatory release studies

Meclofenamic acid sodium salt (Mc) (15% w/w) was added to the film-forming solution as described before for the soy membranes. This resulting suspension containing anti-inflammatory drug was then spray-dried ($0.9XSI_{mp}+15Mc$), as mentioned above. To study the release pattern of this anti-inflammatory, microparticles were placed in a dialysis tube (14 kD cut-off), which was immersed, at 37°C, in ISS – NaCl 9g/l at pH 7.4. The quantification of released anti-inflammatory was assessed as described above for protein membranes (section 2.2.2.).

2.4. SOY THERMOPLASTICS

2.4.1. *Extrusion and Conventional Injection Moulding*
Native soy protein was converted into a thermoplastic material in a co-rotating twin-screw extruder (Berstorff ZE 25(CL) x 40D) in the presence of water and glycerol. The extrusion was carried out at temperatures ranging from 70 to 90°C, a screw speed of 200 rpm and pH 7. The soy protein was also crosslinked with different amounts of glyoxal (0, 0.3 and 0.6% w/w relative to the protein amount) (SI_{tp}, $0.3XSI_{tp}$ and $0.6XSI_{tp}$, respectively).

The extruded thermoplastic materials (in the form of pellets) were moulded into ASTM dumb-bell tensile test bars (2 x 4 mm^2 of cross section) after being conditioned (60°C and 24 hours) until the respective moisture content reached 12 to 14%.

These specimens were moulded using a DEMAG D25 NC IV machine under optimised processing conditions, namely temperatures ranging from 120 to 140°C. Half of the injection moulded specimens were submitted to a thermal treatment at 80°C during 24 hours in an air-circulating oven ($24TTSI_{tp}$). Subsequently, the moulded samples were conditioned at 25°C and 60% RH for at least 1 week before testing.

2.4.2. *Thermoplastics characterization*

Tensile testing and moisture content
The mechanical performance of the produced soy thermoplastics was also assessed by means of tensile experiments. The test speed was always 1 mm/min and a static load cell of 5kN was used. The E-modulus at 0.05-0.25% strain ($E_{0.05-0.25\%}$), the maximum tensile strength (σ_{max}) and the strain at break (ε_r) were evaluated. The tensile tests were performed in a controlled environment (20°C and 55% RH) equivalent to the atmosphere used for conditioning the test specimens.

After testing, specimens were ground using liquid N_2 and weighed into aluminium dishes for subsequent dried for 24 h in a vacuum oven at 40°C. Moisture content (MC) was determined by equation (1).

Free amino group measurement
The free amino group content of the protein samples was determined using the o-phthaldialdehyde (OPA) method [25].

Solubility studies

The protein solubility at the different pHs was determined by Kjeldahl analysis [26] and calculated according to the method applied in [27].

In-vitro degradation

Soy thermoplastics were submitted to in-vitro degradation tests. Pre-weighed dry specimens were immersed up to 14 days in ISS - NaCl 9g/l, at pH = 7.40 ± 0.02 and 37°C. The weight loss of the thermoplastics was evaluated with the method described in section 2.2.2.

3. Results & Discussion

3.1. SOY MEMBRANES

3.1.1. *Morphology*
The membranes obtained from the film-forming solutions of soy protein isolate were flexible, transparent and homogeneous, indicating the molecular level compatibility of the components [28,29].

3.1.2. *Mechanical Properties*
As expected, the crosslinking of the material molecular structure favoured the mechanical strength and reduced the ductility (Table 3). The maximum tensile strength (σ_{max}) increased from 10.5 MPa for SI to 12.6 MPa for $0.9XSI_m$. The strain at break (ε_b) significantly decreased for the $0.9XSI_m$ (24.1%) when compared with the SI_m (41.8%). The E-modulus ($E_{0.05-0.25\%}$) increased upon crosslinking from 299 MPa for the SI_m membrane to 421 MPa for the $0.9XSI_m$.

TABLE 3. Mechanical properties of soy membranes.

Membrane	σ_{max} (MPa)	ε_b (%)	$E_{0.05-0.25\%}$ (MPa)	MC (%)
SI_m	10.5 ± 0.5	41.8 ± 11.9	299 ± 13	8.91 ± 0.10
$0.9XSI_m$	12.6 ± 0.6	24.1 ± 14.9	421 ± 47	7.58 ± 0.07

SI_m – soy protein membrane; $0.9XSI_m$ – crosslinked soy protein membrane

These results confirm the occurrence of crosslinking action of glyoxal on the soy structure [28,29]. Increases in σ_{max} have previously been reported for soy protein isolate films reacted with dialdehyde starch [30].
It is generally assumed [31] that glyoxal crosslinking of proteins results of the reaction of the aldehyde groups of the glyoxal with the free ε-amine groups of lysine or hydroxylysine residues of the protein.

3.1.3. *Water-vapour transmission rate*
No substantial differences were detected between the two membrane materials tested (Table 4). The slight decrease of the WVP of the crosslinked membrane can be attributed to the presence of covalent links that reduced the intermolecular free volume.
In general, the WVP of the membranes (~100 g/m² per day) is smaller than that present by a normal human skin (WVP ~204 g/m² per day) [32]. However, the developed membranes were non-porous, conversely to the normal human skin. So, an eventual production of porous soy membranes should lead to a much higher WVP, which can match that of the human skin.

TABLE 4. Physico-chemical properties of soy membranes.

Membrane	WVP	WL$_{24}$	Q$_s$		
	(g/m^2 per day)	(%)	pH 4.2	pH 7.4	pH 10
SI$_m$	96.96 ± 0.01	21.85 ± 1.53	2.11	3.22	3.95
0.9XSI$_m$	89.04 ± 0.49	19.37 ± 0.43	1.56	2.38	3.10

SI$_m$ – soy protein membrane; 0.9XSI$_m$ – crosslinked soy protein membrane

3.1.4. *Swelling studies*
The membranes showed a substantially higher pH-dependent pattern (Table 4). The swelling of the 0.9XSI$_m$ was significantly lower than that of the SI$_m$ ones over the entire range of pH studied. At alkaline pH, soy membranes presented high swelling degrees due to increased number of negatively charged residues (aspartic acid and glutamic acid) [9]. This net negative charge induces chain relaxation, leading to faster hydrogen-bond dissociation, efficient solvent diffusion and consequent swelling. At pH 4.2 (pI of soy), the membranes presented the minimum-swelling rate. At pI, due to the net neutral charge of the protein [9], the swelling was mainly driven by solvent diffusion, with a minimum of hydrogen-bond dissociation and consequent minimum swelling degrees. In neutral media, the swelling was moderate due to a less net negative charge of the membranes.

3.1.5. *In vitro degradation*
Chain relaxation and consequent solvent diffusion were the main mechanisms responsible for the swelling, hydrolysis and consequent weight loss of the membranes. 0.9XSI$_m$ presented a lower capability to allow chain relaxation due to the crosslinked structure. For this reason, after 24 h of immersion in an ISS at pH 7.4, 0.9XSI$_m$ presented a lower weight loss than the SI$_m$ (Table 4).

3.1.6. *pH-sensitive membranes*
All the membranes presented a noticeable burst effect during the first 12 h (Figure 2), which can be very useful on the initial treatment of an inflammatory process. This burst was controlled by solvent diffusion and resulted in a first-order release of the loaded drug. During this stage, the SI$_m$+20Mc showed cumulative release levels of the entrapped anti-inflammatory drug of ~9% (pH 4.2), ~60% (pH 7.4) and ~90% (pH 10) (Figure 2). After the initial burst of 12 h, the SI$_m$+20Mc tested at pH 7.4 showed a decrease in the release rate of the anti-inflammatory drug (Figure 2). The ~100% release level was only achieved after 240 h of immersion (Figure 2). This type of profile revealed that the process is not solvent diffusion driven, but is mainly related to the degradation of the membrane, which induced the release of the remaining ~40% of entrapped anti-inflammatory drug, following a zero-order profile. Similar release profiles have been reported for other degradable polymeric systems [33-35]. The SI$_m$+20Mc tested at pH 10 (Figure 2) presented a continuous burst effect until 49 h of immersion (Figure 2), when the cumulative release level reached ~100%. By the contrary, the SI$_m$+20Mc tested at pH 4.5 were characterized by a stabilisation of the cumulative release levels after 12 hours of immersion.

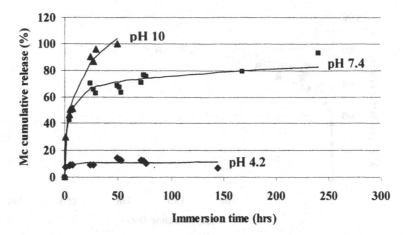

Figure 2. Meclofenamic acid sodium salt cumulative release profiles from SI$_m$.

An inflammation site is characterized by a pH of around 5 [36]. The developed membranes present an initial burst release of about 10% under these conditions[29]. As the inflammation starts to be treated, the pH of the site rises until the normal pH of the human body (pH 7.4) [36]. Consequently, the cumulative release level also tends to increase until about 60% (SI$_m$+20Mc) [29]. After a certain period of time, due to the stabilization of the pH, a decrease in the release rate of the anti-inflammatory drug will be observed (following a zero-order profile) [29]. The ~100% release level will be achieved after some days, indicating the possible end of the inflammatory process [29].

3.2. SOY MICROPARTICLES

3.2.1. *Morphology*

The microparticles obtained by spray-drying present a diameter between 10-100 μm. They tend to have an irregular geometry due to the rapid dehydration and, in some occasions, tend to suffer aggregation.

3.2.2. *Drug release studies*

Like the SI$_m$+20Mc, SI$_{mp}$+15MC also present an observable burst effect during the first 12 h (Figure 3). This burst was controlled by solvent diffusion and resulted in a first-order release of the loaded drug. During this stage, the SI$_{mp}$+15MC showed cumulative release levels of the entrapped anti-inflammatory drug ~50% (pH 7.4), which is slower than the ~60% release presented by the SI$_m$+20Mc tested at the same pH. After the initial burst of 12 h, the SI$_{mp}$+15MC presented a decrease in the release rate of the anti-inflammatory drug (Figure 3). The ~80% release level was achieved after 150 h of immersion (Figure 3). In general, the release rate tended to be slower than in the case of the SI$_m$+20Mc. This feature proves the high efficiency of the spray drying technique on the encapsulation of the anti-inflammatory drug within the microparticles matrix.

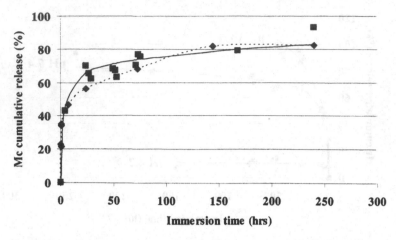

Figure 3. Meclofenamic acid sodium salt cumulative release profiles from soy protein microparticles-SI_{mp} (.....) and soy protein membranes-SI_m (___) tested at pH 7.4.

3.3. SOY THERMOPLASTICS

3.3.1. *Tensile Properties*

The types of interactions between polypeptide chains, the spatial distribution of the crosslinks (intra and intermolecular) and the chemical nature of crosslinking are very important parameters that determine the tensile properties of the studied soy thermoplastics [27,37,38].

Glyoxal Crosslinking

Crosslinking of the thermoplastics with glyoxal resulted in materials with decreased stiffness and ductility (Table 5). This should be associated to the degradation of the crosslinked matrix which can not withstand the thermo-mechanical loads of the moulding cycle. This degradation phenomena is the result of the processing of an already crosslinked structure. These crosslinking has a predominantly intra-molecular nature. The amine groups located on two adjacent chains are apparently too far apart to be bridged by a crosslink. So, the covalent bonds are predominantly introduced within the polypeptide chains and not between them [27]. As a result, the materials remain melt-processable but very sensitive to thermo-degradation. Another evidence of these phenomena is the higher mass loss observed for glyoxal crosslinked thermoplastics during all immersion time, when compared with the non-crosslinked ones (Table 6).

TABLE 5. Mechanical properties of soy thermoplastics

Material	σ_{max} (MPa)	ε_b (%)	E (MPa)	MC (%)
SI_{tp}	22.2 ± 2.3	1.8 ± 0.3	1436 ± 56	5.2 ± 0.1
$0.3XSI_{tp}$	20.7 ± 1.4	1.9 ± 0.2	1241 ± 83	5.4 ± 0.1
$0.6XSI_{tp}$	15.5 ± 2.1	1.3 ± 0.2	1229 ± 38	5.5 ± 0.1
$24TTSI_{tp}$	30.2 ± 3.7	1.1 ± 0.2	2698 ± 269	5.2 ± 0.1

SI_{tp}-soy protein thermoplastics; 0.3-$0.6XSI_{tp}$-crosslinked soy protein thermoplastics; $24TTSI_{tp}$-thermally treated soy protein thermoplastics.

Crosslinking by Heat Treatment

The thermal treatment resulted in the production of highly stiff and brittle soy thermoplastics (Table 5). In this case, the crosslinking of the polymeric matrix occurs after the processing cycle, avoiding the potential degradation of the material. This effect was mainly attributed to the formation of disulfide linkages, hydrogen bonds and also crosslinks (through reaction with the amino groups of soy) under the energetic activation of the heat treatment (Table 6) [14,23]. Conversely to glyoxal crosslinking, heat also introduces linkages between the polypeptide chains (inter-molecular crosslinking). As a result, thermal treated specimens are more resistant to hydrolysis than non-treated thermoplastics, as shown by the mass loss profiles (Table 6).

TABLE 6. Crosslinking degree and mass loss of soy thermoplastics

Material	Free –NH₂ after extrusion (%)	Free –NH₂ after injection moulding (%)	WL$_{14days}$ (%)
SI$_{tp}$	100	86.3	17.53 ± 0.55
0.3XSI$_{tp}$	71.4	58.6	20.86 ± 2.03
0.6XSI$_{tp}$	55.3	49.2	22.03 ± 0.43
24TTSI$_{tp}$	71.6	63.8	11.16 ± 1.07

SI$_{tp}$-soy protein thermoplastics; 0.3-0.6XSI$_{tp}$-crosslinked soy protein thermoplastics; 24TTSI$_{tp}$-thermally treated soy protein thermoplastics.

3.3.2. Solubility

A reduction in the protein solubility with the pH level was observed. The minimum solubility was found between pH 4 and 5 and a subsequent resolubilization of the protein at pHs lower than pH 4. Higher protein solubility (greater than 25%) was observed at pH values greater than 8 as compared to the acidic pH values at which the protein solubility's were 20% (Figure 4). The solubility of these materials is known to vary considerably with pH [27], as referred before in 3.1.4.

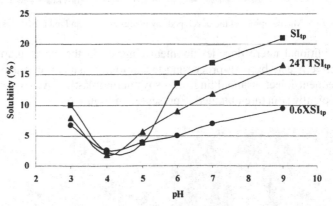

Figure 4. Solubility of soy thermoplastics as a function of pH. SI$_{tp}$-soy protein thermoplastics; 0.6XSI$_{tp}$-crosslinked soy protein thermoplastics; 24TTSI$_{tp}$-thermally treated soy protein thermoplastics.

Glyoxal Crosslinking

Treatment of soy with glyoxal resulted in materials with a decreased solubility and a decrease free amino group content (see section 3.2.) indicating that crosslinking was achieved (Figure 4).

Crosslinking by Heat Treatment

The prolonged heat treatment decreased solubility, less than 15% being soluble after the treatment. This is probably due to the formation of

106

crosslinks through the free amino groups of soy (see section 3.2.) [11,24], to the formation of disulfide bonds and hydrogen bonds [11,24] (Figure 4).

This dependence of solubility on the pH of the immersion solutions and in the degree of crosslinking makes soy thermoplastics an ideal biodegradable polymer that can be tailored in order to be use in pH-triggered biomedical devices, as reported before [29].

4. New Developments

Several developments of different soy structures (membranes, microparticles and thermoplastics) have been made aiming at the production of drug delivery carriers to be used in several tissue regeneration applications (Figure 5). The most promising applications are wound dressings, barrier membranes, oral, nasal and subcutaneous delivery devices.

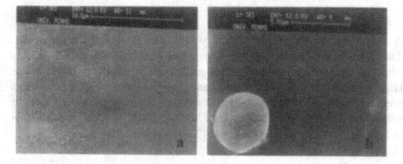

Figure 5. Micrographies of the developed soy membranes (a) and microparticles (b).

An additional interesting development consists in the encapsulation of bioactive agents by melt-processing technologies, namely extrusion and consequent injection moulding, into soy thermoplastics. An example of the typically structure obtained is presented in Figure 6.

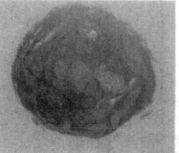

Figure 6. Micrographies of an example of: a) conventionally injection moulded soy carrier; and b) soy bimodal drug delivery device. Core: $SI_{tp}+20Th$; Skin: SI_{tp}. $SI_{tp}+20Th$-soy protein thermoplastic with 20% of encapsulated theophylline (Th); SI_{tp}-soy protein thermoplastic.

Particular attention has been given to the design and processing of bimodal drug delivery carriers. These devices are originally produced by bi-material injection moulding and consist of a core material with an encapsulated active substance and a skin material, which works as a drug diffusion barrier (Figure 6). All the correspondent drug release profiles are under study, as well as their cytocompatibility.

5. References

1. Hubbell, J.A. (1995) Biomaterials in tissue engineering, *Biotechnology* **13**, 565-576.
2. Adriano, K.P., Pohjonen, T., and Tomalla, P. (1994) Processing and characterisation of absorbable polylactide polymers for use in surgical implants, *J. Appl. Biomater.* **5**, 133-140.
3. Cal, Y.L., Rodriguez, A., Vacanti, M., Ibarra, C., Arevalo, C., and Vacanti C.A. (1998) Comparative studies of the use of PGA, calcium alginate and pluronics in the engineering of autologous porcine cartilage, *J. Biomed. Sci.: Polym. Ed.* **9(5)**, 475-487.
4. Tanabe, T., Okitsu, N., Tachibana, A., and Yamauchi, K. (2002) Preparation and characterization of keratin-chitosan composite film, *Biomater.* **23(3)**, 817-825.
5. Mendes, S.C., Reis, R.L., Bovell, Y.P.,Cunha, A.M., van Blitterswijk, C.A., and de Bruijn, J.D. (2001) Biocompatibility testing of novel starch-based materials with potential application in orthopaedic surgery: a preliminary study, *Biomater.* **22(14)**, 2057-2064.
6. Frank, P., and Gendler, E. (2001) Hyalurnic acid for soft-tissue augmentation, *Clin. In Plast. Surg.* **28(1)**, 121-126.
7. Friess, W. (1998) Collagen – biomaterial for drug delivery, *Eur. J. Pharm. And Biopharm.* **45**, 113-136.
8. Hutmacher, D.W. (2001) Scaffold design and fabrication technologies for engineering tissues-state of the art and future perspectives, *J. Biomater. Sci.: Polym. Ed.* **12(1)**, 107-124.
9. Cheftel, J.C., Cuq, J.-L., and Lorient, D. (1985), in O.R Fennema (ed.), *Amino-acids, peptides and proteins*, Marcel Dekker Inc., New York, pp. 245-369.
10. de Graaf, L.A., and Kolster, P. (1998) Industrial proteins as a green alternative for "petro" polymers: potentials and limitations, *Macromol. Symp.* **127**, 51-58.
11. Kolster, P., de Graaf, L.A., and Vereijken, J.M. (1997) in G.M. Campbell (ed.), *Cereals: novel uses and processes*, Plenum Press, New York, pp. 107-116.
12. de Graaf, L.A. (2000) Denaturation of proteins from a non-food perspective, *J. Biotechn.* **79**, 299-306.
13. Gennadios, A., Hanna, M.A., and Kurth, L.B. (1997) Application of edible coatings on meats, poultry and seafoods: A review, *Food Sci. Techn.* **30(4)**, 337-350.
14. Skeist, I. (1990) *Handbook of adhesives*, Von Nostrand Reinhold, New York.
15. Camine, M.E. (1991) Protein functionality modification by extrusion cooking, *J. Amer. Oil Chem. Soc.* **68(3)**, 200-205.

108

16. Winlove, C.P., Parker, K.H., Avery, N.C., and Bailey, A.J. (1996) Interaction of elastin and aorta with sugars and their effects on biochemical and physical properties, *Diabet.* **39(10)**, 1131-1139.

17. Yanauchi, K., Maniwa, M., and Mori, T. (1998) Cultivation of fibroblast cells on keratin-coated substract, *J. Biomater. Sci.: Polym. Ed.* **9(3)**, 259-270.

18. Ulubayram, K., Cakar, A.N., Korkusuz, P., Ertan, C., and Hasirci, N. (2001) EGF containing gelatin-base wound dressings, *Biomater.* **22(11)**, 1345-1356.

19. Schoen, F.J. (1987) Cardiac valve prostheses: Review of clinical status and contemporary biomaterials issues, *J. Biomed. Mater. Res.: Appl. Biomater.* **21(A1)**, 91-117.

20. Olde Damink, L.H.H., Dijkstra, P.J., van Luyn, M.J.A., van Wachem, P.B., Nieuwenhuis, P., and Feijen, J. (1995) Glutaraldehyde as a crosslinking agent for collagen-based materials, *J. Mater. Sci.: Mater. Med.* **6**, 460-472.

21. van Dijk, A. (2001) Tissue Engineering, *Ind. Prot.* **9(1)**, 9-11.

22. Seal, R. (1980) Industrial soya protein technology in R. A. Grant (ed.), *Applied protein chemistry*, Applied Science Publishers Ltd, London, pp. 87-112.

23. Sheard P.R., Mitchell J.R., and Ledward D.A. (1986) Extrusion behaviour of different soya isolates and the effect of particle size *J. Food Tech.* **21**, 627-641.

24. ASTM (1980) *Annual book of ASTM standards*, American Society for Testing Materials, West Conhohocken, Vol. 4.06.

25. Bertrand-Harb, C., Nicolas, M.-G., Dalgalarrondo, M., and Chobert, J.-M. (1993) Determination of alkylation degree by three colorimetric methods na amino acid analysis. A comparative study, *Sc. Alim.* **13**, 577-584.

26. Lynch, J.M., and Barbano, D.M. (1999) Kjeldahl nitrogen analysis as a reference method for protein determination in dairy products, *J. Assoc. Anal. Comm.* **82(6)** 1389-1398.

27. Vaz, C.M., van Doeveren, P.F.N.M., Yilmaz, G., de Graaf, L.A., Reis, R.L., and Cunha, A.M. (2002) Processing and characterization of biodegradable soy thermoplastics: Effects of crosslinking with glyoxal and thermal treatment, *J. Appl. Polym. Sci.*, submitted.

28. Vaz. C.M., de Graaf, L.A., Reis, R.L., and Cunha, A.M. (2002) Effect of crosslinking, thermal treatment and UV irradiation on the mechanical properties and in vitro degradation behaviour of several natural proteins aimed to be used in the biomedical field, *J. Mater. Sci.: Mater. Med.*, submitted.

29. Vaz. C.M., de Graaf, L.A., Reis, R.L., and Cunha, A.M. (2002) pH-sensitive soy protein hydrogels for the controlled release of an anti-inflammatory drug, *J. Biomater. Sci.: Polym. Ed.*, submitted.

30. Rhim, J.-W., Gennadios, A., Weller, C.L., Cezeirat, C., and Hanna, M.A. (1998) Soy protein isolate dialdehyde starch films, *Ind. Crops Prod.* **8**, 195-201.

31. Wong, S. S. (1991) *Protein conjugation and crosslinking*, CRC Press, Boca Raton.

32. Mi, F.-L., Shyu, S.-S., Wu, Y.-B., Lee, S.-T., Shyong, J.-Y., and Huang, R.-N. (2001) Fabrication and characterization of a sponge-like

assymetric chitosan membrane as a wound dressing, *Biomater.* **22**, 165-173.

33. Eliaz, R.E., and Kost, J. (1991) Characterization of polymeric PLGA-injectable implant delivery system for the controlled release of proteins, *J. Biomed. Mater. Res.* **50**, 388-392.

34. Hennink, W.E., Franssen, O., van Dijk Wolthuis, W.N.E., and Talsma H. (1997) Dextran hydrogels for the controlled release of proteins, *J. Control. Rel.* **48**, 107-114.

35. Franssen, O., van der Vennet, L., Roder, P., and Hennink, W.E. (1999) Degradable dextran hydrogels: controlled release of a model protein from cylinders and microspheres, *J. Control. Rel.* **60(2-3)**, 211-221.

36. Friend, D.R. (1998) Review article: issues in oral administration of locally acting glucocorticosteroids for treatment of inflammatory bowel disease, *Alim. Pharmc. Therap.* **12(7)**, 591-603.

37. Vaz, C.M., Fossen, M., van Tuil, R.F., de Graaf, L.A., Reis, R.L., and Cunha, A.M. (2002) Casein and soybean protein-based thermoplastics and composites as alternative biodegradable polymers for biomedical applications, *J.Biomed. Mater. Res.*, accepted.

38. Vaz, C.M., Mano, J.F., Fossen, M., van Tuil, R.F., de Graaf, L.A., Reis, R.L., and Cunha, A.M. (2002) Mechanical, dynamic-mechanical and thermal properties of soy protein-based thermoplastics with potential biomedical applications, *J. Macromol. Sci.: Phys.*, in press.

6. List of Abbreviations

SI_m – soy membrane.

$0.9XSI_m$ – soy membrane crosslinked with 0.9% glyoxal.

$0.9XSI_m + 20Mc$ – soy membrane crosslinked with 0.9% glyoxal and loaded with 20% meclofenamic acid sodium salt (Mc).

SI_{mp} – soy microparticles.

$0.9XSI_{mp}$ – soy microparticles crosslinked with 0.9% glyoxal.

$0.9XSI_{mp} + 15Mc$ – soy microparticles crosslinked with 0.9% glyoxal and loaded with 15% meclofenamic acid sodium salt (Mc).

SI_{tp} – soy thermoplastic.

$0.6XSI_{tp}$ – soy thermoplastic crosslinked with 0.6% glyoxal.

$0.9XSI_{tp}$ – soy thermoplastic crosslinked with 0.9% glyoxal.

$24TTSI_{tp}$ – soy thermoplastics heat treated for 24 hrs at 80 °C.

$SI_{tp} + 20Th$ – soy thermoplastic loaded with 20% theophylline (Th).

RH – relative humidity.

RT – room temperature.

MC – moisture content.

W_0 – initial weight.

W_{0d} – dried weight.

W_e – equilibrium weight.

WL_t – weight loss.

W_t – weight at time t.

Q_s – swelling degree.

Mc – meclofenamic acid sodium salt.

Th – theophylline.

WVP – water vapor permeability.

WVTR – water vapor transmission rate.

110

ISS – isotonic saline solution.
pI – isoelectric point.
$E_{0.05-0.25\%}$ - E-modulus at 0.05-0.25% strain.
σ_{max} – maximum stress.
ε_b – strain at break.
SEM – scanning electron microscopy.
OPA – o-phthaldialdehyde

Characterization of Polymeric Systems and Scaffolds

IN VITRO TESTING OF POLYMERIC SCAFFOLDS

C. MAULI AGRAWAL, PH.D., P.E.
Center for Clinical Bioengineering and Department of Orthopaedics
The University of Texas Health Science Center at San Antonio
San Antonio, TX 28229

1. Introduction

Tissue engineering may be defined as the science of persuading the body to heal or repair tissues that do not do so spontaneously. This is a relative new and exciting field that has experienced tremendous growth in the past decade. In our current understanding, which has developed over the past 10-15 years of research, the most common strategy for tissue engineering is to fill the defect in the tissue with a sponge-like highly porous material. This sponge or scaffold may carry cells and/or biomolecular entities such as growth factors. Needless to say, the scaffold is then critical to the tissue repair or regeneration process because it provides a three-dimensional framework for cells to attach and produce extracellular matrix to form tissue. As the scaffold provides the basic foundation for this tissue engineering approach, its various properties need to be carefully designed and optimized. In an earlier paper we have outlined the essential properties of a scaffold [1]. Briefly, the scaffold should:

1) be biocompatible;
2) biodegradable or can be remodeled;
3) biodegrade in synchronization with the repair or regeneration process;
4) be highly porous;
5) be highly permeable to facilitate diffusion;
6) have the correct pore size range;
7) possess adequate mechanical properties;
8) provide a surface conducive for cell attachment;
9) aid the formation of ECM by promoting cellular functions; and
10) have the ability to carry drugs/growth factors.

Several of these parameters are addressed below in more detail.

2. Biocompatibility

The issue of biocompatibility is complex and difficult to resolve through *in vitro* studies. Basic biocompatibility of the scaffold material can be assessed through laboratory tests such as those specified by standards organizations such as the American Society for

R.L. Reis and D. Cohn (eds.),
Polymer Based Systems on Tissue Engineering, Replacement and Regeneration, 113–123.
© 2002 *Kluwer Academic Publishers. Printed in the Netherlands.*

Testing and Materials (ASTM). However, even the so called "biocompatible" materials may elicit an adverse response *in vivo* if used under certain conditions or indiscriminately. For example, if a large amount of polyglycolic acid (PGA) is implanted in a highly avascular site it may elicit an adverse biologic response. This may be due to the production of large amounts of acidic byproducts of hydrolytic degradation, which the body is unable to clear in an expedient fashion. Thus, biocompatibility may depend not just on the base material itself, but also on the type/design of implant and the site of implantation. The best evaluation of biocompatibility of the scaffold, hence, is through pre-clinical *in vivo* studies.

3. Degradation

A significant number of scaffolds studied in laboratories in recent years have been fabricated from biodegradable polymers [2-11]. It is important that these scaffolds degrade at a rate that is optimum for rapid tissue repair or regeneration. Detailed studies optimizing the degradation rates for rapid tissue healing have not been performed to our knowledge. However, it appears that the degradation rate should be synchronized with the rate of neo-tissue formation. If the scaffold were to degrade very fast it would no longer be able to provide adequate mechanical support to the cells and new tissue, possibly leading to negative results. On the other hand, if the scaffold were to last for a long time, compared to the time required for tissue growth, it may prove to be detrimental or a hindrance to the tissue engineering endeavor. Thus, it is critical to characterize the degradation kinetics of polymeric biodegradable scaffolds.

A variety of changes occur in the scaffold material as a result of degradation. For example, in polylactic acid (PLA) or PGA materials the molecular weight decreases and so does the mass. In fact, hydrolytic scission of the molecular chains of PLA-PGA polymers commences immediately upon contact with water and is reflected in a change in molecular weight of the polymer, corresponding to a decrease in the size of the molecular chains [12]. Usually, the first degradation products are too large at the molecular level (a function of the starting molecular weight) to freely diffuse out and exit the bulk material. Thus, although there is a change in molecular weight, no appreciable change in mass is detected. With time the molecular chains are sufficiently reduced in size by hydrolysis to enable them to diffuse out and cause a significant mass loss [13]. In a study on DL-PLA material, films were implanted in a rabbits model and it was found that polymer degradation proceeded in two stages [14]. Initially only a decrease in molecular weight was detected followed by the polymer experiencing weight loss. Others such as Li and coworkers [15] have shown that the loss of material weight of DL-PLA implants does not begin until after five weeks of implant submersion in saline. Mass loss as a function of time often displays a sigmoidal pattern [12]. Either the physical structure of the implant is initially sufficient to prevent the degradation products to leave the implant, or the number of oligomers and monomers produced initially is very small. As a result, mass does not decrease rapidly although there is a quick decrease in molecular weight. It has been suggested that an exponential increase in degradation occurs once the molecular weight of L-PLA decreases to below 5 kDa [16].

The time period necessary for significant mass loss is a function of many factors including the starting molecular weight of the polymer, the porosity and permeability of the specimen, and dynamic or static conditions.

Vert and his colleagues [15,17] have reported on the development of a degradation differential between the surface and interior of PLA-PGA specimens. This is related to the inability of large sized degradation products to rapidly exit from within the bulk polymer. The above research group used three different copolymers of PLA and PGA and implanted them intramuscularly in rats. Results showed that the degradation proceeded faster in the center of the specimen and slower at the surface. A plausible explanation is as follows: initially, because of the greater availability of water, degradation proceeds rapidly at the surface. However, the degradation products formed at the surface are easily and quickly washed away by the surrounding fluid. The degradation products (with carboxylic end groups) generated within the confines of the specimen take longer to diffuse out due to their high molecular weight and large chain size. Not surprisingly, the concentration of carboxylic end groups increases in the implant center creating an acidic environment which catalyzes ester breakdown, establishes an autocatalytic cycle, and accelerates the hydrolytic scission. Vert and colleagues confirmed this phenomenon in an *in vitro* study [15]. Others have also reported on degradation product buildup and its effects on the rate of degradation [18,19]. These findings imply that larger or denser scaffolds may be subject to greater degradation differentials and may not behave like homogeneous materials.

Scaffolds that are subjected to mechanical stress may degrade at a different rate. It has been shown previously that implants exposed to stresses degrade at a faster rate [20, 21]. The degradation kinetics may also be affected by fatigue loading which might result in accelerated degradation of the polymer due to mechanochemistry. In an *in vitro* study Agrawal et al. [22] have shown that implants fabricated from a 50:50 PLA-PGA copolymer biodegrade at an accelerated rate when subjected to ultrasonic irradiation.

Thus, the degradation of scaffolds should be tested *in vitro* under conditions that best simulate the application. The rate of degradation of a biodegradable scaffold can be monitored using changes in mass and in the molecular weight of the polymer [3,18,19, 22-25]. These techniques are used quite widely and are detailed below:

3.1 MASS

The mass of each scaffold should be measured prior to the *in vitro* study in a dry condition using a balance with a resolution of 1 µg or less. For most *in vitro* degradation studies the scaffolds are immersed in phosphate buffered saline for different periods of time at 37°C. After extraction from the degradation media, the scaffolds should be first carefully dried in a vacuum and their mass re-measured. Because the mass of all the scaffolds in a study may not be exactly the same, it is best to calculate the percent change for each individually.

3.2 MOLECULAR WEIGHT

Changes in the weight average molecular weight of a polymer may be determined as a function of degradation time using gel permeation chromatography (GPC). For PLA-

PGA materials, cross-sections of the scaffolds can be cut, dissolved in chloroform, filtered using a micropore filter (0.45 μm) and analyzed using GPC. Many systems use chloroform as the mobile phase and polystyrene standards. Other techniques to measure molecular changes include measuring changes in intrinsic viscosity.

4. Porosity and Permeability

Pore size and porosity are important parameters for scaffolds. Vacanti et al. [26,27] have reported that desirable properties for a polymeric scaffold for cell transplantation include an open pore network, and a biocompatible and biodegradable polymer fashioned to provide for optimal diffusion of nutrients, oxygen, and wastes. Freed et al. [28] described an ideal scaffold as one where the porosity should be at least 90% in order to provide a high surface area for cell-polymer interactions, adequate space for extracellular matrix formation, and good diffusion during *in vitro* culture. Others such as Mooney et al. [29] and Kim et al. [30] have also espoused the need for high porosity and high surface-area-to-polymer mass ratio, allowing for uniform cell delivery and tissue ingrowth.

Pore size may determine the type of cells that may be excluded from entering the interior of the scaffold. In the past it has been conventional wisdom that the appropriate pore size for bone ingrowth is in the range of 100 to 400 μm. However, recent studies have disputed this range and shown that smaller sized pores may be sufficient [31]. Studies by Mikos and colleagues have indicated that pore sizes in the range of 150 to 710 microns do not have any significant effects on osteoblast behavior on scaffolds [32-34]. The pores allow for the entry of cells and provide space for the formation of extracellular matrix. Irrespective of what the appropriate pore size may be, it is widely accepted that the pores should be interconnected so that the scaffold is highly permeable. High permeability does not depend on porosity alone. For example Agrawal et al. [25] have shown that scaffolds with similar porosity can possess different permeabilities. It is possible to have high porosity with closed pores and little interconnectivity. Such an architecture would result in inadequate diffusion characteristics that would prevent the entry of nutrients to cells within the interior of the scaffold. Additionally, it would prevent degradation products and metabolic waste from exiting the scaffold to the detriment of new tissue formation.

The porosity and permeability of a scaffold can have a significant impact on its degradation characteristics. Agrawal and colleagues have shown that low porosity/permeability scaffolds degrade faster [18,19]. Also, biodegradable scaffolds under dynamic fluid flow conditions degrade slower, maintaining their mass, molecular weight, and mechanical properties longer [19]. It is speculated that this may be due to the lack of buildup of acidic degradation products under flow conditions and the consequent inhibition of autocatalytic degradation.

4.1 POROSITY MEASUREMENTS

Porosity of a scaffold can be measured using the Archimedes' Principle [25]. In this technique, first the dry mass of the scaffold is determined. Then the scaffold is pre-wet by placing it in ethanol under negative pressure following which it is saturated with water using the same procedure; next the scaffold is removed from the water and weighed to determine its wet mass. It is then completely immersed in water and its submerged mass is measured. Percent porosity is then calculated as follows:

$$\%\text{Porosity} = (M_{wet} - M_{dry})/(M_{wet} - M_{submerged}) \qquad (1)$$

Another popular technique for determining the porosity of scaffolds is the use of mercury intrusion porosimetry [35,36]. This procedure forces mercury under pressure into the pores and can be used to determine void volume and surface area. It is based on the principle that the pressure required to force a non-wetting liquid such as mercury into pores, against the resistance of liquid surface tension, is indicative of the pore size. The technique assumes that the pores are cylindrical in shape.

Estimates of porosity may also be obtained from scanning electron microscopy of cross-sections of the scaffolds [37]. The pore to polymer surface area may be ascertained from two-dimensional images using image analysis and then extrapolated to three-dimensions.

4.2 PERMEABILITY MEASUREMENTS

Permeability can be measured using a direct permeation experiment which measures the rate of flow of water through the implant under a known hydrostatic pressure head as described earlier [18,19,25]. This information is then used in conjunction with Darcy's Law to calculate the permeability of the specimen. Darcy's Law can be expressed by:

$$k = Q.L/(h.A.t) \qquad (2)$$

where k is the permeability constant, Q is the quantity of discharge, A is the cross-sectional area of the sample, L is the length of the sample in the direction of flow, h is the hydraulic head, and t is the time.

5. Mechanical Properties

There is considerable evidence that cells respond to mechanical stresses – a phenomenon called mechanotransduction [38-41]. In tissue engineering, the stresses experienced by cells resident in a scaffold are will be determined by the mechanical properties of the scaffold itself and the stress-strain micro-environment it provides for the cells. It is thus important that scaffolds are designed so that their mechanical behavior provides the correct stimuli to cells. However, the desired value of this stress stimulus is still not well defined for the different cell types, leaving tissue engineering specialists in a quandary. Given this lack of information, perhaps the best approach at present is to try and match the mechanical properties of the scaffold to that of the surrounding native tissue.

The need for scaffolds with adequate mechanical properties is strengthened by several studies that have shown that certain cell types exhibit improved ECM formation and tissue regeneration when exposed to mechanical stimuli [42-45]. For example, Freed et al.[43] grew chondrocytes on scaffolds under Earth's gravity and on the Mir Space Station in a gravity free environment. Mir-grown constructs were more rounded, smaller, and mechanically inferior relative to the Earth grown cultures, indicating the positive role of gravity. In a study on articular chondrocytes it was reported that constructs which were not mechanically stressed after cell seeding exhibited lower mechanical integrity, indicating that mechanical stimulation of immature tissue may play a significant role in their becoming mechanically functional [44].

The mechanical properties of scaffolds should ideally be measured under conditions that reflect the type of loading that they will predominantly undergo in their final application. For instance scaffolds destined for articular cartilage repair should be tested under compression. Standard techniques such as those codified by ASTM for testing the compressive properties of porous materials may be used. However, care must be taken to ensure that the standard techniques are applicable and relevant because often the polymeric scaffolds are very fragile in nature due to their high porosity. Other techniques include indentation tests. For example the stiffness of the scaffold implants can be tested by creep indentation using an Automated Stress-relaxation Creep Indentation (ASCI) apparatus [37]. We have tested PLA-PGA scaffolds using this system, first using a tare load of 9.81×10^{-3} N followed by a 29.4×10^{-3} N perpendicular, compressive, step load applied through a 1.5 mm diameter porous, rigid, indenter tip [25]. The maximum creep displacement was determined and used in the Boussinesq-Papkovitch equation, to obtain the elastic modulus of the specimen

$$E = P(1 - v^2)/2aw_0 \qquad (3)$$

where E is the Young's modulus (MPa), P is the load applied (N), v is the Poisson's ratio (assumed to be 0.3), a is the radius of the loading tip (mm), and w_0 is the maximum creep deformation.

Marra et al. [46] have reported on the use of tensile tests to evaluate the mechanical properties of polymer/hydroxyapatite blends. Similarly, Ignjatovic et al. [47] used an Instron instrument to measure the compressive mechanical properties of PLA-hydroxyapatite blends. In another study, Kim and Mooney [48] tested PGA scaffolds under compression using a mechanical tester. In summary, it is important to test the scaffolds under conditions that closely simulate their loading in use.

6. Delivery of Proteins and Drugs

Biodegradable scaffolds for tissue engineering can play an important role in the delivery of growth factors and drugs that may assist or accelerate the tissue repair or regeneration process. In fact, various growth factors have been identified in recent years which enhance cell migration, adherence, and proliferation. Examples are bone morphogenic protein (BMP), fibroblast growth factor (FGF), platelet-derived growth factor (PDGF), insulin growth factor (IGF), transforming growth factor (TGF), and as interleukins. Various studies have proven the efficacy of such factors. Heckman et al.[6], showed that fracture non-unions can be induced to heal with the use of BMP released from PLA carriers. TGFß can increase cartilage production on three-dimensional scaffolds [49]. In conjunction with insulin, TGFß has been shown to triple the rate of chondrocyte turnover on collagen implants and double their glycosaminoglycan content [50]. Another study explored growth factors such as FGF-2 to induce *in vitro* differentiation of avian bone marrow stromal cells into three-dimensional cartilaginous and bone-like tissues [51]. This work indicated that cell expansion in the presence of appropriate growth factors can direct progenitor cells down specific lines of differentiation.

Although, biodegradable scaffolds can serve as delivery vehicles for the sustained release of biomolecules, the incorporation of growth factors or drugs in a safe and efficacious manner is a critical step. For instance the incorporation of proteins in a scaffold can drastically limit the number of fabrication techniques that can be used so that the proteins/drugs are not denatured by heat or solvents utilized in the fabrication. The release kinetics of proteins or drugs are also very important. Agrawal et al. [3] described the release of BMP and a model protein – soybean trypsin inhibitor from a PLA-PGA scaffold which was fabricated using a gel formation technique. The results indicated that there was a burst release in the first few days followed by a more steady, linear pattern thereafter. In a similar study by the same group, the release kinetics of trypsin inhibitor from a scaffold fabricated using a precipitation technique was evaluated [52]. In this case the release kinetics were very different and exhibited a more sigmoidal pattern. In another study this group demonstrated that cyclic stresses on a drug/protein bearing scaffold can significantly alter the protein release profile [53].

Often the *in vitro* drug release studies are performed using water or phosphate buffered saline as the media [3,52-54]. Scaffolds loaded with drugs/proteins are immersed in the media which is then sampled at predetermined periods of time. The protein/drug concentration in this sample is measured using different assays. For proteins often colorimetric assays such as the micro-BCA are used.

7. Scaffold-Cell Interaction

Scaffolds are quite often evaluated *in vitro* with respect to cellular response. The type of cell line used obviously varies depending on the application for which the scaffold is being designed. For example, Kim and Mooney [48] fabricated fiber-based PGA matrices and seeded them with smooth muscle cells. They determined the number of cells on the matrices by measuring the amount of DNA in enzyme-digested samples. Similarly Goldstein et al. [36] seeded PLA-PGA scaffolds using marrow-derived rat osteoblastic cultures. The constructs were evaluated at different time points. Once again, total cell numbers were estimated from fluorometric analysis of DNA content. They also assessed alkaline phosphatase activity which is an indicator of osteoblastic phenotype. Ishaug et al. [33] used a 75:25 PLA-PGA polymer to fabricate specimens utilizing a particulate leaching technique. These were then seeded with osteoblasts from neonatal Sprague-Dawley rat calvaria. Total DNA and alkaline phosphatase activity was measured. Additionally, confocal microscopy was used to visualize cell distribution..

Various other studies have addressed cellular growth on porous scaffolds and have reported varying degrees of success. For example, Freed et al. [4,5,28,55] have seeded both bovine and human articular chondrocytes on tissue engineering scaffolds.

Vunjak-Novakovic and colleagues [4,11] have performed studies to study the effects of hydrodynamic conditions in tissue-culture bioreactors. Their results demonstrated that cell growth rates under static culture conditions are limited by decreased diffusion caused by increasing cell mass and pore occlusion resulting from cartilage matrix regeneration. Cell-scaffold composites that were cultured under a dynamic laminar flow field in rotating vessels were the largest and possessed the highest fractions of glycosaminoglycan and collagen. As a result, constructs produced from these rotating vessels had superior mechanical properties when compared to constructs cultured in static or mixed vessels. Burg et al. [56] compared seeding methods for scaffolds and studied two seeding techniques (static and dynamic) and three proliferation environments – static, dynamic and perfusion bioreactor. They determined that the combination involving dynamic cell seeding followed by bioreactor proliferation yielded the best results. These findings indicate that culture conditions are extremely important when evaluating scaffold-cell interactions.

8. Conclusions

The evaluation and testing of tissue engineering scaffolds is essential to the success of tissue regeneration and repair. Failure to do so can result in negative results which may yield little understanding about where changes or improvements are needed. However, scaffold evaluation should be performed in a systematic fashion. First the physical and mechanical properties, including the architecture should be analyzed. Next it is important to characterize the degradation characteristics of biodegradable scaffolds. This should be followed by *in vitro* cell response studies. Finally, based on the results of these tests, *in vivo* testing may be warranted.

References

1. Agrawal, C.M. and Ray, R.B. (2001) Biodegradable polymeric scaffolds for musculoskeletal tissue engineering., *J Biomed Mater Res* 55(2), 141-50.
2. Yaszemski, M.J., Payne, R.G., Hayes, W.C., Langer, R.S., Aufdemorte, T.B. and Mikos, A.G. (1995) The ingrowth of new bone tissue and initial mechanical properties of a degradable polymeric composite scaffold, *Tissue Engineering* 1(41-52), .
3. Agrawal, C.M., Bert, J., Heckman, J.D. and Boyan, B.D. (1995) Protein release kinetics of a biodegradable implant for fracture non-unions., *Biomaterials* 16(16), 1255-1260.

121

4. Freed, L.E., Vunjak-Novakovic, G. and Langer, R. (1993) Cultivation of cell-polymer cartilage implants in bioreactors, *J Cell Biochem* 51(3), 257-264.

5. Freed, L., Marquis, J.C., Nohria, A., Emmanual, J., Mikos, A.G. and Langer, R. (1993) Neocartilage formation in vitro and in vivo using cells cultured on synthetic biodegradable polymers, *J Biomed Mater Res* 27(1), 11-23.

6. Heckman, J.D., Boyan, B.D., Aufdemorte, T.B. and Abbott, J.T. (1991) The use of bone morphogenetic protein in the treatment of non-union in a canine model, *J Bone Joint Surg (Am)* 73(5), 750-764.

7. Hollinger, J.O. and Schmitz, J.P. (1987) Restoration of bone discontinuities in dogs using a biodegradable implant, *Journal of Oral and Maxillofacial Surgery* 45, 594-600.

8. Klompmaker (1991) Porous polymer implant for repair of meniscal lesions: A preliminary study in dogs, *Biomaterials* 12(9), 810-816.

9. Peter, S.J., Yaszemski, M.J., Suggs, L.J., Payne, R.G., Langer, R., Hayes, W.C., Unroe, M.R., Alemany, L.B., Engel, P.S., and Mikos, A.G. (1997) Characterization of partially saturated poly(propylene fumarate) for orthopaedic application, *J Biomater Sci, Polym Ed* 8(11), 893-904.

10. Peter, S.J., Lu, L., Kim, D.J. and Mikos, A.G. (2000) Marrow Stromal Osteoblast Function on a Poly(propylene Fumarate)/B-Tricalcium Phosphate Biodegradable Orthopaedic Composite, *Biomaterials* 21, 1207-1213.

11. Vunjak-Novakovic, G., Martin, I., Obradovic, B., Treppo, S., Grodzinsky, A.J., Langer, R. and Freed, L.E. (1999) Bioreactor cultivation conditions modulate the composition and mechanical properties of tissue-engineering cartilage, *J Orthop Res* 17(1), 130-138.

12. Fukuzaki, H., Yoshida, M., Asano, M. and Kumakura, M. (1991) In vivo characteristics of high molecular weight copoly(L-lactide/glycolide) with S-type degradation pattern for application in drug delivery systems, *Biomaterials* 12(May), 433-437.

13. Bucholz, B., Accelerated degradation test on resorbable polymers, in Degradation Phenomena on Polymeric Biomaterials, H. Plank, M. Dauner, and M. Renardy, Editors. 1992, Springer Verlag: N.Y. p. 67-76.

14. Pitt, C.G., Gratzl, M.M., Kimmel, G.L., Surles, J. and Schindler, A. (1981) Aliphatic polyesters II: The degradation of poly (DL-lactide), poly (E-caprolactone), and their copolymers in vivo., *Biomaterials* 2(October), 215-220.

15. Li, S.M., Garreau, H. and Vert, M. (1990) Structure-property relationships in the case of the degradation of massive alophatic poly- (alpha-hydroxy acids) in aqueous media; Part 1: Poly-(DL lactid acid), *Journal of Material Science: Materials in Medicine* 1, 123-130.

16. Schakenraad, J.M., Hardonk, M.J., Feijen, J., Molenaar, I. and Nieuwenhuis, P. (1990) Enzymatic activity toward poly (L-lactic acid) implants, *Journal of Biomedical Materials Research* 24, 529-545.

17. Therin, M., Christel, P., Li, S., Garreau, H. and Vert, M. (1992) In vivo degradation of massive poly(alpha-hydroxy acids): Validation of in vitro findings., *Biomaterials* 13(9), 594-600.

18. Athanasiou, K.A., Schmitz, J.P. and Agrawal, C.M. (1998) The effects of porosity on degradation of PLA-PGA implants, *Tissue Engineering* 4, 53-63.

19. Agrawal, C.M., McKinney, J. and Athanasiou, K.A. (2000) Effects of flow on the in vitro degradation kinetics of biodegradable scaffolds for tissue engineering., *Biomaterials* In Press, .

20. Bos, R.R.M., Rozema, F.R., Boering, G., Nijenhuis, A.J., Pennings, A.J. and Jansen, H.W.B. (1989) Bone-plates and screws of bioabsorbable poly (L-lactide) - an animal pilot study, *British Journal of Oral and Maxillofacial Surgery* 27, 467-476.

21. Suuronen, R., Pohjonen, T., Taurio, R., Törmälä, P., Wessman, L., Rönkkö, K. and Vainionpää, S. (1992) Strength retention of self-reinforced poly-L-lactide screws and plates: an in vivo and in vitro study, *Journal of Materials Science: Materials in Medicine* 3, 426-431.

22. Agrawal, C.M., Kennedy, M.E. and Micallef, D.M. (1994) The effects of ultrasound irradiation on a biodegradable 50-50% copolymer of polylactic and polyglycolic acids, *Journal of Biomedical Materials Research* 28, 851-859.

23. Agrawal, C.M. and Athanasiou, K.A. (1997) A technique to control the pH in the vicinity of biodegrading PLA-PGA implants, *Journal of Biomedical Materials Research (Applied Biomaterials)* 38(2), 105-114.

24. Agrawal, C.M., Schmitz, J.P. and Athanasiou, K.A. Can in vitro degradation studies of PLA-PGA be performed at elevated temperatures? 1998: Tissue Engineering.

25. Agrawal, C.M., McKinney, J.S., Huang, D. and Athanasiou, K.A., The use of the vibrating particle technique to fabricate highly permeable biodegradable scaffolds, in STP 1396: Synthetic Bioabsorbable Polymers for Implants, C.M. Agrawal, J. Parr, and S. Lin, Editors. 2000, ASTM.

26. Vacanti, J.P. (1988) Beyond Transplantation. Third Annual Samuel Jason Mixter Lecture, *Arch Surg* 123(5), 545-549.

27. Vacanti, J.P., Morse, M.A., Saltzman, W.M., Domb, A.J., Perez-Atayde, A., Langer, R., Mazzoni, C.L., and Breuer, C. (1988) Selective cell transplantation using bioabsorbable artificial polymers as matrices, *J Pediatr Surg* 23((1 pt 2)), 3-9.

28. Freed, L.E., Vunjak-Novakovic, G., Biron, R.J., Eagles, D.B., Lesnoy, D.C., Barlow, S.K. and Langer, R. (1994) Biodegradable polymer scaffolds for tissue engineering, *Biotechnology (NY)* 12(7), 689-693.

29. Mooney, D.J., McNamara, K., Hern, D., Vacanti, J.P. and Langer, R. (1996) Stabilized polyglycolic acid fibre-based tubes for tissue engineering, *Biomaterials* 17(2), 115-124.

30. Kim, B.S. and Mooney, D.J. (1998) Development of biocompatible synthetic extracellular matrices for tissue engineering. TIBTECH, *Trends Biotechnol* 16(5), 224-230.

31. Itälä, A., Ylänen, H.O., Ekholm, C., Karisson, K.H. and Aro, H.T. (2001) Pore diameter of more than 100 μm is not requisite for bone ingrowth in rabbits., *J Biomed Mater Res (Appl Biomater)* 58, 679-683.

32. Peter, S.J., Miller, M.J., Yasko, A.W., Yaszemski, M.J. and Mikos, A.G. (1998) Polymer concepts in tissue engineering, *J Biomed Mater Res* 43(4), 422-427.

33. Ishaug-Riley, S.L., Crane-Kruger, G.M., Yaszemski, M.J. and Mikos, A.G. (1998) Three-dimensional culture of rat calvarial osteoblasts in porous biodegradable polymers, *Biomaterials* 19(15), 1405-1412.

34. Ishaug-Riley, S. (1997) Bone formation by three-dimensional stromal osteoblast culture in biodegradable polymer scaffolds, *J of Biomed Mater Res* 36(1), 17-28.

35. Nam, Y.S. and Park, T.G. (1999) Porous biodegradable polymeric scaffolds prepared by thermally induced phase separation, *J Biomed Mater Res* 47, 8-17.

36. Goldstein, A.S., Zhu, G., Morris, G.E., Meszlenyi, R.K. and Mikos, A.G. (1999) Effect of osteoblastic culture conditions on the structure of poly(DL-lactic-co-glycolic acid) foam scaffolds, *Tissue Engineering* 5(5), 421-433.

37. Singhal, A.R., Agrawal, C.M. and Athanasiou, K.A. (1996) Salient degradation features of a 50:50 PLA/PGA scaffold for tissue engineering, *Tissue Engineering* 2(3), 197-207.

38. Owan, I., Burr, D.B., Turner, C.H., Qiu, J., Tu, Y., Onyia, J.E. and Duncan, R.L. (1997) Mechanotransduction in bone: Osteoblasts are more responsive to fluid forces than mechanical strain, *Am J Physiol* 273((3 pt. 1)), C810-C815.

39. Sah, R.L., Kim, Y.J., Doong, J.Y., Grodzinsky, A.J., Plaas, A.H. and Sandy, J.D. (1989) Biosynthetic response of cartilage explants to dynamic compression, *J Orthop Res* 7(5), 619-636.

40. Smalt, R., Mitchell, F.T., Howard, R.L. and Chambers, T.J. (1997) Mechanotransduction in bone cells: Induction of nitric oxide and prostaglandin synthesis by fluid shear stress, but not by mechanical strain, *Adv Exp Med Biol* 433, 311-314.

41. Sikavitsas, V.I. (2001) Biomaterials and bone mechanotransduction, *Biomaterials* 22, 2581-2593.

42. Buschmann, M.D., Gluzband, Y.A., Grodzinsky, A.J. and Hunziker, E.B. (1995) Mechanical compression modulates matrix biosynthesis in chondrocyte/agarose culture, *J Cell Sci* 108((Pt 4)), 1497-1508.

43. Freed, L.E., Langer, R., Martin, I., Pellis, N.R. and Vunjak-Novakovic, G. Tissue engineering of cartilage in space. in Proc Natl Acad Sci USA. 1997.

44. Carver, S.E. and Heath, C.A. (1999) Influence of intermittent pressure, fluid flow, and mixing on the regenerative properties of articular chondrocytes, *Biotechnology Bioengineering* 65(3), 274-81.

45. Kim, Y.J., Grodzinsky, A.J. and Plaas, A.H. (1996) Compression of cartilage results in differential effects on biosynthetic pathways for aggrecan, link protein and hyaluronan, *Arch Biochem Biophys* 328(2), 331-340.

46. Marra, K.G., Szem, J.W., Kumta, P.N., DiMilla, P.A. and Weiss, L.E. (1999) *In vitro* analysis of biodegradable polymer blend/hydroxyapatite composites for bone tissue engineering, *J Biomed Mater Res* 47, 324-335.

47. Ignjatovic, N., Tomic, S., Dakic, M., Miljkovic, M., Plavsic, M. and Uskokovic, D. (1999) Synthesis and properties of hydroxyapatite/poly-L-lactide composite biomaterials, *Biomaterials* 20, 809-816.

48. Kim, B.-S.M., David J. (1998) Engineering smooth muscle tissue with a predefined structure, *J. of Biomedical Mat.* 41(2), 322-332.

49. Zimber (1995) TGF-b promotes the growth of bovine chondrocytes in monolayer culture and the formation of cartilage tissue on three-dimensional scaffolds, *Tissue Eng* 1(3), 289-300.

50. Toolan, B.C., Frenkel, S.R., Pachence, J.M., Yalowitz, L. and Alexander, H. (1996) Effects of growth-factor-enhanced culture on a chondrocyte-collagen implant for cartilage repair, *J Biomed Mater Res* 31(2), 273-280.

51. Martin, I., Padera, R.F., Vunjak-Novakovic, G. and Freed, L.E. (1998) In vitro differentiation of chick embryo bone marrow stromal cells into cartilaginous and bone-like tissues, *J Orthop Res* 16(2), 181-189.

52. Athanasiou, K.A., Singhal, A.R., Agrawal, C.M. and Boyan, B.D. (1995) *In vitro* biodegradation and release characteristics of biodegradable implants containing trypsin inhibitor, *Clinical Orthopaedics and Related Research* 315, 272-281.

53. Thompson, D.E., Agrawal, C.M. and Athanasiou, K.A. (1996) The effects of dynamic compressive loading on biodegradable implants of 50-50% polylactic acid-polyglycolic acid., *Tissue Engineering* 2(1), 61-74.

54. Liu, S.-J., Ueng, S.W.-N., Chan, E.-C., Lin, S.-S., Tsai, C.-H., Wei, F.-C. and Shih, C.-H. (1999) *In vitro* elution of vancomycin from biodegradable beads, *J Biomed Mater Res (Appl Biomater)* 48, 613-620.

55. Freed, L.E., Grande, D.A., Lingbin, Z., Emmanual, J., Marquis, J.C. and Langer, R. (1994) Joint resurfacing using allograft chondrocytes and synthetic biodegradable polymer scaffolds, *Journal of Biomedical Materials Research* 28, 891-899.

56. Burg, K.J.L., W.D. Holder, J., Culberson, C.R., Beiler, R.J., Greene, K.G., Loebsack, A.B., Roland, W.D., Eiselt, P., Mooney, D.J., and Halberstadt, C.R. (2000) Comparative study of seeding methods for three-dimensional polymeric scaffolds, *J Biomed Mater Res* 51, 642-649.

CHARACTERIZATION OF DEGRADABLE POLYMERS FOR ORTHOPEDIC APPLICATION

Examining Tyrosine-Derived Polycarbonates for Tissue Engineering of Bone

Sascha D. Abramson, Agnes Seyda, P. Sidney Sit, and Joachim Kohn
Department of Chemistry
Rutgers, The State University of New Jersey
New Brunswick, NJ 08903

1. Abstract

As tissue engineering comes to the forefront as a cutting-edge discipline, one of the principle challenges is the need for new materials that are fully bioresorbable and biocompatible. To this extent, pseudo-poly(amino acid)s such as tyrosine-derived polycarbonates present themselves as a promising new class of degradable polymers. These materials offer a high degree of bone biocompatibility. Variations in pendent chain structure allow the generation of a series of materials with variations in key mechanical and cellular response properties. In addition, their chemical structure provides convenient attachment points for the covalent linkage of bioactive molecules to the polymer backbone. Tyrosine-derived polycarbonates can be shaped by commonly used processing and fabrication methods – another advantage over conventional poly(amino acid)s which are often non-processible .

2. Introduction

One of the important keys to the success of tissue engineering is the ability to construct a matrix which will direct the appropriate cell types to migrate, multiply, and express normal functions such that they replace damaged or diseased tissue. To this end, degradable materials have been designed which could be utilized as scaffold materials for tissue engineering [3]. These materials could provide temporary mechanical stability for the neotissue, but over time would degrade and disappear from the implant site.

Dozens of hydrolytically unstable polymers have been suggested as degradable biomaterials, however, in most cases no attempts have been made to develop them for specific medical applications. Thus, detailed toxicological studies *in vivo*, investigations of degradation rate and mechanism, and careful evaluations of the physicomechanical properties have so far been published for only a very small fraction

125

R.L. Reis and D. Cohn (eds.),
Polymer Based Systems on Tissue Engineering, Replacement and Regeneration, 125–138.
© 2002 *Kluwer Academic Publishers. Printed in the Netherlands.*

of these polymers. An even smaller number of synthetic, degradable polymers has so far been used in medical implants and devices that gained approval by the US Food and Drug Administration (FDA) for use in patients. The simple linear, aliphatic polyesters, poly(glycolic acid) (PGA) and poly(lactic acid) (PLA) and their copolymers (PLGA), are currently the most widely investigated, and most commonly used synthetic, bioerodible polymers for medical applications [4]. These are often regarded as the "gold standard" for synthetic materials for medical applications. With so few degradable polymers available for use as medical implant materials, there is a need for the development of additional implant materials – in particular for use in tissue scaffolds and advanced drug delivery devices.

Since proteins are polymers composed of naturally occurring amino acids, it was an obvious idea to explore the possible use of poly(amino acids) in biomedical applications [5]. Poly(amino acids) were regarded as promising candidates since the amino acid side chains offer sites for the attachment of drugs, crosslinking agents, or pendent groups that can be used to modify the physicomechanical properties of the polymer. Additionally, poly(amino acids) usually show a low level of systemic toxicity, due to their degradation to naturally occurring amino acids.

Despite their apparent potential as biomaterials, poly(amino acids) have found few practical applications. Most are highly insoluble and nonprocessible materials. Since poly(amino acids) have a pronounced tendency to swell in aqueous media, it can be difficult to predict drug release rates. Furthermore, the antigenicity of polymers containing three or more amino acids limits their use in biomedical applications [5].

Attempts have been made to utilize amino acids as monomeric building blocks for biomaterials, while avoiding the unfavorable physicomechanical properties of poly(amino acids). These attempts resulted in the development of a wide range of amino acid derived polymers that do not have the conventional backbone structure found in peptides. Collectively, these materials are referred to as "non-peptide amino acid based polymers" or as "amino acid derived polymers with modified backbones". In spite of their large structural variability, it is possible to identify four main types of non-peptide amino acid based polymers: 1) synthetic polymers with amino acid side chains, 2) copolymers of and non-amino acid monomers, 3) pseudo-poly(amino acids), and 4) block-copolymers containing peptide or poly(amino acid) blocks.

Pseudo-poly(amino acids) are polymers in which α-L-amino acids are linked together by non-amide linkages, such as carbonates or esters. This approach reduces the number of interchain hydrogen bonds due to a reduction in amide linkages in the polymer backbone and results in improved physiomechanical properties over conventional poly(amino acids). In conventional poly(amino acid)s, hydrogen bonding gives rise to secondary structures such as α-helices or β-pleated sheets but also results in high processing temperatures and low solubility in organic solvents.

Pseudo-poly(amino acids) were first described in 1984 [6] and have since been evaluated for use in several medical applications [7-11]. Although a range of different pseudo-poly(amino acids) have been prepared, detailed studies of the physical properties, biological properties, and possible applications of these polymers so far have been conducted for only a select group of new tyrosine-derived polycarbonates, polyiminocarbonates, and polyarylates, which may all be derived from the same monomer (Figure 1). This chapter will focus on tyrosine-derived polycarbonates, which have been extensively study for hard tissue applications.

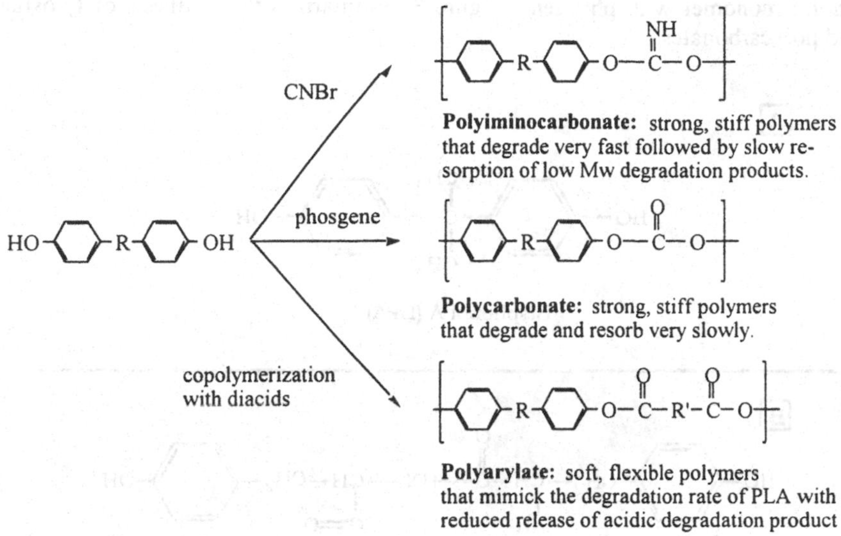
CNBr
phosgene
copolymerization
with diacids
HO
R
OH
NH
O
C
R'
Polyiminocarbonate: strong, stiff polymers that degrade very fast followed by slow resorption of low Mw degradation products.
Polycarbonate: strong, stiff polymers that degrade and resorb very slowly.
Polyarylate: soft, flexible polymers that mimick the degradation rate of PLA with reduced release of acidic degradation product

Figure 1 Synthesis of three different polymer families from the same diphenolic monomers

Tyrosine is the only major, natural nutrient containing an aromatic hydroxyl group. Derivatives of tyrosine dipeptide can be regarded as diphenols and may be employed as replacements for the toxic industrially used diphenols such as Bisphenol A in the design of medical implant materials (Figure 2). The observation that aromatic backbone structures can significantly increase the stiffness and mechanical strength of polymers provided the one rationale for the use of tyrosine dipeptides as monomers.

3. Tyrosine-Derived Polycarbonates

In view of the nonprocessibility of conventional poly(L-tyrosine), which cannot be used as an engineering plastic, variational derivatives were envisioned. The development of tyrosine-based polycarbonates, polyarylates and polyiminocarbonates represents the first time tyrosine-derived polymers with favorable engineering properties have been identified.

As shown in Figure 2, tyrosine dipeptide contains a free amino group and a free carboxylic acid group, which have to be protected during polymer synthesis. Monomer synthesis from 3-(4'-hydroxyphenyl)propionic acid and tyrosine alkyl esters was accomplished by carbodiimide mediated coupling reactions, following known procedures of peptide synthesis [1, 12] in typical yields of 70%. Monomers carrying an ethyl, butyl, hexyl, or octyl ester pendent chain were investigated extensively [13] [1]. These monomers have been named desaminotyrosyl tyrosine alkyl esters (DTR). The R in the abbreviation represents the alkyl pendent chain length: thus the DTE monomer is desaminotyrosyl tyrosine ethyl esters or DTE. These (peptide-like) diphenolic monomers were used as starting materials in the synthesis of polycarbonates. Polycarbonate synthesis can be accomplished by polymerizing the diphenolic monomer with phosgene. Figure 3 summarizes the synthesis of tyrosine-derived polycarbonates.

A

HO—⟨benzene⟩—C(CH₃)(CH₃)—⟨benzene⟩—OH

Bisphenol A (BPA)

B

HO—⟨benzene⟩—CH₂—CH(NH–X₁)—C(=O)—HN—CH(C=O–O–X₂)—CH₂—⟨benzene⟩—OH

protected tyrosine dipeptide

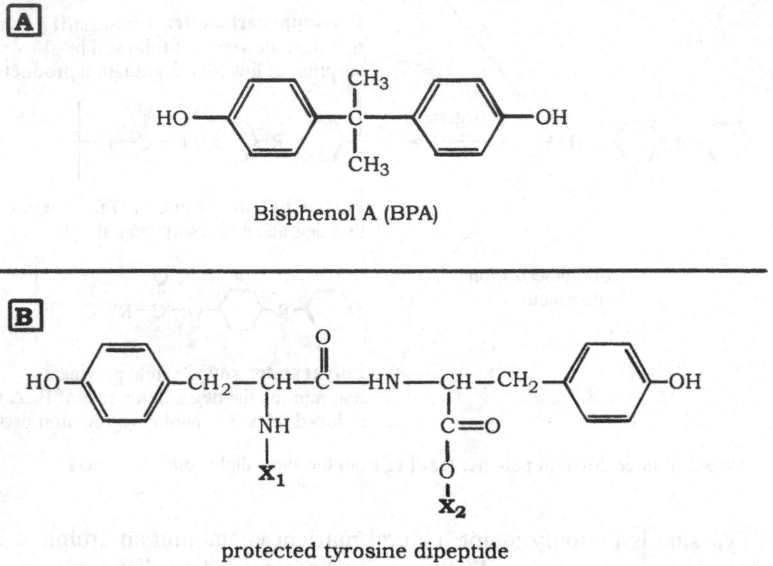

Figure 2: Structures of (A) Bisphenol A, a widely used diphenol in the manufacture of commercial polycarbonate resins; (B) tyrosine dipeptide with specific chemical protecting groups X_1 and alkyl substituents X_2 attached to the N and C termini, respectively.

Tyrosine-derived polycarbonates have important advantages when used in the design of implantable, degradable controlled release systems. First, all members of this series of polymers are amorphous materials with relatively low glass transition temperatures that are a function of the pendent chain length (Table 1). X-ray diffraction exhibited only an amorphous halo, indicative of the lack of crystalline domains. DSC analysis showed a glass transition and decomposition exotherm, but no melting endotherm, further confirming the lack of crystallinity. Contrary to poly(L-tyrosine) which is a non-processible, insoluble polymer, tyrosine-derived polycarbonates are freely soluble in a variety of organic solvents, and are readily processible by conventional solvent casting and thermal processing techniques (extrusion, injection and compression molding) at relatively low temperatures. Typically, they form strong, transparent films. For extrusion and injection molding, processing temperatures are about 70 to 100 °C above the glass transition temperature. Since the thermal decomposition temperature of all four polymers (polymers derived from DTE, DTB, DTH, or DTO monomers) is about 300°C when measured by thermogravimetric analysis (TGA), there is a relatively large gap between the processing temperatures and the thermal decomposition temperature.

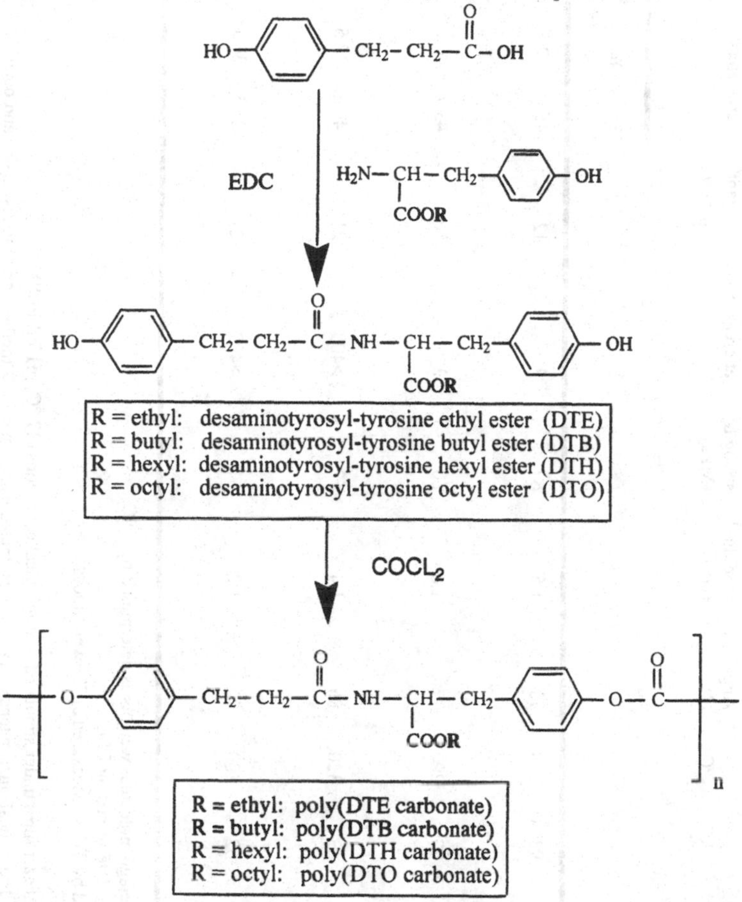

Figure 3 Summary of the chemical synthesis of tyrosine-derived polycarbonates. Reaction scheme couples desaminotyrosine and tyrosine alkyl esters to obtain a diphenolic monomer, which carries an alkyl ester pendent chain (R). The length of the pendent chain was varied from two to eight carbons by choosing one of four available desaminotyrosyl-tyrosine alkyl esters as monomers. EDC = ethyl-3-(3'-dimethylamino)propyl carbodiimide hydrochloride salt [13].

Table 1: Comparison of different physical properties of tyrosine-derived polycarbonates

Polymer	Mw[a] (Mw/Mn)	Tg[b] (°C)	Td[c] (°C)	Contact Angle (°)	Young's modulus[d] (GPa)	Tensile strength[d] (MPa)	Elongation at break (%)	Time constant[e] (weeks)	Surface Energy Parameters[f]		
									γ_c	γ_d	γ_p
poly(DTE carbonate)	176,000 (1.8)	93	290	73	1.5	67	4	12	46.4	42.5	3.5
poly(DTB carbonate)	120,000 (1.4)	77	290	77	1.6	60	3	16	43.7	40.1	2.4
poly(DTH carbonate)	350,000 (1.7)	63	320	86	1.4	62	>400	21	40.6	37.5	1.1
poly(DTO carbonate)	450,000 (1.7)	52	300	90	1.2	51	>400	21	38.5	36.1	0.6

[a]Data from Ertel and Kohn [1]. Weight average molecular weights as determined by GPC.
[b]Glass transition temperature (midpoint) as determined by DSC.
[c]Decomposition temperature as determined by TGA. Measured at 2% weight loss.
[d]Unoriented samples.
[e]Degradation time constant for thin, solvent cast films under simulated physiological conditions (37 °C, pH 7.4 PBS).
[f]Data obtained from Perez-Luna et al. [2]. The critical surface tension (γ_c) was estimated using Zismann's method, and the dispersive and polar components (γ_d and γ_p) were calculated using the geometric mean approximation to the work of adhesion.

The mechanical properties of these polycarbonates depend strongly on the length of the alkyl ester pendent chain. In general, increasing the length of the alkyl ester leads to a decrease in stiffness and an increase in ductility. Unoriented thin film specimens range in tensile modulus from 1.2 GPa for poly(DTO carbonate) to 1.6 GPa for poly(DTE carbonate). Tensile strength and elongation appear to be strongly affected by the temperature at which the tests are being conducted. At room temperature, the polymers with shorter pendent chains (ethyl and butyl) behave as brittle materials that fail without yielding after about 4% elongation. The polymers with longer pendent chains (hexyl and octyl) orient under stress and can be elongated to over 400%. In their oriented state, the ultimate tensile strength at break surpasses 200 MPa. At slightly higher temperatures (40-60 °C), all four polycarbonates could be elongated under stress. Under carefully controlled conditions, elongation of up to 1000% can be achieved, resulting in highly oriented specimens for which an ultimate tensile strength of up to 400 MPa has been observed.

Tyrosine-derived polycarbonates are stiffer and stronger than many other degradable polymers of comparable molecular weights, such as polycaprolactone and polyorthoesters, but are not as stiff as poly(L-lactic acid) or poly(glycolic acid) [14]. Considering the strength and stiffness of these tyrosine-derived polycarbonates, it is conceivable to fabricate load-bearing devices (such as pins for small bone fixation) or load-bearing drug delivery systems which may find application in orthopedics.

Tyrosine-derived polycarbonates provided a convenient model system to study the effect of pendent chain length on the thermal properties and the enthalpy relaxation (physical aging). It is noteworthy that enthalpy relaxation kinetics are not usually reported in the biomedical literature and that a recent study by Tangpasuthadol [15] represents one of the first attempts to evaluate physical aging in a degradable biomedical polymer.

For the tyrosine-derived polycarbonates tested, the enthalpy relaxation process is not sensitive to the length of the pendent chain. This observation suggests that structural relaxation in these polymers is limited by backbone flexibility, and that the fraction of free volume in these polymers is not the limiting factor for polymer mobility. Furthermore, since the enthalpy relaxation time is short at aging temperatures of 15° C below the Tg, a few hours of storage at that temperature is sufficient to bring the physical aging process to completion. The results obtained by dynamic mechanical analysis support the general notion that an increase in the length of the pendent chain results in a more flexible material [15]. This observation is in agreement with the results obtained in a previous study of these polymers [1]. These structure-property correlations can assist in the selection of suitable polymers for specific applications.

Another significant advantage of tyrosine-derived polycarbonates is their high hydrophobicity. Air-water contact angles increased with a corresponding decrease in the surface energy parameters as a function of increasing pendent chain length (Table 1). The increasing hydrophobicity imparted by longer alkyl ester pendent chains is also demonstrated by the equilibrium water content (EWC) which is 4% at 37°C for poly(DTE carbonate) and less than 1% for poly(DTO carbonate).

It is a general observation that, consistent with the fairly stable polymer backbone, all tyrosine-derived polycarbonates degrade relatively slowly under physiological conditions. The mechanism of degradation has been carefully evaluated [1]. Based on the evidence obtained from ESCA (XPS), ATR-FTIR and GPC, an *in vitro* degradation mechanism has been postulated. According to this mechanism, the ester bonds of the alkyl ester pendent chains are cleaved first, followed by the cleavage of the carbonate bonds in the polymer backbone. The amide bonds are not hydrolyzed *in vitro*. Thus, desaminotyrosyl-tyrosine is the final degradation product. For thin solvent cast films of high initial molecular weight, the *in vitro* degradation time constants are listed in Table 2. The kinetic time constants calculated from Ertel's model [1] indicate that poly(DTE carbonate) degrades hydrolytically almost twice as fast as the more hydrophobic poly(DTO carbonate). *In vivo*, the cleavage of the amide bond by enzymatic or cellular mechanisms may lead to additional degradation products. However, no detailed investigations as to the *in vivo* degradation products have been carried out.

To put these degradation properties into perspective, poly(DTH carbonate) and high molecular weight poly(L-lactic acid) exhibit comparable reductions in molecular weight when incubated in physiological buffer solution at 37 °C. The similarity in the degradation profile between tyrosine-derived polycarbonates and high molecular weight poly(L-lactic acid) was also observed *in vivo* [16]. Original extrapolation from a 26 week implantation study in the femur and tibia of rabbits, estimated a resorption time of 2 to 3 years for high molecular weight poly(DTH carbonate) [17]. A more recent implantation study in rabbits has shown that there is no observable resorption of poly(DTE carbonate), poly(DTB carbonate), or poly(DTH carbonate) up to 3 years [18].

4. Medical Applications and Biocompatability of Tyrosine-Derived Polycarbonates

Because of their strength and toughness, tyrosine-derived polycarbonates are recognized as possible candidates for the development of orthopedic implants [19]. Particularly promising is the development of small bone fixation devices such as pins, screws, and plates that can be used to stabilize fractures in small, non-weight bearing bones. Currently, self-reinforced devices made of poly(glycolic acid) are in clinical use in the USA, but these devices have significant clinical side effects due in part to the release of acidic degradation products [20]. Implants made of polydioxanone (Orthosorb®) are also in use, but these implants tend to degrade too fast, losing their mechanical strength within 1 to 2 months [17].

In vitro attachment and proliferation of fibroblasts on tyrosine-derived polycarbonates are a function of the pendent chain length [1]. Consistent with the hypothesis that most cells favor more hydrophilic and molecularly rigid surfaces, poly(DTE carbonate), the most hydrophilic and rigid of the surfaces tested, supported cell growth and proliferation better than the more hydrophobic poly(DTO carbonate). Poly(DTB carbonate) and poly(DTH carbonate) were intermediate in their ability to support cell attachment and growth.

In an *in vivo* pilot study [17], poly(DTH carbonate) pins were fabricated and compared to commercially available Orthosorb® pins made of polydioxanone. The pins were implanted transcortically in the distal femur and proximal tibia of New Zealand White rabbits for up to 26 weeks. In addition to routine histological evaluation of the implant sites, bone activity at the implant/tissue interface was visualized by UV illumination of sections labeled with fluorescent markers and the degree of calcification around the implants was ascertained by backscattered electron microscopy.

The bone tissue response was characterized by active bone remodeling at the surface of the degrading implant, the lack of fibrous capsule formation, and an unusually low number of inflammatory cells at the bone-implant interface. Poly(DTH carbonate) exhibited very close bone apposition throughout the 26 week period of this initial study. A roughened interface was observed which was penetrated by new bone as early as 2 weeks post implantation. Bone growth into the periphery of the implant material was clearly visible at the 26 week time point [17].

Some of the observations made in the transcortical rabbit model were confirmed in an independent study using the canine chamber model [21] where the bone response to poly(L-lactic acid) and two tyrosine-derived polycarbonates, poly(DTE carbonate) and poly(DTH carbonate) were compared. In this study [16], thin polymeric coupons were used to create narrow channels within a polyethylene housing. Upon implantation of the assembled device into the femur of dogs, the ability of bone to grow into the narrow channels lined by coupons made of different materials was evaluated. In addition, the host bone response was histologically evaluated. Test chambers containing coupons of poly(DTE carbonate) and poly(DTH carbonate) were characterized by sustained bone ingrowth throughout the 48 week study period. Histological sections revealed intimate contact between bone and tyrosine-derived polycarbonates. In contrast, bone ingrowth into the PLA chambers peaked at 24 weeks and dropped by half at the 48-week time point. A fibrous tissue layer was found surrounding the PLA implants at all time points.

The presence of a fibrous layer at the coupon-bone interface is characteristic of a mild foreign body response. The fact that such a fibrous layer was not formed at the bone/material interface for poly(DTE carbonate) and poly(DTH carbonate) is an important advantage of these polymers. Intimate contact between bone and implant, even at 48 weeks post-implantation, is a strong indicator of the biocompatibility of the tyrosine-derived polycarbonates.

To augment the previous data, a second implantation study in the rabbit model was undertaken in which four polycarbonates with different pendent chains were studied [18]. The pins were placed transcortically through the distal end of the femur or the proximal end of the tibia. At timepoints as early as 90 days (3 months) and as late as 1090 days (3 years), all the polymeric implants were surrounded by bone without any obvious deleterious effects such as bone resorption or large concentrations of inflammatory cells at the implant site. Poly(DTE carbonate), poly(DTB carbonate), poly(DTH carbonate), and poly(DTO carbonate) were all osteocompatible according to traditional definitions [22, 23].

A fundamental difference was seen, however, amongst the polymers at the bone-pin interface—where some pins were encapsulated with fibrous tissue (an encapsulation response) whereas others exhibited predominantly direct apposition of bone to the implant surface. The capsule lined the entire circumference of the implant and effectively separated the implant from the surrounding bone. In contrast, in those specimens where bone apposition was observed, the circumference of the implant was devoid of an organized fibrous capsule. In this study, the tissue responses could be readily classified as either an "encapsulation response" or as "direct bone apposition." The intermediate case of partial encapsulation (presence of an organized fibrous capsule along only a fraction of the implant circumference) was very rarely observed.

Most striking was the bone response to poly(DTE carbonate) where direct bone apposition to the implant was the defining feature in 73% of the retrieved implants (22 of 30 pins). In contrast, as the length of the pendent chain was increased, less bone apposition was observed and the predominant response was the formation of a fibrous capsule. Particularly noteworthy was the dramatic difference in the predominant biological response elicited by poly(DTE carbonate) and poly(DTB carbonate) as these two polymers have very closely matched chemical structure and material properties. Poly(DTB carbonate) exhibited direct bone apposition to the surface in only 21% of the implant sites. Clearly, in this family of tyrosine-derived polycarbonates, the predominant response elicited at the bone-implant interface was significantly influenced by a relatively minor modification of the polymer structure.

In addition to the results presented above, the rabbit transcortical pin study provided some information about the *in vivo* degradation pattern exhibited by polycarbonates. It was revealed that while poly(DTE carbonate) had exceptional bone biocompatibility, it had a very extended degradation time. Previously, the mechanism of polycarbonate degradation was examined *in vitro* and it had been shown that tyrosine-derived polycarbonates degrade by hydrolysis of their pendent chain and cleavage of the backbone. Cleavage of the backbone predominates in bulk degradation of the polymers, while pendent chain cleavage predominates at the surface [1].

Table 2: Frequencies of Direct Bone Apposition and Encapsulation Responses at the Bone-Implant Interface [18]

	Poly(DTE carbonate)		Poly(DTB carbonate)		Poly(DTO carbonate)	
	bone apposition	encapsulation	bone apposition	encapsulation	bone apposition	encapsulation
short-term 0-180 days (n=10)	60%	40%	30%	70%	20%	80%
long-term 270-1090 days (n=26)	80%	20%	17%	83%	16%	84%
Overall results (n=36)	73%	27%	21%	79%	17%	83%

Bone apposition responses were reported when an organized fibrous tissue layer could not be identified at the light microscopic level at the bone-implant interface.
Encapsulation responses were reported when a fibrous capsule encompassed the implanted pin.

The hydrolysis of the pendent chain exposes a free carboxylate group at the surface. Poly(DTE carbonate) has a faster pendent chain hydrolysis than polycarbonates with longer pendant chains. Thus when incubated, poly(DTE carbonate) forms free carboxylate groups at the surface more quickly than other tyrosine-derived polycarbonates. Figure 4 is a representation of poly(DTE carbonate) hydrolyzing to the polymer having a free carboxylate containing pendent chain.

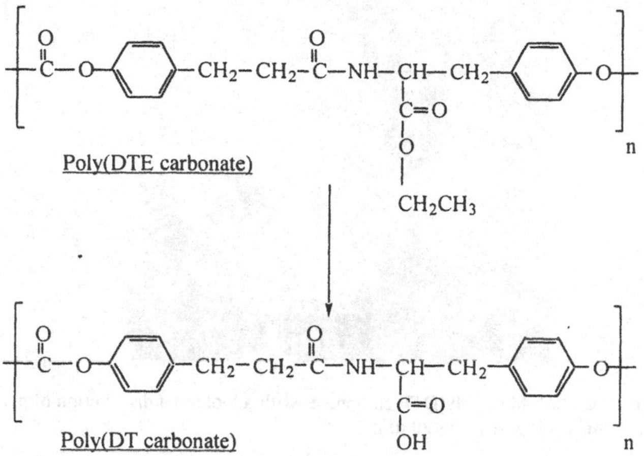

Poly(DTE carbonate)

Poly(DT carbonate)

Figure 4. Hydrolysis of poly(DTE carbonate) to poly(DT carbonate), a polymer with 100% free carboxylate pendent chains

It was hypothesized that the introduction of the free carboxylate pendent chain into poly(DTE carbonate) could increase the rate of degradation. In order to study the effect of the introduction of free carboxylate groups, a series of copolymers were synthesized with DTE and the free carboxylate monomer desaminotyrosyl-tyrosine (DT). Desaminotyrosyl tyrosine benzyl ester (DTBn) was synthesized and polymerized with DTE to yield poly(DTE-co-x%DTBn carbonate) with X representing the percentage of the monomers that were benzylated. The benzyl groups were deprotected using catalytic hydrogenation to yield poly(DTE-co-X%DT carbonate). Polymers were synthesized with 0 - 100% DT content. These polymers proved to have similar mechanical properties and similar processibility. In addition, polymers having a higher molar fraction of free carboxylates degraded at a faster rate than polymers with lower free carboxylate content (unpublished data, Bolikal).

In order to evaluate poly(DTE carbonate) and one member of the poly(DTE-co-X%DT carbonate) family of polymers for guided bone regeneration a delayed healing defect was created in the calvaria of skeletally mature New Zealand White rabbits. Implanted into the defects were porous sponges with a bimodal distribution of pores (larger pores 200-400 μm, small pores <20 μm) (Figure 4). Two polymers were utilized: poly(DTE carbonate) and poly(DTE-co-25%DT carbonate). Each rabbit received two implants in its skull. After 16 weeks there was no statistical difference between the amounts of ingrowth into the scaffold for either polymer. In addition poly(DTE-co-25%DT carbonate) appeared to have disappeared from the site of implantation, while poly(DTE carbonate) showed no signs of resorption [24].

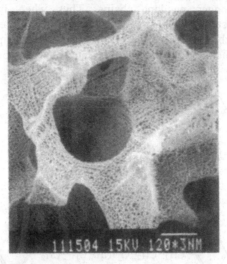

Figure 5: SEM of porous scaffold of poly(DTE carbonate) with a biomodal distribution of pores. Large pores are 200 – 400 µm and the small pores are <20 µm.

In addition to faster degradation than poly(DTE carbonate), free carboxylate copolymers deposit calcium at their surface when incubated in simulated body fluid (SBF) [24]. It has been hypothesized that the free carboxylate groups provide sites of nucleation for calcium phosphate deposition. In this study, poly(DTE carbonate) was used as a control and showed little or no calcium at the surface after incubation in SBF, while poly(DTE-co-25%DT carbonate) and poly(DTE-co-50%DT carbonate) showed measurable calcium deposition within 3 - 7 days.

It is believed that poly(DTE carbonate) has better bone biocompatibility due to its faster surface pendent chain hydrolysis than other tyrosine-derived polycarbonates. The exposure of free carboxylate groups could provide for a site of nucleation for calcium phosphate. Further, it has been theorized that in poly(DTE carbonate) the hydrolysis of the alkyl ester pendent chain occurs before the formation of a fibrous capsule, thus allowing for potential direct bone polymer apposition. In polycarbonates with longer pendent chains, the hydrolysis does not proceed as quickly, thus a fibrous capsule forms prior to exposure of the free carboxylate groups leading to lesser bone biocompatibility in polycarbonates with longer pendent chains.

5. Summary

All currently available tyrosine-derived polymers have been intentionally designed to degrade under physiological conditions. Because of their apparent non-toxicity and high degree of tissue compatibility, these materials are promising candidates for use in degradable medical implants and degradable drug delivery systems.

Tyrosine-derived polycarbonates offer certain advantages over conventional poly(amino acids) such as lower cost of polymer synthesis and processibility by industrially used fabrication techniques. Contrary to petroleum-based polymers, the final degradation products of tyrosine-derived pseudo-poly(amino acids) can be expected to be simple, naturally occurring amino acids. The relatively high mechanical strength of tyrosine-derived polycarbonates has led to their evaluation as orthopedic implants, in addition to their evaluation as drug carries.

6. References

1. S. I. Ertel and J. Kohn, "Evaluation of a series of tyrosine-derived polycarbonates for biomaterial applications", J. Biomed. Mater. Res., **1994**, 28, 919-930.

2. V. H. Perez-Luna, K. A. Hooper, J. Kohn and B. D. Ratner, "Surface characterization of tyrosine-derived polycarbonates", J. Appl. Polym. Sci., **1997**, 63(11), 1467-1479.

3. S. Rimmer, "Biomaterials: Tissue engineering and polymers", in: *Spec. Publ. - R. Soc. Chem.*, Emerging Themes in Polymer Science, Vol. 263, **2001**, 89-99.

4. S. Abramson, J. Kohn and R. Langer, "Bioresorbable and bioerodible materials", in: *Biomaterials Science*, 2nd edition, (B. Ratner, A. S. Hoffman, F. J. Schoen and J. E. Lemons, eds.), **2002**, Academic Press, San Diego, in press.

5. J. M. Anderson, K. L. Spilizewski and A. Hiltner, "Poly-α amino acids as biomedical polymers", in: *Biocompatibility of Tissue Analogs*, (D. F. Williams, ed.), Vol. 1, **1985**, CRC Press Inc., Boca Raton, 67-88.

6. J. Kohn and R. Langer, "A new approach to the development of bioerodible polymers for controlled release applications employing naturally occurring amino acids", in: *Polymeric Materials, Science and Engineering*, Vol. 51, **1984**, American Chemical Society, Washington, DC, 119-121.

7. J. Kohn and R. Langer, "Polymerization reactions involving the side chains of α-L-amino acids", J. Am. Chem. Soc., **1987**, 109, 817-820.

8. J. Kohn, "Design, synthesis, and possible applications of pseudo-poly(amino acids)", Trends Polym. Sci., **1993**, 1(7), 206-212.

9. Q. X. Zhou and J. Kohn, "Preparation of poly(L-serine ester): A structural analogue of conventional poly(L-serine)", Macromolecules, **1990**, 23, 3399-3406.

10. H. Yu-Kwon and R. Langer, "Pseudopoly(amino acids): A study of the synthesis and characterization of poly(trans-4-hydroxy-N-acyl-L-proline esters)", Macromolecules, **1989**, 22, 3250-3255.

11. H. Q. Mao, R. X. Zhuo and C. L. Fan, "Synthesis and biological properties of polymer immunoadjuvants", Polym. J., **1993**, 25(5), 499-505.

138

12. S. Pulapura and J. Kohn, "Tyrosine derived polycarbonates: Backbone modified, "pseudo"-poly(amino acids) designed for biomedical applications", Biopolymers, **1992**, 32, 411-417.

13. K. A. Hooper and J. Kohn, "Diphenolic monomers derived from the natural amino acid α-L-tyrosine: An evaluation of peptide coupling techniques", J. Bioact. Compat. Polym., **1995**, 10(4), 327-340.

14. A. U. Daniels, M. K. O. Chang, K. P. Andriano and J. Heller, "Mechanical properties of biodegradable polymers and composites proposed for internal fixation of bone", J. Appl. Biomaterials, **1990**, 1, 57-78.

15. V. Tangpasuthadol, A. Shefer, K. A. Hooper and J. Kohn, "Evaluation of thermal properties and physical aging as function of the pendent chain length in tyrosine-derived polycarbonates, a class of new biomaterials", *Symposium Proceedings Vol. 394: Spring Meeting of the Materials Research Society*, (A. G. Mikos, K. W. Leong, M. J. Yaszemski, J. A. Tamada and M. L. Radomsky, ed.), San Fransisco, CA, **1995**, Published by: Materials Research Society, Pittsburgh, PA, p. 143-148.

16. J. Choueka, J. L. Charvet, K. J. Koval, H. Alexander, K. S. James, K. A. Hooper and J. Kohn, "Canine bone response to tyrosine-derived polycarbonates and poly(L-lactic acid)", J. Biomed. Mater. Res., **1996**, 31, 35-41.

17. S. I. Ertel, J. Kohn, M. C. Zimmerman and J. R. Parsons, "Evaluation of poly(DTH carbonate), a tyrosine-derived degradable polymer, for orthopaedic applications", J. Biomed. Mater. Res., **1995**, 29(11), 1337-1348.

18. K. James, H. Levene, J. R. Parsons and J. Kohn, "Small changes in polymer chemistry have a large effect on the bone-implant interface: Evaluation of a series of degradable tyrosine-derived polycarbonates in bone defects", Biomaterials, **1999**, 20(23/24), 2203-2212.

19. S. Lin, S. Krebs and J. Kohn, "Characterization of a new, degradable polycarbonate", *Annual Meeting of the Society for Biomaterials*, Scottsdale, AR, **1991**, Published by: Society for Biomaterials, Algonquin, IL, p. 187.

20. O. M. Böstman, "Absorbable implants for the fixation of fractures", J. Bone Joint Surg., **1991**, 73(1), 148-153.

21. J. M. Spivak, N. C. Blumenthal, J. L. Ricci and H. Alexander, "A new canine model to evaluate the biological effects of implant materials and surface coatings on intramedullary bone ingrowth", Biomaterials, **1990**, 11(1), 79-82.

22. H. Plenk, "Prosthesis-bone interface", J Biomed Mater Res (Appl Biomater), **1998**, 43, 350-355.

23. R. Z. LeGeros and R. G. Craig, "Strategies to affect bone remodeling: osteointegration", J Bone Miner Res, **1993**(8), S583-96.

24. H. Levene, V. Phuvanartnuruks, S. Abramson, K. James and J. Kohn, "Chelation of calcium ions by some tyrosine-derived polymers may be related to improved bone biocompatability", *Annual Meeting of the Society for Biomaterials*, Providence, RI, **1999**, Published by: Society for Biomaterials, Minneapolis, MN, p. 13.

DYNAMIC MECHANICAL ANALYSIS IN POLYMERS FOR MEDICAL APPLICATIONS

J.F. Mano[1,2], R.L. Reis[1,2], A.M. Cunha[1]
[1]Department of Polymer Engineering,
University of Minho, 4800-058 Guimarães, Portugal
[2] 3B's Research Group – Biomaterials, Biodegradables and Biomimetics,
University of Minho, 4710-057 Braga, Portugal

Abstract

The Dynamic Mechanical Analysis (DMA) is a powerful thermal analysis technique, which allows to detect phase transitions and relaxation processes in a variety of materials. With this technique, the solid-state rheological properties of viscoelastic materials can be characterised over a wide range of temperature and frequencies. This chapter summarizes the principles and the capabilities of the DMA technique focusing on its uses on polymeric-based systems aimed to medical and environmental applications. The examples presented include the materials that have been investigated in our research group in the last few years, such as starch-based blends, proteins, polyethylenes, and composites thereof, among other materials. These newly developed biomaterials are being proposed for a range of biomedical applications that go from fracture replacement/fixation and tissue engineering scaffolding, to new partially degradable bone cements and hydrogels, carriers for controlled release of drugs and growth factors and new wound dressings and membranes.

1. Introduction

Many polymeric-based materials, both of synthetic or natural origin, have been proposed to be used in biomedical applications, particularly as components for medical devices, fixation or regeneration of tissues, filling of tissue defects, drug delivery systems or in scaffolds for tissue engineering. Some of such materials are becoming more and more complex in structure and composition in order to be able to perform accurately their physiological functions and to avoid a number of undesirable biological responses. Among them we should include copolymers or homopolymer, composites, blends (including interpenetrating polymer networks) and combinations of those compounded with a series of low molecular weight additives or biological material, offering significant property diversification.

In almost any situation the implanted devices are in a solid or gel phase and subjected to complex stress fields, such as the implants for substitution or regeneration of bone,

139

R.L. Reis and D. Cohn (eds.),
Polymer Based Systems on Tissue Engineering, Replacement and Regeneration, 139–164.
© 2002 *Kluwer Academic Publishers. Printed in the Netherlands.*

dental deficiencies, in maxilofacial surgery or in cardiovascular grafts and devices. As polymeric systems are viscoelastic in nature, the knowledge of their solid-state rheological behaviour is very important in the prediction of their mechanical performance in a real service situation [1].

Dynamic mechanical analysis (DMA) is a non-destructive technique widely used in the characterisation of polymer-based systems, including thermoplastics, thermosets, rubbers and composites, which is getting increasing importance in biomaterials research. Basically, it allows to characterise the viscoelastic properties of materials in a wide temperature and frequency ranges, by monitoring the sample's response upon an imposed controlled cyclic loads (with the corresponding development of stresses) or strain. One should alert in this context that, in fact, one imposes loads and the stresses are developed in the material as the result of the testing assembly and the sample's geometry. However, terms such as "application" or "imposition" of stresses are currently used in the literature and have been accepted by the reviewers, although they are not entirely correct.

Several DMA equipments are now commercially available, enabling studies in different mechanical configurations such as in tensile, flexural, compression or shear modes. Moreover, such equipments offer other possibilities of directly investigate other thermal/mechanical materials' features, such as.

(i) The application of a static load gives the creep behaviour of the material if the corresponding deformation (strain) is monitored as a function of time;

(ii) Stress relaxation tests can also be performed, by evaluating the stress as a function of time while the sample is kept at constant stain;

(iii) The application of a continuous increasing constant rate force (or strain) of with the concomitant recording of the developed strain (or stress) is equivalent to the well known quasi-static mechanical tests (tensile, compressive, flexure, etc.);

(iv) The dimensions of a sample (both linear and volumetric) may be register along a specific temperature program like in conventional thermodilatommetry tests. Besides conventional thermal-expansion experiments one may, for example, monitor the swelling in hydrogels or the onset in the temperature axis of the increase of the volume in materials blended with blowing agents, to obtain foams;

(v) Penetration tests are carried out in a heating programme and the softening transitions, such those occurring with the glass transition or melting, are detected.

Progresses are currently made that allow to probe in a sub-micron scale the viscoelastic properties in heterogeneous systems. It should be also mentioned that much more information can be reach if the DMA results are complement with data obtained from other techniques, such as NMR, X-ray diffraction, dielectric spectroscopy and other thermal, spectroscopic or microscopic techniques.

Besides the technological evolutions, the data treatment has also been developed in order to implement phenomenological models that allow to extract valuable informations. Fitting methods using different physical models allows to characterise the viscoelastic parameters in isothermal and isochronal conditions, permitting to correlate the relaxation processes with the molecular dynamics in the material and its microstructure (for example in composites, semi-crystalline systems or blends/composites).

This chapter is an attempt to exemplify and resume the capabilities of the DMA technique in the characterisation of novel polymer-based biomaterials, aimed to be used in a wide range of biomedical applications.

2. Linear viscoelasticity

Perfect elastic solids undergo an instantaneous strain that is a function of the imposed static load. Conversely, in perfect Newtonian liquids the stress is found to be proportional to the strain rate. Elastic solids store all the energy utilised in the deformation induced by the developed stress, which is completely used in an eventual recovery. On the other hand, all the mechanical energy used in the deformation of perfect liquid is dissipated as heat, preventing the recovery of its shape. Polymers feature a hybrid behaviour called viscoelasticity [2,3]. Their mechanical response depends on the time scale of the experiment. For example the creep compliance (ratio between strain and stress), or the relaxation modulus (stress divided by the constant strain applied) varies with time. The total mechanical response of a specimen includes perfect elasticity, total irreversible flow, and a coupling of elastic and viscous components, called anelasticity. The elastic and viscous components of the material's viscoelastic behaviour can be separated upon the study of a specimen response to a cyclic mechanical action.

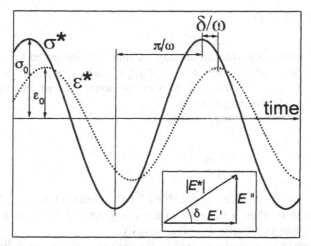

Figure 1: Scheme of the mechanical response (strain, ε) of a viscoelastic material subjected to a sinusoidal stress (σ).

When a specimen is subjected to a sinusoidal load, with a rate defined by a frequency f (in cycles.s^{-1}, or Hz) or an angular frequency $\omega = 2\pi f$ (in rad.s^{-1}), the response (a strain) though sinusoidal, is neither exactly in phase with the developed stress (as it would be for the case for a perfectly elastic solid) nor 90° out of phase (as it would be for a perfectly viscous fluid); thus the strain will lag behind the stress by some phase angle δ, between 0 and 90° (Fig.1). This delay is a result of the time necessary for molecular rearrangements. The sinusoidal stress, $\sigma = \sigma_0 \sin(\omega t + \delta)$, and the strain, $\varepsilon = \varepsilon_0 \sin(\omega t)$, may be written in a complex notation:

$$\sigma = \sigma_0 \exp(i\omega t + \delta) \text{ and } \varepsilon = \varepsilon_0 \exp(i\omega t) \tag{1}$$

where σ_0 and ε_0 are the stress and strain amplitudes and $i = (-1)^{1/2}$.
A full description of the linear viscoelastic response may be provided by the complex modulus, $E^*(\omega)$, defined as

$$E^*(\omega) = \frac{\sigma}{\varepsilon} = (\sigma_0 / \varepsilon_0)\exp(i\delta) = (\sigma_0 / \varepsilon_0)(\cos\delta + i sen\delta) = E' + iE'' \tag{2}$$

The storage modulus, E', is the elastic, or the real component of E^*, which is in phase with σ. The storage modulus is related with the stiffness of the material and compared to the tangent or secant moduli extracted from quasi-static tests [4]. The loss modulus, E'', is the viscous (also called imaginary) component of E^* which is $\pi/2$ out of phase in relation to σ. E'' is associated with the dissipation of energy, as heat, due to internal friction at the molecular level. The dissipation of energy in a complete cycle, per unit of volume, is given by

$$E_{dis} = \oint \sigma d\varepsilon = \pi E'' \varepsilon_0^2 = \pi D'' \sigma_0^2 \tag{3}$$

If no energy is transferred to the surrounding medium the dissipated heat will cause an increase of temperature of the sample: $\Delta T = f E_{dis}/\rho C_p$, where ρ and C_p are the density and the heat capacity of the material. Obviously, some part of the energy will be conducted away to neighbouring material.
The maximum stored elastic energy is given by

$$E_{st,\max} = \int_0^{\varepsilon_0} E' \varepsilon d\varepsilon = E' \varepsilon_0^2 / 2 = D' \sigma_0^2 / 2 \tag{4}$$

The loss factor, $\tan\delta$, is E''/E' and $|E^*|^2 = E'^2 + E''^2$ (inset in Fig.1). The ration of $E_{dis}/E_{st,\max} = 2\pi \tan\delta$ is often a measure of the damping capability of a material and called the specific loss or specific damping capacity.
The rotational/translational mobility within the polymer chains may be of different types, revealed as DMA peaks in E'' or $\tan\delta$ with different temperature/frequency positions. An example is shown in Fig. 2 for an amorphous polymer.
The local motions assigned to low amplitude motions of chain segments or within the lateral groups (β, γ... relaxations in amorphous polymers) are thermally activated processes usually detected at lower temperature (or high frequencies) relatively to the glass transition temperature. The molecular motions associated with the glass transition (α relaxation in amorphous polymers) have a cooperative character and its occurrence is accompanied by drastic changes in the mechanical properties of the material,

especially if it is an unfilled amorphous one. Each relaxation process has its own location in a *f vs T* plot. In this context it is worth to note that the total response of a viscoelastic polymeric material at a given temperature is not the result of a mechanism with a single characteristic time, but rather can be modelled assuming a distribution of times.

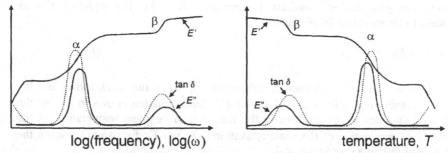

Figure 2: Typical viscoelastic response of an amorphous polymer presenting two relaxation processes: the α-relaxation related with the glass transition and the β-relaxation assigned to localised motions in the polymeric chains. The left graphic corresponds to an isothermal experiment and the right graphic to an isochronal scan.

In a dynamic experiment, a relaxation process is generally detected when its characteristic times are close to the experimental time scale in which the angular frequency ω is the relevant variable. If $\omega\tau<<1$ the frequency is much lower than the material's relaxation rate, τ^{-1}, and the molecular motions assigned to this time are able to fully follow the variation of the periodic field. If no irreversible flow occurs the materials will display typically an elastic behaviour. For the specific case of the glass transition, in this regime the material should be located in the rubbery plateau. When $\omega\tau>>1$, the molecular motions are unable to respond to the rapid change of the field applied. Consequently, the material appears stiffer and elastic and only motions with higher relaxation rates might be able to respond to the applied cycle (glassy state for the glass transition phenomenon). For intermediate values of the $\omega\tau$ product, the material displays a clear viscoelastic behaviour. Thus, the relaxational process appears (i.e., tan δ or E" pass through a maximum) when the value of one of the variables is in the range of the inverse value of the other one (i.e., $\omega\tau\sim1$).

Therefore, the actual parameter that controls the viscoelastic features of a polymeric material is the product $\omega\tau$, and the complete characterisation of the material's dynamics implies the study of the molecular motions throughout the largest range of $\omega\tau$. This can be experimentally carried out by varying either ω or τ. In isothermal experiments the time scale of the material is kept constant (by fixing temperature and pressure) whereas frequency is scanned. In isochronal experiments the time scale of the experiment is constant, by fixing the frequency, and τ is changing by cooling or heating the sample.

The temperature dependence of $\tau=1/(2\pi f)$, where f is here the frequency of maximum E", may present at least two forms [5,6]. In secondary relaxations (β, γ...) or in

144

relaxations assigned to molecular motions in the crystalline lamellae in semicrystalline polymers (α_c process), $\tau(T)$ follows usually the Arrhenius equation:

$$\tau(T) = \tau_0 \exp(E_a / RT) \tag{5}$$

where τ_0 is a pre-exponential factor, E_a is the activation energy and R is the gas constant. The glass-rubber transition follows the Vogel-Fulcher equation that is equivalent to the so-called WLF equation:

$$\tau(T) = \tau_v \exp\left[B / (T - T_v)\right] \tag{6}$$

Figure 3 shows clearly the Arrhenius behaviour (eq. 5) of the β-relaxation and the curvature of the α-relaxation that follows eq. (6). Such equation allows to predict the location of the relaxation processes in the time axis at a given temperature or the characteristic temperature of the same relaxations (T_β and T_α, for example) when the material is probed at a given time scale.

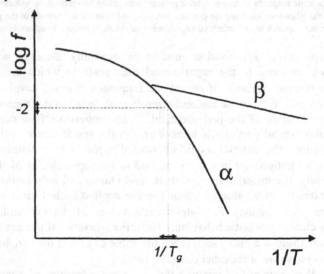

Figure 3: Relaxation plot showing the frequency dependence on temperature for the polymeric system shown in Fig. 2.

The instruments are limited in terms of the available frequency range, usually between 0.01 and 100 Hz. The time scale ranges may be extended by exploiting the time-temperature superposition principle [2,3,5]. Frequency scanning experiments are carried out at different temperatures and the obtained data are shifted along the log f axis in order to construct master curves in thermorheologically simple systems, that allow to estimate behaviour outside the range of the instrument.

3. Relevant information achieved by DMA

This section pretends to briefly describe the potential of DMA for being used in different research areas.

Detection and characterisation of relaxation processes: Conformational mobility within the polymeric chains appear in spectroscopic experiments as relaxation processes. The β transition is often associated with the toughness of a polymer [76]. It should be noticed here that other thermal analysis techniques such as differential scanning calorimetry (DSC) are not sufficiently sensitive to detect such secondary processes. Between T_β and T_α (T_α may be considered a measure of T_g if the experiments are carried at low frequencies) the material is stiff but has sufficient flexibility to not shatter under strain [7]. The study of the dynamics of the α-relaxation may give information about the structure of the material, due to its dependence on the polarity, branching/cross-linking, crystalline degree, composition, etc. Relatively to the last case, the plasticization effect may be particularly well characterised by DMA. Also in semicrystalline polymers, such as in polyethylene, one may detect molecular mobility within the material crystallites, often labelled as α_c-relaxations [2,5]. Such processes are important in several properties of the materials such as creep, processability and drawability.

Measurement of the modulus and damping: The elastic modulus is a basic parameter to characterise the stiffness of a material. The possibility of measure both the storage and loss moduli allows to estimate the damping properties of the material, i.e., to quantify the absorption of mechanical energy of the material is one of the advantages of the DMA technique. The frequency dependence of E' and E'' may be used to estimate the viscoelastic behaviour with time, using appropriate transformations [3]. Moreover, the crossover point between E' and E'' can be related to the molecular weight and molecular weight distribution by the Doi-Edwards theory [7]. The macro-structure of the sample from a same material may also have influence on the measured moduli components. A typical example is the case of porous structures (such as polymeric foams and scaffolds), where the overall mechanical response depends upon the geometrical arrangement, size distribution, interconnectivity, cell wall thickness and anisotropy of the pores [8]. This clearly justifies the high potential of the use of DMA in the characterisation of scaffolds for tissue engineering.

Physical ageing: below T_g all glassy materials are in a non-equilibrium state and undergo structural and physical changes with time. DMA is a suitable technique to characterise this phenomenon as, for example, the glass increases in modulus and decreases in damping [9]. In experiments at moderated frequencies (≈ 1 Hz), where samples that underwent different thermal histories below T_g, are heated across the glass transition region, differences are found in the low temperature side of the tan δ peaks. Lower frequencies should be used in order to have a more complete picture on the change of the viscoelastic response upon ageing in isochronal experiments, due to the competition between the time scales of the dynamic experiment and of the heating process itself (that has equivalent frequencies of about 10^{-2} Hz). In fact, at high frequency experiments the recovery of the thermodynamic equilibrium has already achieved when the dynamic glass transition is manifested.

Interface behaviour in composites and immiscible blends: Usually the interface behaves differently from the bulky phases. Therefore, in heterogeneous systems the interface plays a fundamental role in the overall mechanical behaviour of the material. The effects of the interface characteristics on the viscoelastic properties may be investigated by DMA (*e.g.* refs. [9,10]). The changes in the morphology of the phases in blends, such as the geometry or size of the regions, have also a great effect. For example, in interpenetrating networks (IPNs) the size domains influence strongly the glass transition and the damping properties of the material [11].

Cross-linked materials: a cross-linked polymer is in a rubbery state when well above T_g. Some structural properties may be described from DMA measurements using rubber elasticity theories, such as the molecular mass of the segments between cross-links or the cross-linking density. The thermal behaviour of thermosets or thermoplastic elastomers may be also investigated by DMA. This technique detects, for example, changes in the glass transition temperature and the corresponding distribution of characteristic times and the variation of E' (or G') in the rubbery plateau. Hydrogel systems, which are basically cross-linked polymer networks swollen in a liquid solvent, can also be studied by DMA.

Compatible multi-component systems: In complete or partial miscibile polymeric systems (e.g. polymer/plasticizer mixtures, polymer blends and random or block copolymers) there are the coupling of the dynamics of segments or molecules with different chemical nature. The study of the relaxation processes in such systems may provide valuable information about the structure and the general behaviour of such systems.

4. Applications of the DMA technique in the biomedical field

4.1. INTRODUCTION

In many clinical situations, a biomaterial is subjected to periodic loads. Important examples are implants used in both dental or maxilofacial applications, subjected to the masticatory function; biomaterials used in the orthopaedic field (substitution or repair of bone or cartilage bone cements, bone tissue engineering scaffolds, fixation plates and pins, etc.) that are exposed to the periodic loads of the patient movements; or devices used in cardiovascular applications, that experiment the pulsed blood pressure. Living tissues and, obviously, their biological constituents, present viscoelastic behaviour at certain temperatures and time scales. Interesting enough is the fact that at room or body temperature several natural materials (found in the human body or in other natural sources), including those designed by Nature for performing structural functions, display a clear viscoelastic nature. This provides evidences that Natures has a tendency to develop materials that are able to dissipate certain amounts of mechanical energy. A typical example is wet cortical bone where DMA results showed values of tan δ of ca. 0.04 that results mainly from the viscoelastic properties of type I collagen [12,13]. Such kind of characterisation is extremely useful because they add new specifications for new biomaterials for substitution/regeneration of bone or cartilage tissues. Other structural natural materials such as wood or cork also shows good damping properties at room temperature [8,14]. For the case of cork a peak in tan δ is clearly observed in the temperature axis with a maximum between 20-40 °C (f= 1 Hz) [14]. Another example is protein-based biomaterials. Injection moulded soy-protein thermoplastics presented a broad relaxation process above 40 °C (f= 1 Hz) [15]. The location of this process may be shifted to lower temperature if the protein is plascized, that occurs, for example, in wet conditions. Similarly, soy flour samples also showed a broad relaxation process between −40 and temperatures above 120 °C [16], with tan δ>0.2 at room temperature. Recently, the own viscoelastic properties of living cells have attracted some investigators. Living cells have defined shapes, and can robustly oppose being deformed. This structural integrity is primarily a result of the cells' cytoskeleton − a mesh of protein fibres producing gel-like strucutres. Recent studies reported elastic and frictional properties of a wide variety of living cell types, over a wide range of frequencies and under a variety of biological interventions [17]. It was shown that cells behave as soft glassy materials existing close to a glass transition, implying that cytoskeletal proteins may regulate cell mechanical properties. This kind of evidences leads one to speculate on the existence of some relevant advantages upon biological functions when the materials composing the tissues present a viscoelastic character. Moreover, these observations may strengthen the need to characterise the thermo-rheological properties of materials that are developed with the aim of being used in biomedical applications. DMA studies performed in samples immersed in physiological simulated fluids (saline or SBF solutions), at body temperature, are of particular importance, as they provide more accurate informations about the real rheological behaviour of the material in its clinical use.

4.2. VISCOELASTIC PROPERTIES OF STARCH AND SOME BIODEGRADABLE STARCH / SYNTHECTIC POLYMER BLENDS

Our group has being investigating natural-based biomaterials, mainly from starch and other polysaccharides and proteins, with potential to be used in biomedical applications. These include the use of biomaterials as scaffolds for the tissue engineering of bone and cartilage [18], membranes [19,20], materials for bone fixation and replacement as well as for filling bone defects [21-22], carriers for the controlled release of drugs and other bioactive agents [23,24], and new hydrogels and partially degradable bone cements [25,26]. Furthermore it has been shown in several studies that these polymeric systems exhibit an interesting degradation kinetics [18,27,28], an undoutable biocompatible behaviour [29-31] that is not typical at all of biodegradable systems and tailorable surface characteristics that can be adjusted to favour cell adhesion and proliferation [27,30,32,33]. Before discussing some results obtained in different starch/synthetic polymer blends, a brief presentation on DMA results in a pure thermoplastic starch system will be made.

4.2.1. DMA results in injection moulded potato starch

The design of new thermoplastic polymers from natural origin materials, that are very cheap and from renewable origin, would be a major breakthrough opening new doors for the environmental or biomedical applications of degradable polymers. However it has been difficult, so far, to develop systems that easily processed by melt based techniques, such as injection moulding (that offers the possibility of obtaining complex shape parts from thermoplastics), and at the same type cheap and producible in great quantities. The development of thermoplastic potato starch/flax fibres systems is just another attempt on that direction. After compounding with different fibre content, this material was injected moulded into dumb-bell tensile samples, that were characterised by DMA.

Figure 4 shows the storage and loss moduli obtained in isochronal conditions ($f = 1$ Hz) using the three-point-bending mode (platform with a 20 mm span) of the Perkin-Elmer DMA7e apparatus. A relaxation process appears with a peak near 80 °C, for all materials. The tan δ peak appears at higher temperatures (see Fig. 2). The intensity and position of the E" peak does not vary consistently with the flax content. This is due to fluctuations in homogeneity in the samples, that occurs frequently in natural materials. There is often a need to perform more than one DMA run with different samples in order to have a reliable picture of the rheological characteristic of the materials. The use of statistical tools to handle the results (t-tests, ANOVA tests, Taguchi methods, etc.) has been reported in the description of DMA results [12,34,35].

The origin at the molecular level of the observed relaxation, plotted in Figure 4, must be assigned to motions within the two main structures composing starch: poly(1-4)-α-D-glucan (amylose, essentially linear) and amylopectin, which is branched amylose through (1-6)-α linkages. (More informations about the structure and properties of starch polymers may be found in ref. 36.) Complementary characterisation is useful in the assignment of the observed relaxation processes. For example, the "electric equivalent" of DMA, dielectric relaxation spectroscopy, allowed to detect four relaxations in a broad temperature and frequency ranges [37] in starches from different biological sources:

(i) two local γ-processes around -120 °C, were assigned to motions of the chain backbone and rotation of methylol groups;

(ii) a glass-transition-like β-process around room temperature but located at higher temperature for gelatinised starch and;

(iii) for the case of granular starch, an α-process was detected at 60-80 °C being assigned to gelatinisation.

Note that, in the present study, the starch was extruded in wet conditions before the injection moulding, and thus the final material underwent already gelatinisation. For gelatinised waxy maize starch with low water content, the β-process appears at temperatures above 60 °C (maximum of ε" at ~50 Hz) [37]. As in our case the studied injected moulded samples are also in a highly dry state one could also attribute the observed peak to this process. Note that this relaxation can be shifted to lower temperatures (for example body temperature) with plasticization. The tail in E" at lower temperature is assigned to the γ-processes, that should appear as a peak at lower temperatures.

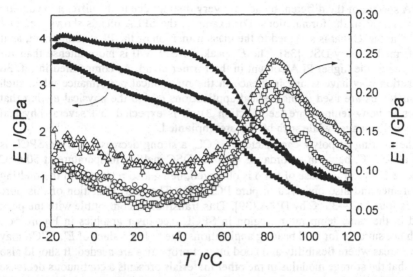

Figure 4 - Storage modulus (solid symbols) and loss modulus (open symbols) against temperature (frequency of 1 Hz) for injected moulded thermoplastic potato starch samples containing different amount of flax fibres: squares (no flax), circles (15% w/w flax fibre) and triangles (25% w/w flax fibre). The samples are bars with ~2 mm thickness and ~5 mm depth. The experiments were carried out under an heating rate of 4 °C.min^{-1}. The authors acknowledge Cláudia Vaz for preparing the samples.

The addition of flax fibre increases the storage modulus that reaches almost 4 GPa at room temperature. Even for the unfilled starch the stiffness is higher than many synthetic polymers. These results show that, in terms of mechanical performance, starch or starch-based materials have good potential to be used in load-bearing biomedical applications, including those related to orthopaedic applications. Further treatments, such as the incorporation of bioactive ceramics and the creation of cross-linkings, could increase the stiffness and the bioactivity.

4.2.2. DMA results on starch blends

The starch-based thermoplastic blends studied in this work were blends of (i) corn starch /ethylene-vinyl alcohol copolymer (SEVA-C) (50/50 wt/wt), (ii) corn starch / cellulose acetate (SCA) (50/50 wt/wt), (iii) corn starch / poly (ε-caprolactone) (SPCL) (30/70 wt/wt), (iv) corn starch / poly (L-lactic acid)-(50/50 wt/wt) (SPLA50) and (v) corn starch / poly (L-lactic acid)- (30/70 wt/wt) (SPLA 70). The as-received materials appear in a granular form. The studied samples were injection moulded into small ASTM dumb-bell tensile pieces (cross-section 2x4 mm^2).

Figure 5 shows the storage modulus, loss modulus and loss factor at 1 Hz against temperature for all the studied blends. Such results complement the information extracted from the thermal properties of such materials, obtained with DSC and thermogravimetry [38].

The DMA spectra of the different blends are very distinct due to the different synthetic polymers present in the formulations. For example, the SPLA blends shows clearly a peak in E" at 58 °C that is assigned to the glass transition of the PLA component, as it was confirmed also by DSC [38]. The E" peak for SPLA70 is more intense than for SPLA50 due to the higher PLA content in the former blend. As commented in ref. 38 this relaxation may have strong influence in the mechanical performance when such kind of materials are used in biomedical applications due to the physical ageing that undergoes at body temperature (see Section 3). It is expected that several physical properties will changes while such blends are implanted.

Due to the melting of poly(ε-caprolactone), PCL, a strong decrease of E' in SPCL is detected at ~50 °C, being in accordance with the DSC melting temperature at 50.6 °C [38]. Above 50 °C an increase of tan δ is observed, being also related with the melting of the polymer and also observed in pure PCL [39]. The glass transition of this pure polymer is found at ~-35 °C by DMA [39]. This transition is compatible with the peak observed in the same temperature region in SPCL (see inset graphics in Figure 5c). Although not suitable for load bearing applications due to low values of E', SPCL may be used in areas where flexibility and good impact properties are needed. It should also be noted that the storage modulus in the other materials presents a continuous decrease below 40 °C. This behaviour was also found in thermoplastic proteins plasticized with glycerol [15] and assigned to a "molecular lubricant" effect of the plasticizer.

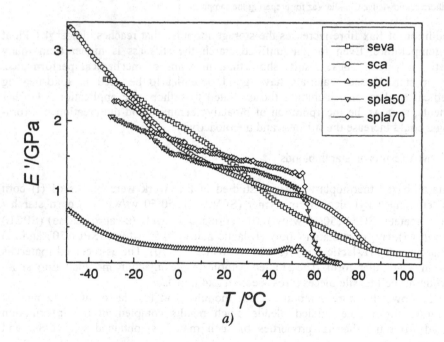

a)

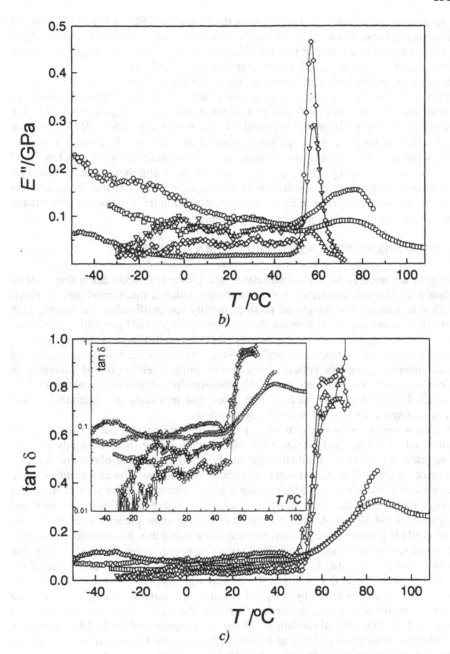

Figure 5 – Temperature dependence of the storage modulus (*a*), loss modulus (*b*) and loss factor (*c*) at 1 Hz for the studied starch-based materials. All the experiments were carried out in a Perkin-Elmer DMA7e apparatus under a three-point bending mode (platform of 15 mm span), at 4 °C.min⁻¹. The inset graphics in Figure 5c shows in more detail the tan δ in a logarithm scale, enabling a better perception of the smaller loss factor values.

The relaxation process observed at ~75 °C in the E'' traces of SEVA-C and SCA should be assigned to molecular motions within the starch fraction. A more complete study of this relaxation may be found in ref. 40. This relaxation has the same origin as the one observed, and already discussed, in pure starch at ~80 °C (Figure 5).

Finally, an important feature observed in Figure 5 is the relaxation observed at ~30 °C for SEVA-C. This process can be meaningful in some biomedical applications due to the ability of this material to absorb part of the mechanical energy associated to the imposed cyclic load (at frequencies around 1 Hz) at body temperature. DMA results on several ethylene-vinyl alcohol copolymers pointed out for a α-relaxation, assigned to the dynamic glass transition of the copolymer, at temperatures around but below 50 °C (E'' peak at 3 Hz) [41]. The glass transition of such copolymers decreases with decreasing ethylene content. As the ethylene content in the copolymer in SEVA-C is lower (around 60% mol) than in the copolymers studied in ref. 41, one can attribute the peak at ~30 °C to the α-relaxation of the copolymer.

4.3. HYDROGELS AND ELASTOMERIC BIOMATERIALS

Hydrogels are used in both pharmaceutical and biomedical application due to their permeability of small molecules, soft consistency, reduced mechanical and frictional irritation to tissues, low interfacial tension, facility for purification and mainly high equilibrium water content that make them, in terms of physical properties, similar to living tissues. Applications of hydrogels include controlled release systems of enzymes, drugs, hormones, anticoagulants, etc., artificial skin, contact lenses, blood contact materials, interface between bone and an implant, etc. (a list of references in this context can be found in reference [25]). Non-swollen cross-linked systems include materials for dentistry uses such as soft lines and maxillofacial materials, or soft membranes or valves for cardiovascular applications.

Soft liner systems are primarily used in patients presenting mucosal deficienties. As commented by Saber-Sheikh et al, [42], other functions for these materials include: aiding retention where the residual ridges are bilaterally undercut, obturation for cleft palates or surgical defects, maxillofacial prosthetics and, occasionally mouth guards. Examples of the use of the DMA technique in the characterisation of soft liners are reported for commercial materials based on silicone, acrylics, fluoroelastomer and phosphazines [34,42-44]. Alternative materials for the same purpose were studied by Water et al [45], which highlighted their potential clinical use. Kalachandra et al. [43], performed compression isochronal DMA experiments on cured materials in dry and wet environment, enabling to compare the moduli and the damping properties of the different materials in both conditions. DMA experiments performed while the sample is immersed in a simulated physiological solution are scarcely reported, but may be extremely significant if one pretends to predict the viscoelastic performance of a biomaterial in physiological conditions. Murata et al performed the DMA experiments in isothermal conditions (37 °C) and compared the studied commercial materials in terms of tensile storage and loss moduli [34]. They also investigated the sensitivity of such materials on the changing of the dynamic viscoelasticity with frequency, between 0.01 to 100 Hz. Changes of properties over a 1 or 3 year period were also monitored

[34,44], indicating that DMA may be a suitable technique to evaluate the long term transformations upon the mechanical properties in biomaterials. Such studies allowed to conclude that the changes of the viscoelastic properties could depend strongly on the type of material tested, all commercially available.

The sol-gel transition in gel systems can be characterised by DMA, by monitoring the strong increase of the storage modulus in the temperature axis. Some DMA works have been reported results on a variety of materials, such as for block-copolymers of polyoxyethylene and polyoxypropylene [46], that have been employed for the topical administration of drugs, poly(vinyl alcohol) in a dimethyl sulphoxide/water solvent [47], hydroxyethylated starch aqueous systems [48] or cellulose in an ammonia/ammonium thiocyanate solvent [49].

Soft matrices for cell cultures are currently in development, to be used, for example in skin substitutes for wound healing. Collagen and its derived matrices have been suggested in this area, due to their compatibility. The characterisation of collagen gel solutions and collagen matrices, cross-linked with different amounts of glutaraldehyde was carried out using DMA [50].

Several formulation of hydrogels have been developed for drug delivery systems. Recently, new hydrogels were designed [25] by blending a biodegradable starch-based blend (SEVA-C, presented in 4.2.2) with copolymers of acrylic acid and acrylamine which are known to have adjustable swelling kinetics with applications for drug (bioactive agents) release. These hydrogels have proved to have properties that allow for their use as in-situ polymerizable partially degradable bone cements [25,51]. Flexural DMA experiments were carried out on different of these formulations, in both the hydrogel (24 h immersed in water) and xerogel (dry state) forms [25]. For the xerogels the influence of the cross-linking was clearly detected by the higher decrease of E' across T_g in the thermoplastic as compared with the cross-linking material. Strong variations in T_g are found for the different compositions of the copolymer, due to the different T_g of poly(acrylic acid) and poly(acrylamide) and the plasticization of absorbed water was clearly confirmed. The effect of the composition was also discussed for the hydrated hydrogels, allowing to characterise such materials in a more realistic situation taking into account the potential of such materials for clinical uses.

An interesting example of using the DMA technique in hydrogel systems were reported by Cauich-Rodriguez et al [52] for blends of poly(vinyl-alcohol-vinyl acetate) with poly(acrylic acid) or poly(vinyl pyrrolidone). The xerogels were characterised under tensile isochronal experiments allowing to study the influence of composition and thermal treatment on tan δ. The swelling of the hydrogels was followed under dynamic measurements where the storage modulus was monitored against time while the sample is immersed in water. Finally the authors were also able to measure both the storage modulus and tan δ as a function of frequency and E' as a function of temperature of the hydrogels.

4.4. CURABLE SYSTEMS FOR DENTAL APPLICATION AND AS BONE CEMENTS FOR ORTOPHAEDIC SURGERY

Resins systems that are able to polymerise under heat, visible light and microwave radiation may be employed in orthopaedic surgery and dental applications for the fixation of joints, prosthesis to act as a stress distributor between the artificial implant and the bone, as well as filling self-curing materials for bone and dental cavities.

Whilst the kinetics of the evolution of the curing process may be investigated by DMA (see, for example, ref. [53] for applications in polymeric coatings), this technique has been mainly used to characterise the final cured biomaterial. Usually, the commercial available materials used in this field contain poly(methyl methacrylate) (PMMA), due to its biostability, and good mechanical properties. It is in fact the only material for anchoring cemented arthroplasties to the contiguous bone. However, the formulations can be complex, containing initiators, cross-linking substances, ceramic fillers and other additives, and the resulting materials can appear with very different viscoelastic properties. For example, Vaidyanathan and Vaidyanathan [54], compare different commercial denture base resins and concluded that both the microwave and thermal cure resins have lower flexural storage and loss moduli and presented lower glass transitions relative to the visible light cure resin system. Papadogiannis et al [55], also reported DMA studies on chemical- and light- cured composites for dental applications. Pascual et al [56] introduced a higher molecular weight and more hydrophilic monomer to the current methyl methacrylate monomer in order to produce an improved cement, in terms of decrease of maximum polymerisation temperature and decrease of residual monomer content. These new formulations were also characterised by DMA, showing a strong dependence on the new monomer content. The DMA technique was also intensely used in the characterisation of surgical Simplex-P radiopaque bone cement, modified by tricalcium phosphate, hydroxyethyl methacrylate and ethylene glycol dimethacrylate [57]. The same group also grafted methyl methacrylate onto ultrahigh molecular weight polyethylene fibres in order improve the interfacial bonding between the grafted fibres and the acrylic bone cement matrix. This procedure allowed to produce cements with enhanced mechanical properties, as confirmed with DMA analysis [58].

New partially biodegradable and bioactive materials with potential to be used as bone cements were developed by combining the biodegradable character of the starch/cellulose acetate blend (see section 4.2.2) with the biostability of PMMA and poly(acrylic acid) and the bioactive and stiff character of hydroxyapatite [26]. All formulations were characterised by DMA, enabling to monitor both moduli components. The glass transition of the different materials did not correlate with the solid/liquid fraction or the hydroxyapatite content. The dispersion of the results when one repeats the same experiment were also discussed and the flexural dynamic moduli of the materials were compare with the quasi-static mechanical results. A broad relaxation process observed at 10 °C in all formulations may be attributed to the β-relaxation of the copolymer and can provide interesting damping capabilities of such materials in some clinical applications.

4.5. DEGRADABLE BIOMATERIALS

Many degradable polymers have been proposed to be used in a variety of applications in the biomedical field. Good examples are poly(lactic acid) - PLA, poly(ε-caprolactone) - PCL and polyglycolic acid (PGA). There is still, however, a need for the development of novel degradable polymer biomaterials exhibiting an adequate balance of mechanical properties and degradation kinetics.

4.5.1. Classic degradable biomaterials

Blends of PLA and PCL were studied by DMA, which permitted to conclude about the immiscibility of such blends [59]. However, DMA studies on PCL and cellulosic esters also allowed to deduce that PCL could be partially miscible with cellulose acetate butyrate [60]. Block copolymers based on segments of poly(D,L-lactic-glycolic acid) and PCL have also been intensively studied in order to obtain systems with improved degradation rate and permeability; DMA provided a valuable method to observed the relaxation processes in such materials [61]. PLA can also be blended with plasticizers in order to improve the flexibility and impact properties. Different plasticizers were used (glycerol, citrate ester, polyethylene glycol, PEG monolaurate and oligomeric lactic acid) by Martin and Avérous and the materials were characterised by DMA that allowed to conclude about the plasticization efficiency and compatibility extent [62], by looking at the glass transition. As commented before, the T_g of PLA systems are usually ~20 °C above body temperature and thus implanted materials undergo structural relaxation. It was shown that such process could be investigated by DMA, providing additional information on the typical differential scanning calorimetry results [63]. However, a complete description of structural relaxation, and of the glass transition dynamics in general, is still to be carried out. Phenomenological approaches should be applied in order to achieve the distribution of relaxation times and their temperature dependence. Complementary dielectric spectroscopy results should be helpful as their provide the dynamics of molecular motions throughout a broad frequency range [64,65].

A composite system of poly(DL-lactic acid), as the amorphous matrix, reinforced with hydroxyapatite (HA) particles and/or semi-crystalline poly(L-lactic acid) fibres were characterised by DMA [66]. It was found that HA incorporation in the PDLLA matrix almost doubled the storage modulus but the further inclusion of the PLLA fibres did not improve the E' at body temperature. However, the three-component system showed an higher T_g in comparison with the matrix alone and damping capabilities that are extended in a broader temperature range, ranging from ~30 to ~80 °C. Tetraethoxysilane/PCL hybrid biomaterials were also studied by DMA [39]; it was possible to conclude that by changing triethoxysilane PCL end-groups by hydroxyl ones, the glass transition increases and a small increase in E' is observed. The same effect is also seen when the number of functional end-groups per PCL chains increases. The effect of the composition and curing was also investigated by DMA in that work [39].

4.5.2. Blends with biopolymers

Blends of synthetic and natural origin polymers represent a modern way to develop new biomaterials. Synthetic polymers may present a wide and tailored range of mechanical properties and can be processed by a variety of techniques. On the other hand, as commented in Section 4.2., natural origin polymers are obtained from renewable sources and present good biocompatibility [29-31]. Cascone presented a series of DMA studies in partially biodegradable polymer blends in the form of films or hydrogels [67]. Poly(vinyl alcohol), poly(acrylic acid), PMMA, ethylene/vinyl alcohol copolymers and polyurethanes were used as synthetic components and collagen, gelatin, hyaluronic acid, starch and dextran as biological components. The miscibility and the effect of treatment by glutaraldehyde vapour, which induce cross-linking, or thermal dehydration were factors that were especially inspected in that work.

In Section 4.2.2. the DMA spectra of biodegradable blends of starch and a series of synthetic polymers were presented. As in other natural-based biomaterials, these blends have a dominant hydrophilic character, fast degradation rate and, in some cases, unsatisfactory mechanical properties, especially under wet environments. The effect of the moisture in SEVA-C was investigated by DMA [40]; a clear plasticization effect due to the presence of water was detected, particularly for the relaxation process at ~30 °C and a consequently decrease of the storage modulus above ~20 °C as the moisture level increases. This work also includes the study of the effect of the degradation of SEVA-C in physiologic fluids on the solid-state rheological performance.

4.5.3. Effect of the β-radiation

It should relevant to briefly discuss on the effect of the sterilisation in the general performance of materials for biomedical applications. Such procedure was carried out in three of the starch-based biomaterials mentioned in Section 4.2.2. using β-radiation. Different doses of radiation were used to investigate the possibility of using this sterilization technique also has a treatment to tailor the surface and bulk properties (namely mechanical) of these polymers [68]. DMA was used in this study to explore this issue. It is interesting to notice that no clear changes were detected in both E' and tan δ for the irradiated samples as compared to as-processed samples [68]. One may then conclude that, at least no deterioration of the viscoelastic behaviour of such materials occurs in the studied materials.

4.5.4. Chemical and physical modification of starch-based blends

Such starch-based may be chemically modified in order to reduce the water uptake ability and the degradation rate and to improve their mechanical performance. Such changes of properties were achieved in SEVA-C, SCA and SPCL by chain cross-linking [27]. The DMA results showed a tendency for a higher E' in the modified materials. Moreover, the relaxation peak at ~80 °C for SEVA-C and SCA (Figures 5b and 5c) seems to shift to higher temperatures with the cross-linking reactions [27]. Physical methods, such as adding stiffer fillers or changing the processing conditions or methods can also improve the mechanical properties of the studied starch-based materials. The introduction of hydroxyapatite (HA) increases the stiffness of injection moulded SEVA-C pieces, as reported in a previous work using the DMA technique [69]. The position of the relaxation at ~30 °C of SEVA-C do not change with the inclusion of HA but the relaxation peak at ~80 °C broadens, suggesting the presence of a new relaxation at ~60 °C [69]. This process can be assigned to molecular motions in lower molecular mass starch-based chains that results from the scission or other degradation processes in the starch fraction that occur during the processing of the composites. The work in ref. 69 also included a DMA study on SEVA-C materials with porous structure that was induced by processing the thermoplastic with different contents of distinct blowing agents. The obtained injected moulded parts present a very compact and oriented surface layer and a inner porous structure. Such kind of procedure has been used in our research group in order to produce scaffolds for hard tissue engineering. The DMA results on the foamed materials showed that their viscoelastic properties did not change significantly with respect to the bulk material, indicating, for example, that the inducing of porosity did not decrease considerably the flexural stiffness of the materials. This is a good example that indicates that DMA is also a suitable technique to investigate the mechanical performance of materials for tissue engineering applications.

Beside adding a ceramic filler to the starch-based blends, it was found that further improvement of the mechanical performances of SEVA-C could be reach by using non-conventional processing techniques, such as *Shear controlled orientation in injection moulding* (SCORIM) [70,71]. This processing technique relies on the application of shear stress fields to the melt/solid interfaces during the packing stage by means of hydraulically actuated pistons. After the filling of the cavity mould, the molten polymer is continuously sheared as the solidification progressively occurs from the mould wall to the moulding core part inducing a significant improvement of anisotropy, especially in semi-crystalline systems. More details on the SCORIM processing of biomaterials and the obtained results may be found in some of our previous works [70-73]. The application of the SCORIM process typically doubled the flexural storage modulus of SEVA-C as compared with the same material processed by conventional injection moulding. The simultaneous use of SCORIM and the filling of SEVA-C with HA particles increased even more the stiffness. For example if one takes the E'(37 °C) at 1 Hz of the unfilled SEVA-C processed by conventional injection moulding, that yield ~ 1 GPa (see Fig. 5a), the work reported in ref. 68 proved that this values could reach almost 4 GPa with the inclusion of 50% (wt/wt) of HA. Further increase of HA and the addition of appropriate coupling agents could enable, in principle, to produce composite with modulus above 7 GPa (in the bounds of the lower limit for human cortical bone). The SCORIM processing of SEVA-C/HA composites also broaden the relaxation peak of starch (at ~80 °C) in the low temperature side, in comparison with the pure SEVA-C material [71], being consistent with the results reported in ref. 69.

4.6. NON-DEGRADABLE POLYMER SYSTEMS FOR HARD TISSUE REPLACEMENT

The development of new hard tissues replacement materials from synthetic polymers have been hindered by the mechanical performance required for the implant. Polyethylene (PE), has a bio-inert character and can be easily sterilised, being the most important polymer used in hard tissue replacement. Ultra-high-molecular-weigh PE (UHMWPE) has been used in several orthopaedic applications, such as in hip joint implants and accetebular cups. The high viscosity of this material prevents its easy processability by injection moulding.

158

In our group the SCORIM processing technique has been used to improve the mechanical properties of injection-moulding PE grades. The viscoelastic properties of PE processed with SCORIM were compared with those on the same material processed by conventional injection moulding methods [73]. In this study isothermal DMA experiments were carried out at different temperatures. The construction of master-curves at a single temperature from the all set of results allowed to predict the viscoelastic behaviour of the materials in an extremely broad time scale. The shift factors assigned to the master curve and the fitting of the obtained master curves using known models allowed to access to relevant structural and rheological informations. The flexural storage modulus of the SCORIM processed PE is approximately twice the one of conventional processed PE, in the studied frequency/temperature range [73]. The flexural results for a frequency of 1 Hz are shown in Figure 6, as a function of temperature.

The relaxation peak observed in Figure 6 (centred at ~50 °C) is assigned to the molecular motions within the crystalline fraction of PE and is usually labelled as α_c-relaxation. This process has been assigned to screw-like motions within the lamellae structure of polyethylene, involving also some mobility in the inter-spherulitic amorphous phase. The values of the α_c-relaxation's activation energies scatter strongly in the different studies that have been carried out. Typical values vary from ca. 100 to higher than 200 kJ.mol^{-1}. More details on the features of this process may be found in refs. [73,74]. Figure 6 shows that the α_c-relaxation is already active at 1Hz at the physiological temperature, providing an important dissipation mechanism during cyclic deformation in clinical situations. Even in physiological conditions we do not expect a significant changes in the position of this process, due to the hydrophobic character of PE. At higher temperatures (or lower frequencies) the intensity of this process is even enhanced with the SCORIM processing.

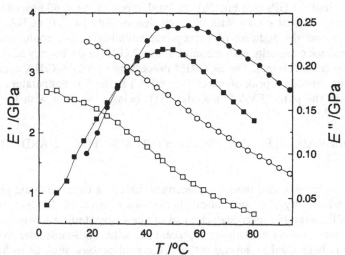

Figure 6- Flexural storage (open symbols) and loss (filled symbols) moduli for an high-density PE in the α-relaxation region. Square: material processed by conventional injection moulding. Circles: material processed by SCORIM. The results were extracted from ref. 73.

Hydroxyapatite (HA) reinforced PE composites have been proposed as so-called bone analogue materials [75], combining the ductile and viscoelastic polymeric matrix with a stiff bioactive inorganic phase. DMA was used to characterise such composites, focusing on the effect of HA content, temperature and particle size on the storage modulus and tan δ [76]. E' was found to increase concomitantly with the quasi-static modulus as the HA content increases. Moreover, the filler-matrix interface showed to have an effect on the overall viscoelastic performance of these composites [76].

It was found that the mechanical properties of such PE/HA composites could be improved using the SCORIM processing [77]. For example, at 25 °C the storage modulus at 1 Hz of the conventionally injection moulded and processed SCORIM PE/HAs composites are 2.1 and 3.2 GPa, respectively, while the quasi-static tensile moduli are 1.9 and 5.9 GPa, respectively. The stiffness could be even enhanced by using coupling agents, that improves the interfacial interaction between the polymer and the ceramic phases. As observed in Figure 6 for pure PE, the E'' peak of the α_c – relaxation also slightly shifts to higher temperatures for the composites when the SCORIM is used instead of the conventional moulding [77].

5. General conclusions

This work outline the usefulness of the dynamic mechanical analysis technique (DMA) in the characterisation of biomaterials aimed to be used in different applications. It is manly based on the fact that (i) many living tissues (even living cells) or biological molecules display frequently a viscoelastic character in physiological conditions and (ii) the general performance of an implanted biomaterial will depend on its viscoelastic properties. The literature overview on this subject allowed to conclude that this technique is being increasingly applied on the biomedical area due to the enhanced perception of researchers on the valuable informations that are extracted by DMA and due to the improved equipments that are now commercially available.

There are currently apparatus that allow to perform experiments while the material is immersed in a fluid, at controlled temperature, which allows to reproduce realistic environment. Besides dynamic experiments, modern equipments permit to run conventional static or quasi-static experiments (creep, stress-relaxation, stress-strain, penetration, dilatometry...) or even more unusual tests, such as thermally stimulated recovery/creep, that has been scarcely used in the biomaterials area but has proved to be a powerful method, complementary of DMA, for characterising relaxation processes [78,79].

The list of applications of DMA in different materials presented in this work is not exhaustive but rather pretends to exemplify the use of this technique in different areas, showing that it is a valuable tool in laboratories aimed to develop or characterise materials for biomedical applications.

References
1. Jones, D. S. (1999) Dynamic mechanical analysis of polymeric systems of pharmaceutical and biomedical significance, *Internat. J. Pharmac.* **179**, 167-178.
2. McCrum, N. G., Read, B. E. and Williams, G. (1991) Anelastic and Dielectric Effects in Polymer Solids, Dover, New York.
3. Ferry, J. D. (1980) Viscoelastic Properties of Polymers, 3rd Ed., Wiley, New York.
4. Mano, J. F., Viana, J. C. (2001) Effects of the strain rate and temperature in stress-strain tests: study of the glass transition of a polyamide-6. *Polym. Testing* **20**, 937-943
5. Strobl, G. (1996) The Physics of Polymers, Springer-Verlag, Berlin Heidelberg.
6. Richert R., Blumen A., *Eds.* (1994) Disordered Effects on Relaxational Processes, Springer-Verlag, Berlin.
7. Menard, K .P. (1999) Dynamic Mechanical Analysis, A Practical Introduction, CRC Press, Boca Raton.
8. Gibson, L. J., Ashby, M. F. (1997) Cellular Solids. Structure and Properties Cambridge University Press, 2nd Ed., Cambridge
9. Wetton, R. E. (1986) Dynamic mechanical thermal analysis of polymers and related systems, in Dawkins, J. V. (ed.), *Developments in Polymer Characterization – 5*, Applied Science Publishers, London, pp. 179-221.
10. Harris, B., Braddell, O. G., Almond. D. P., Lefebvre, C., Verbist, J. (1993) Study if carbon-fibre surface treatments by dynamic-mechanical analysis, *J. Mat. Sci.* **28**, 3353-3366
11. Kim, S. C., Sperling, L. H., eds (1997) IPNs around the world, Science and Engineering, John Wiley & Sons, Chichester
12. Yamashita, J., Furman, B. R., Rawls, H. R., Wang, X., Agrawal, C. M., (2001) The use of dynamic mechanical analysis to assess the viscoelastic properties of human cortical bone, *J. Biomed. Mater. Res. (Appl. Biomater.)* **58**, 47-53
13. Yamashita, Li, X., J., Furman, B. R., Rawls, H. R., Wang, X., Agrawal, C. M. (2002) Collagen and bone viscoelasticity: a dynamic mechanical analysis, *J. Biomed. Mater. Res. (Appl. Biomater.)* **63**, 31-36.
14. Mano, J. F. (2002) The viscoelastic properties of cork, *J. Mat. Sci.*, **37(2)**, 257-263.
15. Vaz, C. M., Mano, J. F., Fossen, M. van Tuil, R. F., de Graaf, L. A., Reis, R. L., Cunha, A. M. (2002) Mechanical, Dynamic-Mechanical and thermal Properties of Soy Protein-Based Thermoplastics with potential Biomedical Applications, *J. Macrom. Sci.-Phys.*, **B41(1)**, 33-46.
16. Yildiz, M. E., Kokini, J. L. (2001) Determination of William-Landel-Ferry constants for a food polymer system: Effect of water activity and moisture content *J. Rheol.* **45**, 903-912.
17. Fabry, B., Maksym, G. N., Butler, J. P., Glogauer, M., Navajas, D., Fredberg, J. J. (2001) Scalling the microrheology of living cells, *Phys Rev Lett* **87(14)** art. no. 148102.
18. Gomes, M. E., Ribeiro, A. S., Malafaya,P B., Reis, R. L., Cunha, A. M. (2001) A New Approach Based on Injection Moulding to Produce Biodegradable Starch Based Polymeric Scaffolds, *Biomaterials* **22**, 883-889

19. Vaz, C. M., Fossen, M., Cunha, A. M., Reis, R. L. (2000) Casein and Soybean Thermoplastic Proteins as Alternative Biodegradable Polymers for Biomedical Apllications, 6th World Biomaterials Congress, Hawai, USA, Apr., p. 429.
20. Silva, R. M., Mano, J. F., Reis, R. L. (2002) Development of chitosan membranes with controlled swelling properties for biomedical applications, ESB2002, European Society for Biomaterials, Barcelona, Spain, Sept.
21. Reis, R. L., Cunha, A. M. (2000) New Degradable Load-Bearing Biomaterials Composed of Reinforced Starch Based Blends, J. of Appl. Med. Polym. 4, 1-5
22. Sousa, R. A., Mano, J. F., Reis, R. L., Cunha, A. M., Bevis, M. J. (2002) Mechanical performance of starch based bioactive composites molded with preferred orientation for potential medical applications. Polym. Eng. Sci., in press
23. Malafaya, P. B., Elvira, C., Gallardo, A., San Román, J., Reis, R. L. (2001) Porous Starch-Based Drug Delivery Systems Processed By A Microwave Treatment, J. of Biomat. Sci. – Polym. Ed. 12, 1227-124
24. Vaz, C. M., De Graff, L. A., Reis, R. L., Cunha, A. M (2002) pH-Sensitive Soy Protein Hydrogels for the Controlled Release of an Anti-Inflamattory Drug, J. of Biomat. Sci. – Polym. Ed., submitted
25. Elvira, C., Mano, J. F., San Román, J., Reis, R. L. (2002) Starch Based Biodegradable Hydrogels With Potencial Biomedical Applications As Drug Delivery Systems, Biomaterials 23, 1955-1966
26. Espigares, I., Elvira, C., Mano, J. F., Vasquez, B., San Roman, J., Reis, R. L. (2002) New Biodegradable And Bioactive Acrylic Bone Cements Based On Starch Blends And Ceramic Fillers, Biomaterials 23(8), 1883-1895
27. Demirgöz, D., Elvira, C., Mano, J. F., Cunha, A. M., Piskin, E., Reis, R. L. (2000) Chemical modification of starch based biodegradable polymeric blends: effects on water up-take, degradation behaviour and mechanical properties, Polym. Degrad. Stability 70, 161-170
28. Vaz, C. M., Cunha, A. M., Reis, R. L. (2001) Degradation Model of Starch-EVOH/HA Composites, Mat. Res. Innovat., 4, 375-380
29. Mendes, S. C., Bovell, Y. P., Reis, R. L., Cunha, A. M., de Bruijn, J. D., van Blitterswijk, C. A. (2001) Biocompatibility Testing of Novel of Starch-Based Materials with Potential Application in Orthopaedic Surgery, Biomaterials 22, 2057-2064.
30. Gomes, M. E., Reis, R. L., Cunha, A. M., Blitterswijk, C. A., de Bruijn, J. D. (2001) Cytocompatibility and response of osteoblastic-like cells to starch based polymers: effects of several additives and processing conditions, Biomaterials 22, 1911-1917
31. Marques, A. P., Reis, R. L., Hunt, J. A. (2002) In Vitro Evaluation Of The Biocompatibility Of Novel Starch Based Polymeric And Composite Material, Biomaterials 6, 1471-1478.
32. Leonor, I. B., Ito, A., Onuma, K., Kanzaki, N., Reis, R. L. (2002) In–Vitro Bioactivity of Starch Thermoplastic/Hydroxylapatite Composite Biomaterials: An In Situ Study Using Atomic Force Microscopy, Biomaterials, in press
33. Leonor, I. B., Sousa, R. A., Cunha, A. M., Zhong, Z., Greenspan, D., Reis, R. L. (2002) Novel Starch Thermoplastic/ Bioglass® Composites: Mechanical

Properties, Degradation Behaviour And In-Vitro Bioactivity, *J. Mat. Sci.: Mat. in Med.*, *in press*

34. Murata, H., Taguchi, N., Hamada, T., McCabe, J.F. (2000) Dynamic viscoelastic properties and the age changes of long-term soft denture lines, *Biomaterials* **21**, 1421-1427.

35. Rodríguez-Pérez, M.A., Almanza, O., del Valle, J.L., González, A., de Saja, J. A. (2001) Improvement of the measurements process used for the dynamic mechanical characterisation of polyolefin foams in compression, *Polym. Test.* **20**, 253-267.

36. Reis, R. L., Cunha, A. M. (2001) Starch Polymers in Encyclopedia of Materials: Science and Technology, Elsevier Science Ltd, p. 8810.

37. Butler, M. F., Cameron, R. E. (2000) A study of the molecular relaxations in solid starch using dielectric spectroscopy, *Polymer* **41**, 2249-2263

38. Mano, J. F., Koniarova, D., Reis, L. R., (2002), Thermal properties of thermoplastic starch/synthetic polymer blends with potential biomedical applicability, *J. Mater. Sci. Mater in Med.*, *in press*

39. Tian, D., Blacher, S., Dubois, Ph., Jérôme, R. (1998) Biodegradable and biocompatible inorganic-organic hybrid materials. 2. Dynamic mechanical properties, structure and morphology, *Polymer* **39**, 855-864.

40. Mano, J. F., Reis, R. L., Cunha, A. M. (2000) Effects of moisture and degradation time over the mechanical dynamic performance of starch based biomaterials. *J. Appl. Polym. Sci.* **78**, 2345-2357.

41. Cerrada, M. L., Pereña, J. M., Benavente, R., Pérez, E. (2000) Viscoelastic processes in vinyl alcohol-ethylene copolymers. Influence of composition and thermal treatment. *Polymer* **41**, 6655-6661.

42. Saber-Sheikh, K., Clarke, R. L., Braden, M. (1999) Viscoelastic properties of some soft lining materials. I- effect of temperature. *Biomaterials* **20**, 817-822.

43. Kalachandra, S., Minton, R. J., Takamata, T., Taylor, D. F. (1995) Characterization of commercial soft liners by dynamic mechanical analysis. *J. Mater. Sci. Mat. Med.* **6**, 218-222.

44. Saber-Sheikh, K., Clarke, R. L., Braden, M. (2000) Viscoelastic properties of some soft lining materials. II- ageing characteristics. *Biomaterials* **20**, 2055-2062.

45. Waters, M., Jagger, R., Willianms, K., Jerolimov, V. (1996) Dynamic mechanical thermal analysis of denture sift lining materials. *Biomaterials* **17**, 1627-1630.

46. Brown, A. F., Jones, D. S., Woolfson, A. D. (1997) Investigation of the thermorheology of poloxamers using oscillatory rheometry. *J. Pharm. Pharmacol.* **49**, 26.

47. Watase, M., Nishinari, K. (1989) Effects of the degree of saponification and concentration on the thermal and rheological properties of poly(vinyl alcohol)-dimethyl sulfoxide-water gels. *Polym. J.* **21**, 567-575.

48. Jauregui, B., Munoz, M. E., Santamaria, A. (1995) Rheology of hydroxyethylated starch aqueous systems. Analysis of gel formation. *Int. J. Biol. Macromol.* **17**, 49-54.

49. Frey, M. W., Cuculo, J. A., Khan, S. A. (1996) Rheology and gelation of cellulose/ammonia/ammonium thiocyanate solutions. *J. Polym. Sci. B: Polym. Phys.* **34**, 2375-2381.

50. Sheu, M-T, Huang, J-C, Yeh, G-C, Ho, H-O (2001) Characterization of collagen gel solutions and collagen matrices for cell cultures. *Biomaterials* **22**, 1713-1719.

51. Elvira, C., I. Espigares, I., Mano, J. F., Vazquez, B., San Roman, J., Reis, R. L. (2001) New Starch Based Partially Biodegradable In-situ Polymerazible Bioactive Bone Cements, 27th Annual Meeting of The Society For Biomaterials, St. Paul, Minnesota, USA, Apr., p. 271

52. Cauich-Rodriguez, J. V., Deb, S., Smith, R. (1996) Dynamic mechanical characterization of hydrogel blends of poly(vinyl-alcohol-vinyl acetate) with poly(acrylic acid) or poly(vinyl pyrrolidone). *J. Mat. Sci. Mat. Med.* **7**, 349-353.

53. Skrovanek, D. J. (1990) The assessment of the cure by dynamic mechanical thermal analysis. *Prog. Org. Coatings* **18**, 89-101.

54. Vaidyanathan, J., Vaidyanathan, T. K. (1995) Dynamic mechanical analysis of heat, microwave and visible ligh cure denture base resins. *J. Mater. Sci. Mat. Med.* **6**, 670-674.

55. Papadogiannis, Y., Lakes, R. S., Petrou-Americanos, A., Theothoridou-Pahini, S. (1993) Temperature dependence of the dynamic viscoelastic behavior of chemical- and light-cured composites. *Dent. Mater.* **9**,118-122.

56. Pascual, B., Gurruchaga, M., Ginebra, M. P., Gil, F. J., Planell, J. A., Vázquez, B., San Románm J., Goñi, I (1999) Modified acrylic bone cement with high amounts of ethoxytriethyleneglycol methacrylate, *Biomaterials* **20**, 453-463

57. Yang, J-M, Li, H-M, Yang, M-C, Shih, C-H (1999) Characterization of acrylic bone cements using dynamic mechanical analysis, *J. Biomed. Mater. Res. (Appl. Biomater.)* **48**, 52-60

58. Yang, J-M, Huang, P-Y, Yang, M-C, Lo, S.K. (1997) Effect of MMA-*g*-UHMWPE grafted fiber on mechanical properties of acrylic bone cements, *J. Biomed. Mater. Res. (Appl. Biomater.)* **38**, 361-369

59. Tsuji, H., Ikada, Y. (1996) Blends of aliphatic polyesthers. I. Physical properties and morphologies of solution-cast blends of poly(DL-lactide) and poly(ε-caprolactone). *J. Appl. Polym. Sci.* **60**, 2367-2375.

60. Vasquez-Torres, H., Cruz-Ramos, C.A. (1994) Blends of cellulosic esters with poly(ε-caprolactone): characterisation by DSC, DMA and WAXS. *J. Appl. Polym. Sci.* **54**, 1141-1159.

61. Penco, M., Sartore, L., Bignotti, F., D'Antone, S., Di Landro, L. (2000) Thermal properties of a new class of block copolymers based on segments of poly(D,L-lactic-glycolic acid) and poly(ε-caprolactone). *Eur. Polym. J.* **36**, 901-908.

62. Martin, O., Avérous, L. (2001) Poly(lectic acid) : plasticization and properties of biodegradable multiphase systems. *Polymer* **42**, 6209-6219.

63. Celli, A., Scandola, M. (1992) Thermal properties and physical ageing of poly(L-lactic acid). *Polymer* **33**, 2699-2703.

64. Starkweather, H. W., Avakian P., Fontanella, J. J., Wintersgill, M. C. (1993) Internal motions in polylactide and related polymers. *Macromolecules* **26**, 5084-5087.

65. Maquet, V., Blacher, S., Pirard, R., Pirard, J. P., Jerome, R. (2000) Characterization of porous polylactide foams by image analysis and impedance spectroscopy. *Langmuir* **16**, 10463-10470.

66. Nazhat, S. N., Kellomäki, M., Törmälä, P., Tanner, K. E., Bonfield, W. (2001) Dynamic mechanical characterization of biodegradable composites of hydroxyapatite and polylactides. *J. Biomed. Mater. Res. (Appl. Biomater.)* **58**, 335-343.

67. Cascone, M. G. (1997) Dynamic-mechanical properties of bioartificial polymeric materials. *Polym. Int.* **43**, 55-69.

68. Oliveira, A. L., Mano, J. F., San Román, J., Reis, R. L.(2001) Study of the effect of β-radiation sterilisation on the properties of different starch based polymers. European Society for Biomaterials – 2001 Conference. London (U.K.), 12-14 September.

69. Mano, J. F., Vaz, C. M., Mendes, S.C., Reis, R.L., Cunha, A. M. – Dynamic Mechanical Properties of Hydroxylapatite Reinforced and Porous Starch-Based Degradable Biomaterials. *J. Mater. Sci.: Mater. in Medicine* **10**, 857-862 (1999).

70. Sousa, R. A., Kalay, G., Reis, R. L., Cunha, A. M., Bevis, M. J. (2000) Injection Molding of a Starch/EVOH Blend Aimed as an Alternative Biomaterial for Temporary Applications, *J. of Appl. Polym. Sci.* **77**, 1303-1315

71. Sousa, R. A., Mano, J. F., Reis, R. L., Cunha, A. M., Bevis, M. J. (2002) Mechanical performance of starch based bioactive composites molded with preferred orientation for potential medical applications. *Polym. Eng. Sci.*, **42**, 1032-1045.

72. Kalay, G., Sousa, R. A., Reis, R. L., Cunha, A. M., Bevis, M. J. (1999) The Enhacement of the Mechanical Properties of a High Density Polyethylene, *J. of Appl. Polym. Sci.* **73**, 2473-2483.

73. Mano, J. F., Sousa, R.A., Reis, R. L., Cunha, A. M., Bevis, M. J. (2001) Viscoelastic behaviour and time-temperature correspondence of HDPE varying the degree of orientation induced by processing. *Polymer* **42**, 6187-6198.

74. Mano, J. F. (2001) Cooperativity in the crystalline α-relaxation in polyethylene. *Macromolecules* **34**, 8825-8828.

75. Bonfield, W., Wang, M., Tanner, K. E. (1998) Interfaces in analogue biomaterials, *Acta. Mater.* **46**, 2509-2518.

76. Nazhat, S. N., Joseph, R., Wang, M., Smith, R., Tanner, K. E., Bonfield, W. (2000) Dynamic mechanical characterization of hydroxyapatite reinforced polyethylene: effect of particle size. *J. Mat. Sci. Mat. Med.* **11**, 621-628.

77. Sousa, R. A., Mano, J. F., Reis, R. L., Cunha, A. M., Bevis, M. J. (2002) Mechanical behaviour of polyethylene/hydroxyapatite bone-analogue composites moulded with an induced anisotropy, in Bioceramics14, *Ed.* S. Brown, I. Clarke, P. Williams, Trans Tech Publications , Zurich, Switzerland, p. 469-472.

78. Alves, N. M., Mano, J. F., Gómez-Ribelles, J. L. (2002) Molecular mobility in polymers studied with Thermally Stimulated Recovery. II- study of the glass transition of a semicrystalline PET and comparison with DSC and DMA results. *Polymer*, **43**, 3627-3633.

79. Alves, N. M., Mano, J. F., Gómez-Ribelles, J. L. (2001) Molecular Mobility in a Thermoset as seen by TSR and DMA near T_g. *Mat. Res. Innovat.*, **4**, 170-178.

Surface Modification, Functionalization and Osteoconductivity

MACROMOLECULAR DESIGN OF NEW SYNTHETIC BIODEGRADABLE MEDICAL POLYMERS

Functional Polyesters and Copolymers with Cell-adhesion Peptide Sequences

F. RYPÁČEK

Institute of Macromolecular Chemistry, Academy of Sciences of the Czech Republic
Heyrovský Sq. 2, 162 06 Prague 6, Czech Republic

1. Introduction

The use of polymeric biomaterials in health care, medicine and pharmaceutics presently comprises variety of applications, ranging from contact lenses, prosthetic devices, and wound healing materials, to drugs delivery systems and, increasingly also, materials for cell therapies and tissue repair and engineering.[1, 2] Though, the use of polymers as an artificial extracellular matrix and a support for tissue regeneration promises exciting perspectives for medicine, at the same time, it also poses a major challenge to the fundamental science interfacing the polymer chemistry and cell biology. The materials sought as templates for implanted cells, or as a support for tissue regeneration, must fulfill rather complex requirements, among which two issues become increasingly important: (a) a controlled biodegradability of the polymer matrix, and (b) polymer-cell and/or polymer -tissue interactions.[1, 3] In the past decade, significant advances have been made in understanding the ways in which cells interact with their natural environment of extracellular-matrix proteins.[4] This knowledge has laid the foundation for the rational design of polymers that could play an active role in tissue regeneration and repair. New polymers and their supramolecular constructs, tailored to provoke a physiological response such as growth and differentiation of cells in a controlled way, are being sought.[5, 6]

A range of problems related to the specific requirements of a given application have to be addresses by the polymer chemist, when designing the biomaterial.[3, 7] The mechanical properties, such as strength, stiffness, or pliability of the polymer supports shall be well adjusted to match those of the tissue that is to be substituted, and they have to be retained at least for the period until the newly formed or healed tissue can take over. Other physical properties relate to the microscopic architecture of the polymer matrix. The matrix should exhibit a porous structure providing for interconnected channels that allow for the ingrowth of cells, in some cases also for the ingrowth of blood vessels, vascularization of the implant, as well as for the permeation of nutrients, metabolites and humoral factors controlling the growth and differentiation of cells.

R.L. Reis and D. Cohn (eds.),
Polymer Based Systems on Tissue Engineering, Replacement and Regeneration, 167–182.

For many applications, the biomaterial has to have capacity to be safely eliminated and gradually replaced by either the newly formed native extracellular matrix (ECM) or the organized regenerated tissue after it fulfils its supporting role. This scheme anticipates that polymers will be degradable by biological mechanisms present during the tissue remodeling, i.e. predominantly by enzyme-catalyzed hydrolysis, or by chemical reactions effective under mild conditions of the tissue, which mostly are limited again to hydrolytic reactions.

The requirement of controlling the interactions at biomaterial-tissue interfaces calls for a rational design of the molecular structure and pattern at the biomaterial surface. The term surface used in this respect is not limited to a flat solid surface, it could be also any diffuse interface with no apparent discrete boundary, at which the biomaterial and macromolecules of cell membranes come in a contact and can interact at a molecular level.

2. Biodegradable Polymers with Pendant Functional Groups

The synthetic absorbable polymers most often utilized for three-dimensional porous scaffolds in tissue engineering are aliphatic polyesters derived from polymerization of lactones, such as polylactide (PLA), polyglycolide (PGA), polycaprolactone (PCL). Polylactones can be prepared in a wide range of structural variations either through the copolymerization of lactones themselves, and/or through copolymerization with other cyclic monomers, such ethers or carbonates (e.g. dioxanone, ethylene oxide, trimethylene carbonate). The family of polymers thus available can span a wide range of properties, from loosely associated hydrophilic gels to semi-crystalline high-strength materials. In common, these polymers exhibit sterically unhindered aliphatic ester bonds in polyester parts of the polymer chains, that make these chains prone to hydrolysis even under mild conditions in living tissues. In addition, they are mostly well processable by conventional polymer processing techniques and offer ample ways in which they can be processed in a desired shape, microstructure, and mechanical parameters. On the other hand, the polyesters offer only limited possibilities for chemical modification and creation of functional groups that could be used for binding of bioactive ligands and/or molecular recognition moieties. The functional groups can be introduced to polyesters through copolymerization with functional monomers. The following paragraphs will focus on the reviewing the chemical approaches to the synthesis of biodegradable functionalized copolymers with at least partial polyester backbone.

2.1. STATISTICAL COPOLYMERS

2.1.1. Polyesters with Carboxyl Side Chains

There has been significant effort to develop polyesters with functional polar groups in side chains, such as carboxyl or amine groups. Polyesters based on dicarboxy-hydroxyalkanoic acids, such as malic acid, were investigated in attempt to provide for functional polyesters with carboxylic side chains. Vert et al.[8-11] and Lenz et al.[8, 11] prepared poly(β-malic acid) by ring-opening polymerization of benzyl malolactone and subsequent debenzylation of the resulting intermediate polymer. Poly(α-malic acid) has been prepared by ring-opening polymerization of malide dibenzyl ester, followed by removal of protecting groups through debenzylation.[12]

By copolymerization of malolactonic and 3-alkylmalolactonic acid esters different degradable materials for biomedical field can be prepared, exhibiting hydrophilic, functional and hydrophobic moieties with a random or block constitution, capable to form biodegradable associating networks, macromolecular micelles constituted by degradable amphiphilic block copolymers of poly(β-malic acid) as hydrophilic segments and poly(β-alkylmalic acid alkyl esters) as hydrophobic blocks.[13] Bioactive poly(β-hydroxyalkanoate)s derived from malic acid were prepared and tested for bone repair and muscle regeneration. The functionalized and hydrolyzable polyesters were obtained by anionic copolymerization of three malolactonic acid esters and subsequent chemical modifications of the resulting terpolymer. The benzyl-protecting groups were turned into carboxyl groups and allyl groups into sulfonate groups. The resulting polymer bore carboxylate, sulfonate, and sec-butyl pendent groups which were aimed at interacting with heparan binding growth factors.[14]

Functional biomaterials based on polyesters with carboxyl groups of malic acid were prepared by ring-opening copolymerizations of L-lactide with 3-(s)-[(benzyloxycarbonyl)methyl]- and 3-(s)-[(dodecyloxycarbonyl)methyl]-1,4-dioxane-2,5-diones. The solution-cast films of these copolymers were either alkali-treated, or the benzyl groups were removed by catalytic hydrogenolysis before casting, to form a carboxyl-functionalized surface. The cell-binding Arg-Gly-Asp tripeptide (RGD) was immobilized on the surface of the cast film with dicyclohexylcarbodiimide as coupling agent. The RGD-immobilized films thus prepared exhibited improved cell attachment compared with the original ones.[17]

2.1.2.   *Polyesters with Amine Side Chains*

Poly(L-serine ester) is an example of polyesters with amine pendant groups. The poly(amino serinate)s were prepared by ring-opening polymerization of N-protected L-serine β-lactones.[15] By mixing a carboxyl-bearing polyacid, such as poly(β-malic acid), with a polybase, such as poly(amino serinate), degradable polyelectrolyte complexes were prepared.[16]

2.1.3.   *Copolymers of a-hydroxy and a-amino acids*

Another approach to the synthesis of functionalized fully biodegradable polymers is based on combining the properties of two classes of important synthetic biodegradable polymers, such as poly(hydroxy acid)s and poly(α-amino acid)s. The α-amino acid component introduces peptide bonds into the polymer backbone chain, thus bringing the potential of degradation by of the polymer chain by peptidases and proteinases. In addition, the amino acid residues can provide for functional groups and poly(α-amino acid) blocks or peptide sequences can add to biospecificity of polymer interactions. Polymers of α-hydroxy and α-amino acid units, with the backbone chain containing both the ester and peptide bonds, can be prepared either as statistical or block copolymers.

Polydepsipeptides: Polymerization of morpholine-2,5-derivatives. Statistical copolymers of hydroxy acids and α-amino acids are called poly(depsipeptides). Sequential polydepsipeptides have been prepared on a small scale by polymerization of tetra-, tri- or di-depsipeptide activated esters, which were synthesized via a multi-step synthetic route.[18-20] However, facile synthesis of various types of high-molecular-weight polymers containing both hydroxy acid and α-amino acid derived units was achieved by ring-opening polymerization of the cyclic monomers, morpholine-2,5-dione derivatives, and/or their copolymerization with lactones, such as lactide, glycolide, e-caprolactone, etc. The first successful polymerizations were performed with p-dioxanone and unsubstituted or alkyl-substituted 2,5-morpholinedione.[21] These copolymers showed accelerated in vivo absorption and excellent strength retention. Alternating poly(glycine-D,L-lactic acid)was obtained by ring-opening polymerization of 6-methyl-2,5-morpholinedione.[22] Alternating and random copolymers containing lactic acid or e-caprolactone in combination with various amino acids, including functionalized ones, were described by Feijen's group. [22-24] The effect of the amino acid component in the ester-amide copolymers on the polymer properties and their biodegradability was evaluated.[25-27]

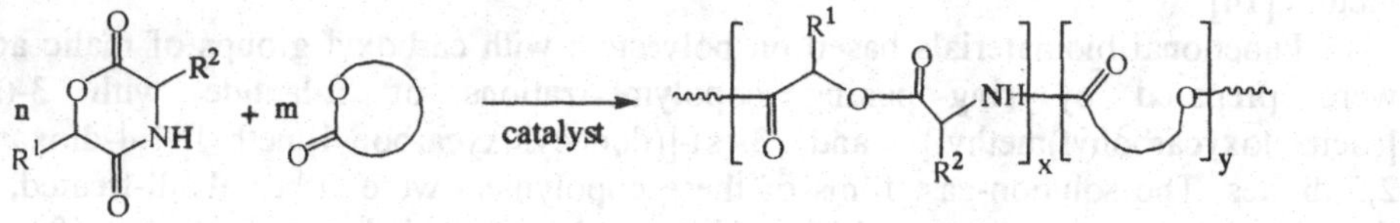

R^1 : H - glycolic acid
CH$_3$ - lactic acid

R^2 : H - glycin
CH$_3$ - alanin
CH$_2$-CH$_2$-COO-Bzl - benzyl glutmate

Figure 1. Synthesis of functional poly(ester-amides) by copolymerization morpholine-2,5-dione derivatives and lactones.

Monomers for these polymerizations are substituted morpholinediones, which are prepared as cyclic heterodimers of α-hydroxy and α-amino acids. Their synthesis is based on cyclization of N-(2-halogenacyl)-amino acids prepared by the reaction of the amino acid with bromopropionyl bromide or bromopropionyl chloride.[24] The polymerization of morpholine-2,5-diones is typically carried out in bulk, with tin(II)-octanoate as a catalyst. The synthesis of polydepsipeptides by copolymerization of morpholine-2,5-dione derivatives with lactones is schematically depicted in Figure 1.

The side-chain substituents (R^2) of polymers can be derived from trifunctional amino acids, such as aspartic or glutamic acids for carboxyl, lysine for amine, and cystein for thiol groups. The functional groups have to be temporarily protected. Typically, benzyl ester or benzyloxycarbonyl groups were used for protection of carboxylic or amine functions, respectively.[23] The yield of polymerization of substituted morpholinediones and the molecular weight of obtained depend on the size and structure of the substituents. While the polymerization of morpholine-2,5-dione and 3-methylmorpholine-2,5-dione proceeds with good to moderate yields giving polymers with reasonable molecular weights ($>10^4$ g/mol), with increasing size of the substituents the reactivity of depsipeptides decreases. This is particularly important with respect to the polymerization of depsipeptides derived from protected functional amino acids, such as benzyl aspartate and glutamate esters, which cannot be used in ring-opening homopolymerization. Nevertheless, these derivatives can be copolymerized with lactones, such as glycolide, lactide and ε-caprolactone, giving rise to poly(ester-amide)s with pendant protected functional groups, which can be selectively deprotected. The polymers in a molecular-weight range above 10^4 g/mol can reasonably be produced with depsipeptide content in a range up to 20 mole %.[23, 26] Functionalized poly(ester-amide)s were used for preparation of biomaterials with specific cell-adhesion properties. The copolymer poly(lactic acid-co-lysine) contains functional amino groups to which cell adhesion molecules -RGD peptides- were grafted [28] and the cell interactive properties of bioactive surfaces were evaluated.[29]

Poly(ester-amides): Copolymerization of lactones and NCAs. An alternative approach to the synthesis of statistical α-hydroxy acid/α-amino acid copolymers is based on direct ring-opening copolymerization of α-amino acid *N*-carboxyanhydrides (NCA) and lactones.[30] Ring-opening polymerization of α-amino acid *N*-carboxyanhydrides, when initiated by protic nucleophiles, such as primary amines or alcohols, propagates through a nucleophilic end group (amine or carbamate). [31] Similarly, the ring-opening polymerization of lactones can be initiated by protic nucleophiles, such as alcohols, water or amines. We found that the Sn(II)octanoate-catalyzed polymerization of lactide can equally be initiated by both alcohols and amines.

Under well-controlled conditions, the number-average polymerization degree of the resulting PLA corresponded well with the monomer/initiator (nucleophile) ratio and the M_w/M_n ratio was within the range of 1.1-1.3. The NMR data showed that both types of initiating nucleophiles were incorporated in the polymer chain in a molar ratio well corresponding to the polymerization degree. [32] Consequently, when both types of monomers, i.e. NCA and lactone, are present, the amine end group of the terminal amino acid residue may attack either another NCA molecule or the lactone monomer, thus giving rise to a peptide or an amide linkage. On the other hand, the reactivity of NCA is sufficient to propagate the polymer growth by reaction with nucleophilic hydroxy group of the terminal lactic acid residue as well, thus forming an ester bond while exposing the amine end group of the added amino acid residue for the next propagation step. The structures and reactions that could be involved in the ring opening copolymerization of lactones and NCAs are schematically shown in Figure 2.

The copolymerization of NCAs and lactones was studied with the aim to evaluate (i) the effect of NCA content in the reaction mixture on properties of the copolymers, (ii) the possibility of controlling the molecular weights of the copolymers, and (iii) the effect of amino acid structure on the copolymerization. The polymerizations were carried out in solution (dioxane) at 60 °C. Decanol was used as a "co-initiator" with Sn(II) octanoate to speed up the initiation reaction and control the molecular weight of polymers. In the presence of co-initiator, copolymers with a narrow molecular-weight distribution were obtained, while in its absence, the yields of polymers were lower for the same reaction times. With extended reaction times, the polydispersity of resulting copolymers increased, due to more pronounced transfer reactions.[30]

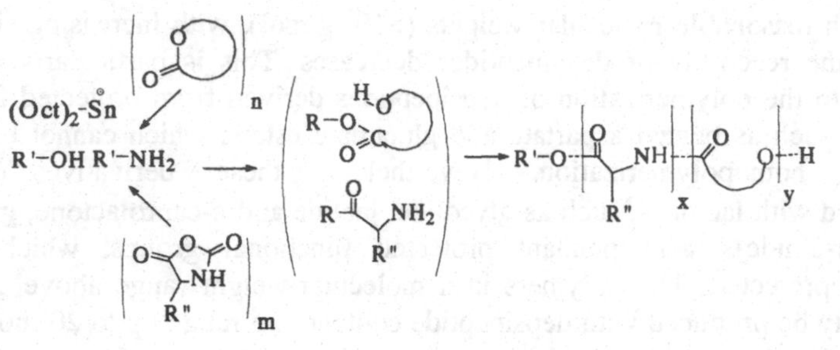

172

Figure 2. Synthesis of functional poly(ester-amides) by ring-opening copolymerization of lactones and *N*-carboxyanhydrides of α-amino acids

Three types of α-amino acids differing in the structure of side chains were used with emphasis on fuctionalizable amino acids. The side chain of both γ-benzyl L-glutamate and β-benzyl L-aspartate contain an ester bond, while phenylalanine, although having similar steric requirements for the aromatic ring, is without a side-chain ester. All three types of amino acid NCAs formed copolymers with both the lactones studied, i.e., lactide and ε-caprolactone. The composition of copolymers was determined by NMR spectra. The ^{1}H NMR and ^{13}C NMR spectra revealed that both types of structural units and both types of bonds, ester and amide, were present in the copolymers. If decanol was used as an initiator, its incorporation to one end of the copolymer chain could be detected by NMR spectra. Both types of structures, lactic acid and amino acid, were found as terminal units. With all amino acids, with increasing the concentration of NCA in the reaction mixture the mole fraction of amino acid residues in the copolymer increased proportionally (Table 1).

The effect of the amino-acid side-chain structure on the course of copolymerization was evaluated by comparing the copolymerization of aspartate and glutamate derivatives with phenylalanine, an aromatic amino acid with similar steric requirements, however, without the side chain ester bond. When amino acids with ester protecting groups in side chains were used, such as benzyl glutamate and aspartate, with the increasing content of NCA in the reaction feed the yield of the copolymerization and the average molecular weight obtained decreased (Table 2).

The lower molecular weight and higher polydispersity of copolymers containing glutamate and/or aspartate esters may indicate possible side-chain reactions involving the pendant ester groups. It is worth noticing that the copolymers with phenylalanine, containing an inert side chain, were obtained with molecular parameters very closely approaching those calculated from the composition of the feed, with the molecular-weight distribution well controlled by the M/I ratio.

TABLE 1. Copolymerization of L-lactide with BzlGlu-NCA: the effect of NCA content in the feed on molecular parameters of copolymers [a]

Bzl Glu-NCA [b] mol %	y [c] mol%	yield %	M_n [d] (theor)	M_n (GPC)	M_w/M_n (GPC)
0	0	96	7060	6440	1.26
8	8	92	7040	4000	1.37
15.5	15.1	88	6980	3800	1.6

[a] Polymerization conditions: M/I (decanol) = 50 (mol/mol), dioxane, 60 °C, 72 h; [b] BzlGlu-NCA in the feed; [c] BzlGlu residues in the copolymer (from NMR); [d] calculated from the feed composition and the copolymer yield.

TABLE 2. The effect of amino acid structure and the monomer-to-initiator ratio (M/I) on the properties of copolymers [a]

Amino acid	M/I [b]	Yield (%)	y (mol %)	M_n [c] (theor)	M_n (GPC)	M_w/M_n (GPC)
BzlGlu	50	88	15.1	6990	3140	1.6
	100	86.3	16.2	13600	4670	1.36
BzlAsp	50	86.4	14.2	6780	4510	1.13
	100	85.6	12.7	13300	7230	1.12
Phe	50	85	14.6	6270	7630	1.1
	100	83	12.8	12120	1348	1.1

[a] 16 mol% of NCA in the polymerization mixture ; dioxane, 60 °C, 72 h; [b] mole ratio of monomers to co-initiator (decanol); [c] calculated from the feed composition and the yield of copolymer.

In addition to L-lactide copolymers, analogous copolymers of e-caprolactone. were prepared using the same method. With respect to lactide, the polymerization with e-caprolactone was significantly slower. Under the same conditions (dioxane, 60 °C, 72 h) the conversion of e-caprolactone copolymerization reached typically about 60 - 70 % of that for lactide copolymers.[30] The copolymers with free carboxylic acid side chains can then be prepared by the debenzylation of their protected precursors. Catalytic hydrogenation using a Pd catalyst was used to prepare copolymers with functional side chains.

2.2. BLOCK COPOLYMERS OF HYDROXY ACIDS AND α-AMINO ACIDS

The functional polymers are needed to control the surface properties of polymer biomaterials through binding of biospecific molecules. On the other hand, as it follows from the above discussion, introduction of functional groups into polyesters through synthesis of statistical copolymers, such as polydepsipeptides, brings about significant changes in the properties of resulting copolymers. The molecular parameters of the copolymers are affected by the fraction of amino acid component as well as by its structure. Often, the molecular weights of functionalized copolymers are far bellow those of polyesters alone and, generally, the copolymers do not reach the parameters needed for processing them to high strength materials. Consequently, the range of materials that can be prepared from statistical functional copolymers becomes narrow and we loose much of the advantages of polyesters -such as processability, strength and mechanical qualities. Therefore, methods providing for polyester biomaterials with functionalized surfaces only without significant modification of bulk properties become important. To this end, several groups investigate modification of polymer surfaces by biologically specific molecules or specifically designed block copolymers.

2.2.1. Surface Modification by Amphiphilic Block Copolymers

The microphase separation of incompatible copolymer blocks leads to formation of supramolecular structures due to association of similar blocks and exclusion of incompatible polymer segments. Amphiphilic block copolymers can organize at interfaces. They may form polarized molecular monolayers at the water-air interface, that can be transferred to solid surfaces as Langmuir-Blodgett films. Such copolymer layers can be used as model surfaces, which afford a great deal of control over the surface structure and topography. Valuable information about the ability of copolymers to self assembly and its relation to the molecular parameters of copolymers can be derived from measurements of the dependence between surface pressure (π) and area (A) of the film during its compression (π/A-isotherm). Block copolymers can form micelles in selective solvents. When the micelles are formed in a solution containing a polymer compatible with one of the copolymer blocks, discrete domains rich in the other copolymer segments can be obtained in a continuous polymer matrix in cast films.[33] Block copolymers will adsorb from a selective solvent onto the surfaces that are made of polymer materials compatible with one of the copolymer blocks.[34, 35] By application of these principles, functional biomaterials with functional groups assembled in surface domains [33] and copolymer brush layers at the material surface can be designed. [34]The examples of behavior we can observe with block copolymers are schematically shown in Figure 3.

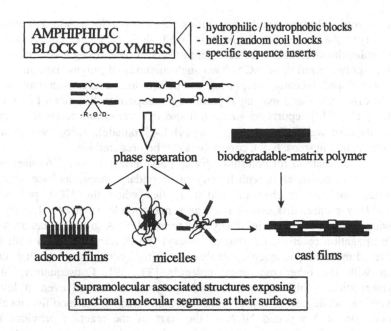

Figure 3. Supramolecular assemblies of amphiphilic block copolymers

Our approach is based on using block copolymers that could be applied to the polyester surface. Therefore, such block copolymers should contain polyester blocks, that could provide for good anchoring the copolymer molecules at the surface to the polyester bulk, and one or more hydrophilic blocks, containing the functional groups. Through the microphase separation of blocks of copolymers at the polymer/air or the polymer/water interface, the functional groups will become enriched at the material surface.

Suitable amphiphilic block copolymers can be prepared based on polylactone and poly(α-amino acid)s and/or functionalized polyethylene oxide blocks. In the present discussion we will focus on block copolymers with polylactide (PLA) and poly(amino acid), poly(AA), blocks. The synthesis of block copolymers is carried out in two steps. In the first step, poly(amino acid) with a protic end group, that can later be used as a macromolecular co-initiator in the ring opening polymerization of lactones. Two controlled-polymerization processes are needed for the synthesis. Poly(AA)s can be prepared by polymerization of *N*-carboxyanhydrides of α-amino acids (NCA) .

2.2.2.   Synthesis of Poly(a-amino acid) Blocks: Living Polymerization of NCAs

NCA polymerization can provide for both homopolymers and statistical copolymers of α-amino acids. Although the synthesis of NCA and its polymerization, first described by Leuchs already in 1906 [36], is well known and investigated for last four decades, neither of the common polymerization mechanisms produce polymers with narrow molecular-weight distribution (see Ref.[31]for a thorough review on NCA chemistry). An increasing interest in well defined block copolymers and their supramolecular structures boosted recent efforts in developing a controlled process for "living" polymerization of NCA. Two such methods of polymerization of NCAs have been developed recently, reporting on poly(amino acid)s with narrow molecular-weight distribution and making it possible to prepare well defined block copolymers. Deming et al.[37] reports on successful use of transition metal-amine initiators for preparation of well-defined poly(γ -benzyl L-glutamate). Also, through a controlled polymerization in steps, block copolymers can be prepared [38].

Another method for controlled living polymerization of NCAs uses *N*-acyl-NCA derivatives as co-initiators with tertiary or secondary amines as base catalysts. In the activated-monomer mechanism, which applies when the NCA polymerization is initiated by a base, that cannot react as a nucleophile, such as a tertiary amine, the growing polymer chain carries *N*-acylated terminal NCA-group. It has been known that the propagation reaction, i.e. that of *N*-acyl- NCA terminal group with NCA anion (activated monomer) is much faster than the initiation reaction, i.e. that of the NCA-anion with the other monomer molecule [31, 39]. Consequently, in a typical polymerization, a high-molecular-weight polymer is formed even at low monomer conversions while new chains are still being initiated. In the modified mechanism, the introduction of *N*-acylated NCA at the start of the reaction provides for defined concentration of initiating species from which polymer chains start to grow. The growing polymer chains have one acyl-protected end and a "living" end-group formed by the NCA ring N-acylated with the polymer chain (Figure 4).

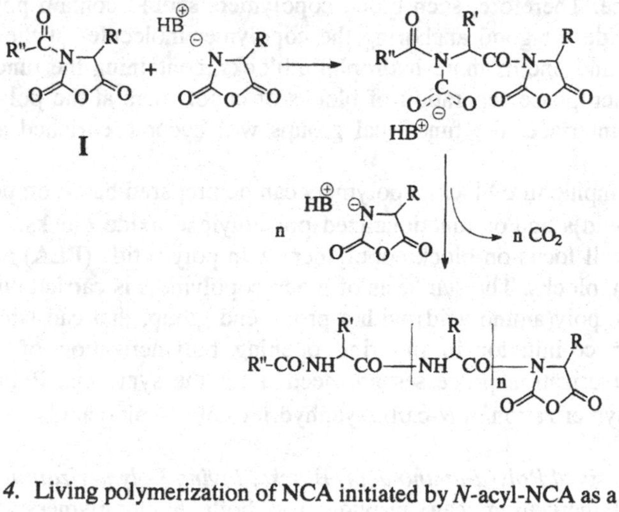

Figure 4. Living polymerization of NCA initiated by *N*-acyl-NCA as a co-initiator.

Thus the propagation is limited to only one type of reaction. The polymerization degree could be controlled by the amount of N-acyl-NCA initiator added.[40] The polymerization can be terminated by reacting the *N*-acyl-NCA end group with amino alcohols or mono-protected diamines, providing for poly(AA) with either hydroxyl or amino terminal groups which, subsequently, could be used as macroinitiators for ring-opening polymerization of lactide or other lactones.[41] The feasibility of preparing polyglutamates and polyaspartates with a controlled molecular-weight distribution by using this concept was evaluated.[42]

2.2.3. Synthesis of Poly(hydroxy acid) Blocks: Living Ring-opening Polymerization of Lactones (using lactide as an example)

In the second step, functional poly(AA)s are to be used as macromolecular co-initiators in the ring-opening polymerization of lactone. A bulk polymerization cannot be effectively applied because of insolubility of co-initiators in the lactone melt and/or because of side reactions due to high temperature. A living polymerization of lactide in solution using Sn(II)alkoxides, such as 2,2-dibutyl-2-stanna-1,3-dioxepane [43] and Sn(II)butoxide [44] has been reported, suggesting that also in Sn(II)octanoate catalyzed polymerizations of lactide, the initiation actually takes effect through Sn(II)alkoxide formed in situ by reaction of Sn(II)octanoate with the alcohol added as a co-initiator. We investigated the feasibility of using Sn(II)octanoate as a catalyst in a solution polymerization, under mild conditions, which would enable us to use poly(α-amino acid)s with protected side chains as co-initiators.[32]

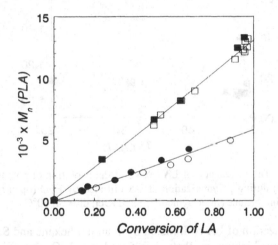

Figure 5. The relation of M_n of PLA (or a PLA block in block copolymers) to the monomer conversion after the initiation by PEO (open symbols) and 1-aminohexane (filled symbols); the lines show theoretical dependencies based on the monomer/co-initiator ratios: ● ○ [LA]$_0$/ [I] = 40; ■ □ [LA]$_0$/ [I] = 90.

The feasibility of incorporating amine-ended precursors in block copolymers by using their amine group for co-initiating the ring-opening polymerization of lactones becomes important with poly(AA)s. In model studies, the kinetics of LA polymerization in dioxane solution, using either hydroxyl-terminated (PEO) or amine-terminated (1-amino hexane, AH) as co-initiators with Sn(II)Oct as a catalyst was investigated in detail and the living character of polymerization was documented in both cases.[32] In both polymerizations, solid evidence was obtained that the polymerization started by addition of the monomer to the hydroxy or amine group of the co-initiators. The polymerization degree increased linearly with the monomer conversion and in accordance with the monomer/co-initiator ratio (Figure 5).

The values of M_n of the copolymer determined by SEC and NMR analysis, respectively, were in a good agreement. The polydispersity of the polymer product in the range 1.03-1.08 was maintained behind the point, when the polymerization reached the equilibrium conversion of lactide (about 97 %, at 60 °C) (Figure 6).

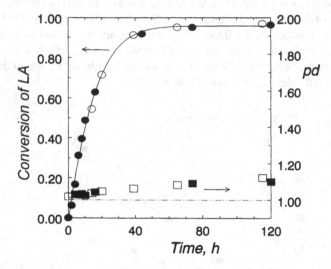

Figure 6. The conversion of LA (α) and the evolution of polydispersity index (pd) during polymerization of LA initiated by PEO (open symbols) and 1-amino hexane (filled symbols); Sn(II)Oct, dioxan, 60°C.

The polymerization of LA co-initiated by 1-amino hexane and Sn(II)Oct followed essentially the same kinetics as that co-initiated by PEO, providing PLA with a polydispersity below 1.1 and the polymerization degree corresponding to the ratio of $([LA]_{eq}-[LA]_t)/[AH]$ through the all course of polymerization (data with filled points in Figures 5 and 6). The NMR analysis confirmed the presence of an amide bond in the polymer and the ratio of LA structural units to hexyl groups was found corresponding to M_n values obtained from SEC.

2.2.4. Block Copolymers with Adhesion-peptide Sequences

The feasibility of using either hydroxyl- or amine-terminated polymers to co-initiate the living polymerization of LA with Sn(II)Oct as a catalyst, can conveniently be used to prepare block copolymers of PLA with poly(amino acids). By selection of suitable *N*-acyl-NCA in the preparation of poly(AA) blocks and through capping reactions, PLA-*b*-polyAA copolymers containing a defined peptide sequences at the ends of poly(AA) blocks can be prepared. The strategy used in the synthesis of block copolymers, end-capped with the Arg-Gly-Asp tripeptide - a known amino acid sequence of fibronectin, is depicted in Figure 7.

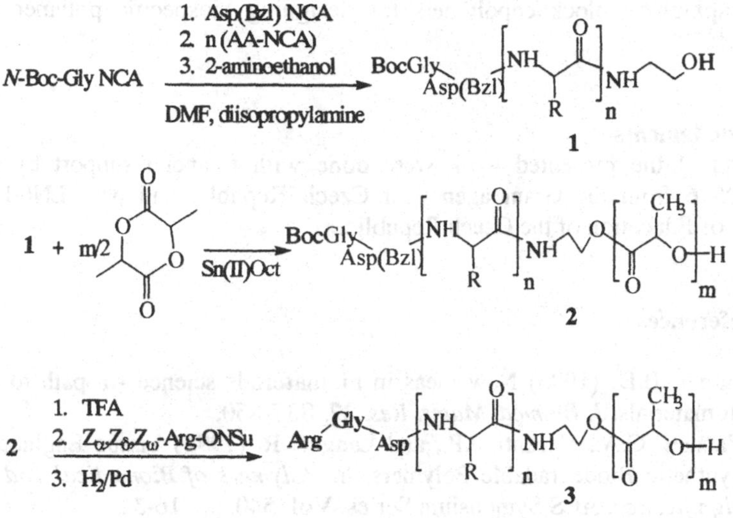

Figure 7. Strategy of the synthesis of polylactide/poly(amino acid) block copolymers, carrying cell-adhesion peptide sequences.

AB-type block copolymers were prepared by ring-opening polymerization of lactide using poly(AA) derivatives **1** as macroinitiators and Sn(II) octanoate as a catalyst. The polymerization of N-carboxyanhydrides using N-Boc-glycine NCA as a co-initiator assures that virtually all polymer chains start with a protected glycine residue. By using NCA of β-benzyl L-aspartate as the first amino acid monomer, the aspartate amino acid residue is introduced next to initiating glycine. The poly(AA) chain then can be extended by continuing the polymerization with NCA of chosen amino acid (AA). In the presented scheme, the polymerization was terminated by the reaction of living N-acyl-NCA group with 2-aminoethanol, providing for **1**. Analogously, a semi-protected diamine can be used for termination, providing for an amine terminal group. The ring-opening polymerization of lactide in a dioxane solution provided for PLA-*b*-poly(AA) block copolymers. Arg-Gly-Asp sequence was completed by a capping reaction as indicated. Hydrogenation on a Pd catalyst was used to remove all protecting groups in polypeptide side chains.[45]

3. Conclusions

Biodegradable polymers with functional groups can be prepared through copolymerization of lactones and functional hydroxy acid and α-amino acid derivatives. Statistical copolymers with pendant carboxylic and or amine groups can be prepared. Block copolymers containing poly(hydroxy acid) and poly(α-amino acid) blocks can be prepared by stepwise application of controlled ring-opening polymerizations of NCAs and lactones. Biospecific peptide sequences can be introduced into the polymer structure either by biding through pendant functional groups in the side chains or as a peptide segment in block copolymers. Strategy of using amphiphilic block copolymers for designing biospecific polymer surfaces is presented.

Acknowledgments

The parts of the presented work were done with financial support by grants No. 203/99/0576 from the Grant agency of Czech Republic and No.: LN00A065 from Ministry of Education of the Czech Republic.

4. References

1. Ratner, B.D. (1993) New ideas in biomaterials science - a path to engineered biomaterials, *J. Biomed. Mater. Res.* **27**, 837-850.
2. Vacanti, C.A., Vacanti, J.P., and Langer, R. (1994) Tissue Engineering Using Synthetic Biodegradable Polymers, in *Polymers of Biomedical and Biological Significance*, ACS Symposium Series. Vol. 540, pp. 16-34.
3. Vacanti, J.P., Langer, R. (1999) Tissue engineering: The design and fabrication of living replacement devices for surgical reconstruction and transplantation, *The Lancet* **354**, SI32-SI34.
4. Adams, J.C. (2001) Cell-matrix contact structures, *Cellular and Molecular Life Sciences* **58**, 371-392.
5. Langer, R., Cima, L.G., Tamada, J.A., and Wintermantel, E. (1990) Future directions in biomaterials, *Biomaterials* **11**, 738-745.
6. Hubbell, J.A. (1999) Bioactive biomaterials, *Curr. Opin. Biotechnol.* **10**, 123-129.
7. Thomson, R.C., Shung, A.K., Yaszemski, M.J., and Mikos, A.G. (2000) Polymer scaffold processing, in R.P.Lanza, R.Langer, and J.Vacanti (eds.), *Principles of Tissue Engineering*, Academic Press, London, pp. 251-262.
8. Lenz, R.W., Vert, M. (1981). US. Pat. 4 265 247. *Chem. Abstr.* 95, 116201g.
9. Guerin, P., Vert, M., Braud, C., and Lenz, R.W. (1985) Drug carriers: Optically active poly(β-malic acid), *Polym. Bull.* **14**, 187-192.
10. Braud, C., Vert, M. (1985) Poly(β-malic acid) derivatives - a new type of polyvalent polymeric drug carriers., *Polym. Prepr.* **24**, 71-72.
11. Arnold, S.C., Lenz, R.W. (1985) Synthesis of stereoregular poly(alkyl malonates), *Makromol. Chem., Macromol. Symp.* **6**, 285-303.
12. Ouchi, T., Fujino, A. (1989) Synthesis of poly(α-malic acid) and its hydrolysis behavior in vitro, *Makromol. Chem.* **190**, 1523-1530.
13. Cammas, S., Bear, M.M., Moine, L., Escalup, R., Ponchel, G., Kataoka, K., and Guerin, P. (1999) Polymers of malic acid and 3-alkylmalic acid as synthetic PHAs in the design of biocompatible hydrolyzable devices, *Int. J. Biol. Macromol.* **25**, 273-282.

14. Jeanbat-Mimaud, V., Barbaud, C., Caruelle, J.P., Barritault, D., Cammas-Marion, S., Langlois, V., and Guerin, P. (2000) Bioactive functionalized polymer of malic acid for bone repair and muscle regeneration, *J. Biomater. Sci. Polym. Ed.* **11**, 979-991.

15. Zhou, Q.X., Kohn, J. (1990) Preparation of poly(L-serine ester): a structural analogue of conventional poly(L-serine), *Macromolecules* **23**, 3399-3406.

16. Rossignol, H., Boustta, M., and Vert, M. (1999) Synthetic poly(beta-hydroxyalkanoates) with carboxylic acid or primary amine pendent groups and their complexes, *Int. J. Biol. Macromol.* **25**, 255-264.

17. Yamaoka, T., Hotta, Y., Kobayashi, K., and Kimura, Y. (1999) Synthesis and properties of malic acid-containing functional polymers, *Int. J. Biol. Macromol.* **25**, 265-271.

18. Nissen, D., Gilon, C., and Goodman, M. (1975) Polydepsipeptides. 4. Synthesis of the alternating polydepsipeptides poly(Ala-Lac) and poly(Val-Lac), *Makromol. Chem., Suppl.* **1**, 23-53.

19. Mathias, L.J., Fuller, W.D., Nissen, D., and Goodman, M. (1978) Polydepsipetides. 6. Synthesis of sequential polymers containing various ratios of L-alanine and L-lactic acid, *Macromolecules* **11**, 534-539.

20. Yoshida, M., Asano, M., Kumakura, M., Katakai, R., Mashimo, T., Yuasa, H., Imai, K., and Yamanaka, H. (1990) A New Biodegradable Polydepsipeptide consisting of (L-alanyl)$_n$-L-lactyl Sequences (n=0,1,2, and 3), *Makromol. Chem., Rapid Commun.* **11**, 337-343.

21. Shakaby, S.W., Koelmel, D.F. (1983) Eur.Pat.Appl. 86,613.

22. Helder, J., Kohn, F.E., Sato, S., Van den Berg, J.W.A., and Feijen, J. (1985) Synthesis of poly[oxyethylidenecarbonylimino-(2-oxyethylene)] [poly(glycine-D,L-lactic acid] by ring opening polymerization, *Makromol. Chem., Rapid Commun.* **6**, 9-14.

23. In't Veld, P.J.A., Dijkstra, P.J., and Feijen, J. (1992) Synthesis of biodegradable polyesteramides with pendant functional groups, *Makromol. Chem.* **193**, 2713.

24. In't Veld, P.J.A., Dijkstra, P.J., van Lochum, J.H., and Feijen, J. (1990) Synthesis of alternating polydepsipeptides by ring-opening polymerization of morpholine-2,5-dione derivatives, *Makromol. Chem.* **191**, 1813-1825.

25. Helder, J., Dijkstra, P.J., and Feijen, J. (1990) In vitro degradation of glycine/DL-lactic acid copolymers , *J. Biomed. Mater. Res.* **24**, 1005-1020.

26. Ouchi, T., Nozaki, T., Ishikawa, A., Fujimoto, I., and Ohya, Y. (1997) Synthesis and enzymatic hydrolysis of lactic acid-depsipeptide copolymers with functionalized pendant groups, *J. Polym. Sci. Part A:Polym. Chem.* : **35**, 377-383.

27. Ouchi, T., Nozaki, T., Okamoto, Y., Shiratani, M., and Ohya, Y. (1996) Synthesis and enzymatic hydrolysis of polydepsipeptides with functionalized pendant groups, *Macromol. Chem. Phys.* **197**, 1823-1833.

28. Barrera, D.A., Zylstra, E., Lansbury, P.T., and Langer, R. (1993) Synthesis and RGD peptide modification of a new biodegradable copolymer: Poly(lactic acid-co-lysine), *J. Am. Chem. Soc.* **115**, 11010-11011.

29. Cook, A.D., Hrkach, J.S., Gao, N.N., Johnson, I.M., Pajvani, U.B., Canizzaro, S.M., and Langer, R. (1997) Characterization and development of RGD-peptide-

182

      modified poly(lactic acid-co-lysine) as an interactive, resorbable biomaterial, *J. Biomed. Mater. Res.* **35**, 513-523.

30. Rypáček, F., Štefko, I., Machová, L., Kubies, D., and Brus, J. (1998) Synthesis of ester-amide copolymers from lactones and α-amino acid *N*-carboxyanhydrides, *Polym. Prepr.* **39**, 126-127.

31. Kricheldorf, H.R. (1987) *Alpha-Aminoacid-N-Carboxy-Anhydrides and Related Heterocycles*, Springer Verlag,Berlin, Heidelberg, New York.

32. Kubies, D., Machová, L., and Rypáček, F. (2000) Solution Polymerization of Lactones Promoted by Tin(II) 2-Ethylhexanoate., 660.

33. Kubies, D., Rypáček, F., Kovářová, J., and Lednický, F. (2000) Microdomain structure in polylactide-*block*-poly(ethylene oxide) copolymer films, *Biomaterials* **21**, 529-536.

34. Elbert, D.L., Hubbell, J.A. (1998) Self-assembly and steric stabilization at heterogeneous, biological surfaces using adsorbing block copolymers, *Chem Biol.* **5**, 177-183.

35. Meiners, J.C., Quintel-Ritzi, A., Mlynek, J., Elbs, H., and Krausch, G. (1997) Adsorption of block-copolymer micelles from a selective solvent, *Macromolecules* **30**, 4945-4951.

36. Leuchs, H. (1906) *Ber. Dtsch. Chem. Ges.* **39**, 857.

37. Deming, T.J. (1997) Transition metal-amine initiators for preparation of well-defined poly(gamma-benzyl L-glutamate), *J. Am. Chem. Soc.* **119**, 2759-2760.

38. Deming, T.J. (1997) Facile synthesis of block copolypeptides of defined architecture, *Nature* **390**, 386-389.

39. Bamford, C.H., Block, H. (1961) The initiation step in the polymerization of N-carboxy-alpha-amino-acid Anhydrides. Part I. Catalysis by tertiary bases, *J. Chem. Soc.* 4989-4991.

40. Dvořák, M., Rypáček, F. (1996) The Feasibility of Preparing Poly(α-Amino Acids) With Narrow Molecular-Weight Distribution Via Controlled Living Polymerization of *N*-Carboxyanhydrides, 532-535.

41. Rypáček, F., Dvořák, M., Kubies, D., and Machová, L. (1999) Functionalised Polymers of α-Amino Acids and the Method of Preparation Thereof, PCT/CZ99/00016.

42. Rypáček, F., Dvořák, M., Štefko, I., Machová, L., Škarda, V., and Kubies, D. (2000) Poly(amino acid)s and Ester-Amide Copolymers: Tailor-made Biodegradable Polymers, in R.A.Gross and C.Scholz (eds.), *Biopolymers From Polysaccharides and Agroproteins*, ACS Books,Washington,D.C., pp. 258-275.

43. Kricheldorf, H.R., Lee, S.R., and Bush, S. (1996) Polylactones .36. Macrocyclic polymerization of lactides with cyclic Bu_2Sn initiators derived from 1,2-ethanediol, 2- mercaptoethanol, and 1,2-dimercaptoethane, *Macromolecules* **29**, 1375-1381.

44. Kowalski, A., Libiszowski, J., Duda, A., and Penczek, S. (2000) Polymerization of L,L-dilactide initiated by tin(II) butoxide, *Macromolecules* **33**, 1964-1971.

45. Rypáček, F., Machová, L., Kotva, R., and Škarda, V. (2001) Polyesters with functional-peptide blocks: Synthesis and application to biomaterials, *Polym. Mater. Sci. Eng.* **84**, 817-818.

SURFACE TREATMENTS AND PRE-CALCIFICATION ROUTES TO ENHANCE CELL ADHESION AND PROLIFERATION

A.L. OLIVEIRA[1,2], I.B. LEONOR[1,2], C. ELVIRA[3], M.C. AZEVEDO[1,2], I. PASHKULEVA[1,2], R.L. REIS[1,2]

1. Department of Polymer Engineering, University of Minho, Campus de Azurém, 4800-058 Guimarães, Portugal, Tel: + 351 253 510245, Fax: + 351 253 510249

2. 3B's Research Group (Biomaterials, Biodegradables & Biomimetics), University of Minho, Campus de Gualtar, 4710-057 Braga, Portugal, Tel: + 351 253 604490 (Ext. 5497); Fax: + 351 253 604492

3. Institute of Polymer Science and Technology, CSIC, c/Juan de la Cierva 3, 28006 Madrid, Spain.

Abstract

When coated with a bone-like apatite layer, biodegradable polymers have a great potential to be used as bone-repairing materials, since they can exhibit not only mechanical properties analogous to the natural bone but also a bioactive character. Presently available methods to produce such type of coatings are usually difficult to control on what concerns to the calcium-phosphate (Ca-P) layer composition, resorbability, and ability to generate strong bonds with substrates. On the other hand the presently available methodologies are not so effective on coating 3D architectures for being used as tissue engineering scaffolds. These are some challenges addressed in our work. In that perspective, our research group is developing several biomimetic coating methodologies, inspired in natural physiological processes, to coat the surface of starch based biodegradable polymers with tailored apatite layers that will be able to bond to living bone. The different biomimetic approaches that are being proposed go from adaptations of the traditional biomimetic methodology (performed for untreated and surface modified materials using chemical and physical means) to innovative sodium silicate gel treatments or a novel autocatalytic methodology. To understand the mechanisms of apatite formation, particularly in the earlier stage of nucleation, the atomic force microscopy (AFM) has been used as an extremely powerful tool, since it allows for in-situ studies of the surface, simulating the chemical environments founded in-vivo.

The recent developments on tissue engineering in the field of orthopaedic research allow for creating an engineered living tissue. However, it is required the attachment, proliferation and differentiation of living cells on the surface of appropriate polymeric scaffolds. In this case, surface modifications have an important role, since they can improve the cell adhesion and proliferation at the surface of materials. Therefore,

183

R.L. Reis and D. Cohn (eds.),
Polymer Based Systems on Tissue Engineering, Replacement and Regeneration, 183–217.
© 2002 Kluwer Academic Publishers. Printed in the Netherlands.

plasma and chemical induced graft polymerisation of active groups on starch-based blends is another approach proposed by our group in order to achieve this goal. Preliminary cell adhesion and proliferation tests, carried out in materials on which acrylic monomers have been grafted, have shown remarkable improvements with respect to original starch based samples.

1. Introduction

Bioinorganic materials constitute a source of inspiration for the field of biomaterials. Mineralized tissues such as oyster shells, coral, ivory or pearls are just a few wonderful examples of what we can learn with the vast variety of bioinorganic materials engineered by organisms [1]. Who better than nature to design complex structures and control intricate processing routes that lead to the final shape of living creatures? Can a material scientist design and process materials with structures that perform as well as those found in living creatures? The answer is of course not or not yet in an optimistic looking forward view. The enormous diversity of the phenomenon of biomineralization among living organisms implies that much can be learned about mineral formation from these organisms, in particular the control of crystal formation. Furthermore, many of the mineralized tissues formed by organisms have advantageous mechanical properties [2-4]. The study of these natural biomineralized structures has generated a growing awareness in materials science that the adaptation of biological processes may lead to significant advances in the controlled fabrication of superior "smart" materials [5-8].

1.1. THE BONE

Bone is probably the most complex example of a biomineralized material [4]. Understanding the inherent complexity of the molecular systems controlling its biological synthesis is a main challenge to many materials scientists. In fact, to copy the structure, property and performance (function) of this elegant structure, would represent a major breakthrough in the field of biomaterials. Bone is formed by cell-mediated processes on which hard tissue is synthesised as a composite material, consisting of an *organic matrix* reinforced with a *mineral phase*. The organic matrix is mainly constituted by collagen (almost 90%), the most abundant animal protein in mammals, accounting for up to 30% of all proteins [9]. Collagen molecules, after being secreted by cells, assemble into characteristic fibers, which are responsible for the functional integrity of tissues such as bone, cartilage, skin and tendon. In order to withstand the physical stresses to which they are exposed, covalent cross-links are formed between adjacent collagen molecules. The mineral phase in bone is a calcium phosphate mineral, largely in the form of a partially amorphous hydroxyapatite (HA, $Ca_{10}(PO_4)_6(OH)_2$), and contains not only Ca^{2+}, PO_4^{3-} and OH^- ions, but also small amounts of CO_3^{2-}, Mg^{2+}, Na^+ or trace amounts of elements such as F^- and Cl^- [10]. Other forms (e.g. with substitutions of the hydroxyl groups) are possible, and as much as 30% of the skeletal mineral is actually present in the form of "amorphous" calcium phosphates, such as: $CaHPO_4 \cdot 2H_2O$ and $Ca_3(PO_4)_2 \cdot 3H_2O$ [9,11]. This mineral phase is involved in both the biomechanical and metabolic functions of tissue. Essentially, the mineral has two functions [12]. First, it provides structural stability to the skeleton (protecting and supporting vital organs) and, secondly, it acts as a storage site for calcium, phosphate, sodium, magnesium, carbonate and other ions. Concerning this latter function, calcium phosphate mineral can either provide body fluids with these ions, in order to maintain the biologically required levels, or else to act as a detoxifying depository to store unwanted ions like lead and strontium.

There are two kinds of bone in the organism [9]: the cortical bone (the outer, denser envelope of most bones, which plays a major role in the support function) and the trabeculae or cancellous bone (which has a porous structure and is metabolically more active). The process of bone formation, as it occurs in embryogenesis, begins when mesenchymal stem cells (MSCs- progenitor cells that can differentiate into bone or cartilage-forming cells) start to differentiate into chondrocytes and secrete a cartilaginous matrix [13]. After vascularization, a new set of MSCs begin to differentiate into osteoblasts (bone-forming cells) that proliferate to secrete a collagen bone matrix. The predominant collagen matrix secreted by the osteoblasts undergoes mineralization. As the osteoblasts are separated by the mineralizing matrix, they are entrapped in spaces called lacunae. These entrapped osteoblasts are now called osteocytes, and they gradually lose the ability to produce matrix. At tissue level bone undergoes remodelling, being continuously resorbed and rebuilt (or formed) [9,12,13]. The primary cells involved are the osteoblasts that form bone, and osteoclasts that digest bone. The process begins with resorption of old bone, including the matrix and expired osteocytes. When breakdown and formation of bone are not in balance (i.e. more bone is broken down than it is being formed), there is bone loss [13]. Most diseases of the skeleton, like osteoporosis or arthritis, are due to such an imbalance, resulting in systemic or local bone loss [14-16].

The normal physiological process for the formation of bone, is called mineralization and is characterized by the deposition of calcium phosphate (Ca-P), in the form of mineral crystals upon collagen [14,17]. Only a few years ago [18] mineralization was visualized as being primarily a physicochemical phenomenon in which mineral nucleation was achieved by extracellular, non-living chemical structure(s) in the matrix (e.g., collagen) that serve as templates upon which the first mineral crystals were formed. Initial crystals could then serve as nuclei for further mineral propagation [18]. In this perspective both initiation and propagation were controlled by non-living chemical factors residing in the matrix. In recent years, however, the prevailing opinion has shifted to a view that envisions cells as being more importantly involved in mineral initiation (through matrix vesicles) while mineral propagation remains primarily extracellular and physico-chemical [9,12,13,19,20,]. Nevertheless, there is an increasing awareness that mineral propagation is also regulated by cells by creating the matrix and ionic milieu in which mineralization may or may not progress.

The mechanism of cell-mediated mineralization is best visualized as a cascade requiring the interaction of many different factors that either promote or retard/inhibit this phenomenon [19]. The whole mechanism and the specific role of each factor in particular have not yet been completely elucidated, and different perspectives have been proposed [12-14]. The most widely accepted mechanism is based on the biphasic hypothesis [12,13,18]. During initiation (phase 1), the first mineral is formed inside the matrix vesicles (MV). These are small, membrane-bound vessels of cellular origin [13]. Active transport (ion pumps) may be used to raise the amount of calcium and phosphate in the vesicles to levels above supersaturation, creating favourable ionic conditions for deposition of nascent mineral within the protected microenvironment of the MV membrane. Initially the mineral is in the form of amorphous calcium phosphate, octacalcium phosphate and/or brushite, with later conversion to hydroxyapatite. The membrane of the vesicles provides a protected microenvironment in which Ca^{2+} and PO_4^{3-} can be concentrated, localized and interact to form the first, unstable, nuclei of mineral. Alkaline phosphatase activity (and the activity of other vesicle phosphatases) functions to promote calcification during this phase, as do the Ca-binding phospholipids and proteins of the MV. Annexins may serve as Ca^{2+} ion channels, promoting Ca^{2+} transport to the initial site of mineralization, beneath the MV membrane.

Once these nuclei are transformed into HA, mineral propagation begins with the penetration of the MV membrane by crystals, and their exposure to the extracellular fluid (ECF) (phase 2) [12,13,18]. Crystal exposure is promoted by proteinases and lipases in MVs that speed vesicle breakdown. In the absence of apatite, ECF contains insufficient Ca^{2+} and PO_4^{3-} to initiate mineral deposition, but a sufficient amount is present to support crystal proliferation. The preformed apatite crystals from the MV act as templates for new crystal proliferation. Molecules at the outer MV surface such as collagen type X may serve as a bridge for mineral to spread into the adjacent collagenous matrix.

In summary, cells control all phases of mineralization by [19]:

    i.   releasing to selected sites in the matrix, MVs which are equipped by structure and composition to promote calcification;
    ii.   creating and modifying the local matrix composition in order to induce a faster propagation or delimit the spread of mineral;
    iii. regulating through endocrine means the ionic milieu at the mineralization front, thus providing an uninterrupted supply of Ca^{2+} and PO_4^{3-}, at a permissive pH, for mineral propagation.

The processes by which the mineralized tissues are formed can be a source of information for the development of new materials for biomedical applications, capable of mimicking the living tissues, i.e. biomimetic materials. In case of bone replacement, this new concept can lead to innovative ideas, inspired in mineralized tissues, easily found in nature. However, if the principles by which living organisms produce mineralized tissues are understood, how might they be used for the development of new biomaterials? If biomimetic is interpreted as the reproduction of the entire sequence of biomineralization steps, then it is clear that any developed process would be extremely complex and would lead to unbearable costs. A less literal use of the term biomimetic should then be applied. If a materials scientist can be inspired by a biological prototype to apply its principles to his research area, then a biomimetic result has been realized. This is the type of research approach we try to use in our research group.

1.2. BONE-LIKE APATITE COATINGS

When considering an ideal material to replace and mimic bone, synthetic calcium phosphates (currently designated as Ca-Ps or apatites) can be an obvious answer, since they can replicate the structure and composition of bone mineral. Therefore they are widely used for a number of orthopaedic and dental applications, either in bulk as bone grafts or as coatings [21,22]. However, the mechanical properties of Ca-Ps are far from being close to those of human bone, which limits their applications. In fact Ca-Ps are too stiff and very brittle. Nevertheless, coating the surfaces of orthopaedic materials with an apatite layer can elicit favourable chemical and biological responses on the surfaces, which allows for the reproduction of the reactions occurring in the natural calcified tissues, without loosing bulk properties of the materials [22-27]. In fact, Ca-Ps have a biocompatible behaviour with most of the cell types such as osteoblasts, osteoclasts, firoblasts, and periodontal ligament cells being found in the calcified tissues. Furthermore, Ca-Ps disclose osteoconductive properties allowing for the formation of bone on its surface by attachment, migration, proliferation, and differentiation of bone-forming cells [25,28]. Therefore, these types of coatings have great potential for bone fixation applications, for being used as fillers of bone defects, or scaffolds in tissue engineering. The later can be loaded with osteogenic biological molecules or can serve as beds for the seeding of living cells that will stimulate bone formation [27].

In the history of bone replacement, various "from the shelf" orthopaedic implant materials have been introduced, particularly metallic materials like stainless steel or titanium alloys [29]. Because a stable fixation of these implants to the bone was found to be critical to the long term stability in applications like joint replacement, an input was made for the development of different Ca-P coating methodologies to induce bone bonding ability and therefore, to create more stable interfaces [30,31].

The plasma-spraying technique is the major method commercially available method used for coating Ca-P on metallic implants [32-35]. Reproducibility and economic efficiency of the process are outstanding advantages [33,34]. However, this method presents some crucial drawbacks affecting the long-term performance and lifetime of the implant. The most significant are the poor coating-substrate adherence [34] and lack of uniformity of the coating in terms of morphology and crystallinity [35,36]. Since plasma-spraying is a high-temperature and line-of-slight process, there are also some aspects that were not solved yet, such as the deteriorating effect of intense heat on substrates, non-uniformity in coating density, wide range of band strength and limitation in coating implant devices with complex shapes [37].

Other currently used approaches are sputter coating techniques that are able to increase the bond strength between the coating and the substrates [38-40]. However, the drawback inherent to this technique is that the deposition and the process itself are very slow. By using a magnetically enhanced variant of radiofrequent sputtering, this problem can be solved but the endurance and the Ca/P ratio of the coating require further *in vitro* and *in vivo* studies before it can be applied routinely to produce crystalline pure Ca/P ceramic coatings on implant surfaces [41]. Other techniques are available such as: dip coating sintering [42], chemical vapour deposition [37], sol-gel deposition [43,44], ion implanting [45] laser deposition [46-48] and electrochemical processes like electrophoretic deposition [49], electrocrystallization [50] and anode oxidation [51]. Despite of all the investigations carried out, the produced coatings can suffer from at least one of the following problems [22]: lack of coating adherence to the substrate, thickness non-uniformity, poor structure integrity, and non-stoichiometric composition of the coatings. In fact, each of the above mentioned techniques has its own technical limitations, and so far, an optimal technique for producing physiologically stable and interfacially adherent apatite coatings has yet to be developed. Thus, there is a demand to develop a method being able to form an apatite layer with properties similar to those of bone calcium-phosphates on various complex shaped materials, that is capable to enhance biocompatibility as well as bioactivity, when engineering bone implants or designing tissue engineering scaffolds.

1.2.1. Biomimetic approach

The Ca-P minerals found in natural hard tissues are fabricated in a physiological environment at low temperatures from moderately supersaturated mineralizing solutions [52]. In the recent years, there has been an increasing interest in the so-called biomimetic preparation of calcium phosphate coatings on implant materials. Rapid progress has been made in the development of these coatings and several methodologies have emerged. This type of approach is particularly suitable to coat polymeric materials [53-55], as it can be carried out at low temperature reaction conditions.

A calcium phosphate coating was first grown on a substrate by a biomimetic process by Kokubo *et al* in 1990 [56]. For this purpose, a bioactive CaO-SiO$_2$-based glass was used. Silicate ions containing silanol groups (Si-OH) were supposed to be released from that bioactive glass and adsorbed on the substrate surface to induce apatite formation. The bioactive glass was used in the form of particles that were set in contact with the substrates to induce apatite nucleation on their surface in a simulated body fluid (SBF) with ion concentrations nearly equal to those of the human blood plasma and at body temperature. This solution was developed also in 1990 by the same author [57]. After one week (the time established for apatite nuclei to form into a layer), the apatite grew subsequently by immersing the substrates in 1.5 SBF with ion concentrations 1.5 times higher than those of SBF. By this method, an apatite layer has been coated on ceramics, metals and polymers, such as alumina glass, zirconia, titanium, polyethylene, polymethylmethacrylate, ethylene-vinyl alcohol copolymer, polyethylene terephthalate (PET), poly(ethersulfone) (PES), polyamide (PA 6,6), poly(vinyl alcohol) and silicone [54-56]. It was also shown that this apatite layer could be formed not only on flat surfaces but also on curved surfaces of small particles, long fibres and woven cloths of various materials [53,55,58].

The adhesion strength of the apatite layer to polymer surfaces by this process was, however, not sufficient for clinical applications [53]. Surface modifications, were then applied, prior to the biomimetic treatment, to improve the adhesion strength by increasing the amount of polar groups on the substrate surface [59-63]. These groups were found to act as favourable nucleating sites for apatite formation in the surface of the polymers. Various methods [59-63] for surface modification of different polymers were reported by the group of Kokubo. Chemical treatments with sodium hydroxide (NaOH) [59] or hydrocloric acid (HCl) [60] solutions, before biomimetic coating, have shown to improve the adhesion strength and to reduce the induction periods for apatite formation, being strictly dependent of the type of polymer. Surface modifications induced by a glow-discharge treatment [61] and ultraviolet irradiation (UV) [62,63] were also tested for the same polymers and similar effects were described.

Another way of tailoring the apatite formed by the herein described biomimetic methodology is by the side of the solution, i.e. changing the composition of the SBF [64,65]. Kim *et al* [64] have reported that different apatite layers can be produced on polyethylene terephthalate (PET) substrates in solutions where the ion concentrations were changed from 0.75 to 2.00 times those of SBF. Increasing the ionic activity product has then resulted in lower Ca/P ratios of the apatites. The same author [65] have also reported that by increasing the carbonate ion content, apatites with composition and structure nearly identical to those of bone apatite could be produced. In fact, this solution is known to be deficient in relation to the HCO_3^- content, when comparing to the human blood plasma [65,66]. Therefore, the japanese group is now proposing a new revised SBF (R-SBF) with an ion composition closer to the human blood plasma (higher amounts of HCO_3^-), to replace conventional SBF [66].

1.2.2. Other pre-mineralization routes

There is an enormous amount of published work using other different biomimetic routes for the formation of apatite layers on the surface of different materials [62,67-71]. Some of them are based in surface modifications, by chemical and physical means that are claimed to induce direct bioactivity on the surface of the materials [68-71]. Other methodologies use nucleating agents to induce the formation of the bioactive layer [62,67].

Bonelike apatite was successfully formed for the first time on organic polymers by a biomimetic process using sodium silicate solution as a nucleating agent, instead of bioactive glass particles, in a work by Miyaji *et al* [67,72]. A dense apatite layer was formed not only on limited surfaces but also on whole surfaces of fine PET fibers constituting a fabric. Therefore, this method enabled the apatite coating on various materials with complex shapes. The "traditional" biomimetic process is not so effective on coating materials with complex shapes, since the apatite nuclei are formed only on the material surface which is facing the glass grains [72]. On the other hand it was also possible to reduce the incubating periods for apatite formation to only 6 hours [67,72].

The grafting technique has also become very popular as a way to immobilize functional groups at the surface of polymers that can encourage apatite formation. Oyane *et al.* [73,74] developed a different methodology in which ethylene-vinyl alcohol copolymer (EVOH) substrates were modified by grafting silane coupling agents like tetraethoxysilane (TEOS) at its surface, in order to produce silanol groups. Their apatite forming ability was examined in SBF and 1.5 SBF. Only in the later case it was possible to observe the formation of an apatite layer after 21 days [74]. Nevertheless, the Ca/P molar ratio of the apatite formed in 1.5 SBF was much lower than that of the apatite found in the natural bone. Therefore, the same authors tried also to incorporate calcium ions in the silicate phase, by using a calcium silicate solution prepared with the coupling agent [75,76]. The release of these ions accelerated the apatite nucleation by increasing the ionic activity product of the apatite locally near the surface [76]. Recently, Kim *et al.* [68] proposed a different methodology to incorporate silanol groups at a polyethylene substrate via photografting of vinyltrimethoxysilane and hydrolysis. The substrate modified in this way formed a dense and homogeneous bone-mineral-like apatite layer in a SBF solution with ion concentration 1.5 times, after 7 days [68].

Tretinnikov *et al.* [69] developed a bioactive polymer by using also surface modification by grafting to immobilize organic compounds. An organophosphate polymer was chemically bound onto a polymeric film by surface graft polymerization of a phosphate-containing monomer [69]. As the phosphate group is one of the building blocks of HA and has a high affinity toward calcium ions, polymeric materials modified by surface grafted water soluble organophosphate polymer induce bone deposition of Ca and PO_4 ions in the form of a hydroxy carbonate apatite (HCA) layer [69]. It is then expected that covalent immobilization of organophosphates will open the way for developing bioactive bone-bonding polymers. Another approach was used by Mucalo *et al.* [70] to coat apatite in cotton substrates, by means of grafting phosphate groups at the surface of the substrates using a phosphorylation methodology and subsequent SBF immersion. Yokogawa *et al* [71] found that soaking in $Ca(OH)_2$ treated phosphorylated chitin fibers, also lead to the deposition of an apatite layer, after SBF immersion.

An interesting approach is the Langmuir Blodgett (LB) technique, a type of supramolecular assembly which can be used to produce an organic template with a specific head group for the nucleation of calcium phosphate crystals [77]. The LB process produces a thin organic film with a very organized structure, fulfilling the requirements of an organic template for a controlled mineralization process. Costa *et al* [77], investigated apatite formation in SBF using a ω-tricosenoic acid monolayer film that was prepared by the LB method, using calcium carboxylate as a functional group. They reported that the nucleation of apatite was induced by carboxyl groups that reduced the interfacial energy between the nucleus and substrate. The morphology of the apatite crystals grown was strongly affected by the structure of the monolayer. It was further indicated the crystal orientation between the apatite and organic material took place *in vitro*, possibly due to the similar mechanisms observed in the biogenic materials [77].

Besides the use of SBF as a mineralizing environment, other biomimetic solutions have been developed for the formation of an apatite layer. Taguchi *et al.* [78] have developed an apatite formation process using hydrogels – an alternating soaking process – to form large amounts of apatite in SBF for a considerable short period of time (42 hours). The soakings have alternated every hour between $CaCl_2$ and H_2PO_4 solutions. The same methodology was also effectively applied by Furuzono *et al.* [79] to coat a silk fabric proposed as a biomaterial for bone replacement. Nevertheless, this methodology does not seem very practical since it requires a constant change of solutions every hour. Kim *et al.* [80] have used a simple method of coating thin films of low crystalline apatite crystals by using a filtrated solution containing calcium and phosphate ions to coat a poly(lactide-co-glicolide) co-polymer sponge (PLGA). The material was coated by this process within 24 hours. The work of Yuan *et al.* [81], indicates that an apatite layer was formed in the surface of poly(L-lactic acid) (PLLA) after two weeks without the aid of a nucleating agent, by immersing the substrates in a solution inspired in SBF but with 1.5 times its ion concentration. Rhee *et al.* [82,83], have shown that an apatite layer could be formed in the surface of a collagen membrane after one week with immersion in 1.5X SBF solution, to which citric acid was also added. This same solution is able to induce the formation of an apatite layer, after the same period, in the surface of a cellulose cloth [84]. The results therefore suggest that citric acid has a nucleating ability and can accelerate the nucleation of apatite on the non-bioactive studied substrates.

The works described above are just some examples that can well illustrate the variety of possibilities that can be explored when using/developing a biomimetic approach. All these surface modifications and pre-mineralization routes described before can play an important role in enhancing the biocompatibility and bioactivity of a biomaterial. In fact cells are sensitive to several surface properties such as roughness, energy, chemistry and even to more subtle characteristics such as relative crystallinity.

It is possible to find several advantages of the biomimetic approach over the methodologies previously described. In fact, a biomimetic coating is expected to show higher bone-bonding ability due to its similarity to the mineral of bone. On the other hand, the adhesion to the substrate can be enhanced by means of several methodologies or by different surface pre-treatments. Tailored apatite coatings with different Ca/P ratios and crystallinities are possible to obtain. Another very important advantage is that no adverse effect of heat on substrate occurs, since these methodologies work at operating temperatures, which allow them to be applied to a range of different materials such as biodegradable polymers. It is also the simplest and more cost effective of the approaches available to create a biological-like apatite layer. Taking advantage of all this, different biomimetic approaches are being developed by our group to be able to effectively coat a range of bioinert and biodegradable polymers processed into a large variety of shapes for bone related applications.

1.3. STARCH BASED POLYMERS

Starch-based biodegradable polymers are particularly interesting for bone replacement. Besides being biodegradable, inexpensive (when compared to other biodegradable polymers) and available in large quantities [85-88], starch-based polymers can be converted into complex geometries that exhibit interesting mechanical properties, by using standard equipment developed for the processing of synthetic polymers or by means of using distinct innovative methodologies. Furthermore, in addition to their processing versatility, they exhibit a biocompatible behaviour, already demonstrated by in vitro [89-91] and in vivo studies [92]. Therefore, they are under consideration for a wide range of biomedical applications like bone replacement/fixation [93-95], novel hydrogels and partially degradable bone cements [96], drug delivery carriers [97] or temporary scaffolds for tissue engineering applications [97-99].

In case of bone related applications like tissue replacement/fixation, or tissue engineering scaffolds to be applied in load-bearing sites, these systems must exhibit mechanical properties that match those of human bone, associated to degradation kinetics adequate to the healing of the tissues to be replaced or fixed [100]. It was reported by our group that the physical and mechanical properties of these materials could be optimized by controlling the morphologic developments within the moulds, using non-conventional processing routes [101]. On the other hand, the incorporation of bone-like inorganic fillers, such as hydroxylapatite (HA) or bioactive glasses is another interesting approach, allowing for the development of degradable composites that can combine an attractive range of mechanical properties with the so-desirable bone-bonding behaviour [102,103]. Since the essential condition for materials to bond to living bone is the formation of a biologically active bone-like apatite layer on their surfaces different coating methodologies are also being developed through biomimetic processes for producing such type of layers on the proposed materials [104]. Finally, a novel emerging application is also currently being proposed, on which starch based polymers would serve as temporary scaffolds for the transplanted cells to attach, grow and maintain differentiated functions in a range of tissue engineering applications [97-99]. In fact it was already possible to develop distinct porous architectures based on these starch based biodegradable blends [97,105] by means of using different and innovative processing routes either based on melt-processing technologies [105] or on microwave baking [97] and subsequently to produce biomimetic coatings on these materials that are not only aimed at enhancing cell adhesion and proliferation but also tissue ingrowth.

2. Results & Discussion

When coated with a layer similar to the natural mineral of bone, starch based biodegradable polymers have a great potential to be used as bone-repairing materials, since they can exhibit not only mechanical properties analogous to the natural bone but also a bioactive character [106], i.e. can stimulate the formation of new bone. One of the main goals addressed by the 3B's research group of University of Minho is to induce a bioactive behaviour on the surface of these promising biodegradable polymers, via different surface modifications and biomimetic routes, trying to overcome all the difficulties arising from the pH changes and continuous degradation of the polymeric surfaces. Another important research approach is to develop adequate ways to modify the surface of the polymers by means of using specific hydrophilic monomers able to directly enhance cell adhesion and proliferation, for bone engineering applications.

2.1. STANDARD BIOMIMETIC METHODOLOGY

As discussed above, although very popular and effective, the "traditional" biomimetic process, using bioactive particles as nucleating agents, still present some difficulties on what concerns to the adhesion of the apatite layer to polymeric surfaces as well as on coating materials with complex shapes [72]. Therefore, an adaptation of this biomimetic methodology was made by Reis et al. [106], in which the samples were rolled on a bed of wet bioactive glass particles before immersion in an SBF solution. The methodology was successful on coating different types of polymers and shapes like a high molecular polyethylene, a biodegradable starch poly (ethylene vinyl alcohol) blend (SEVA-C) and a polyurethane foam. However, the same problem associated with some lack of coating adhesion was still also observed (although better results than for the original method could be obtained). To overcome this problem, different surface treatments were experimented by Oliveira et al. [104], like potassium hydroxide (KOH), acetic anhydride ($CH_3(CO)_2O$, Ac_2O), UV radiation and overexposure to ethylene oxide sterilization (EtO), on SEVA-C substrates before the biomimetic process. In this process the samples were: (i) involved in a dispersion of a bioactive glass and then soaked in SBF (37 °C), in order to form apatite nuclei; (ii) soaked in another solution with ion concentrations 1.5x SBF, after 7 days, for making apatite nuclei to grow. The influence of surface pre-treatments was studied: (i) over the induction and growing periods of the apatite layer formation; (ii) on the substrate/Ca-P coating adhesive strength. A biodegradable blend of starch with poly(ethylene vinyl alcohol) (SEVA-C) was selected for this study.

The surface modifications were identify by FTIR (ATR) analysis that showed differences between non-treated SEVA-C blends and samples treated with KOH and $(CH_3CO)_2O$. No remarkable changes were observed for the UV and sterilization methods. Table 1 shows the assignment of the most characteristic FTIR signals of SEVA-C treated and non-treated samples.

TABLE 1. Most significant FTIR (ATR) band position for SEVA-C blends. SEVA-C treated with KOH and SEVA-C treated with (CH3CO)₂O.

Sample	Peak assignement (cm^{-1})			
	- OH strech	C-H stretch	C=O stretch	C-O stretch
SEVA-C	3500-3100 Broad	2920-2850 two bands	1700. 1664 weak band	1110. 1020 Weak
SEVA-C (KOH treatment)	3500-3100 medium (several peaks)	2950-2880 two bands	1680 Weak	1150. 1075. 1025 (3 bands) Intense
SEVA-C (Ac₂O treatment)	3500-3100 Weak	2940-2860 two bands	1640 Broad	1108. 1048 Intense

The most important changes were observed on the region from 3500 to 3100 cm^{-1} assigned to the OH vibration of the starch and vinyl alcohol hydroxy groups, on which non-treated SEVA-C exhibits a broad band which intensity is reduced when treated with KOH and Ac₂O. SEVA-C samples treated with KOH exhibited also 3 very intense bands at 1150, 1075, 1025 cm^{-1} which are assigned to C-O stretching vibrations. The decrease of -OH signal intensity and the appearance of the third band in the 1100 cm-1 region suggests that hydroxyl groups of SEVA-C are forming molecular complexes by hydrogen bonding interactions with KOH, being also possible the formation of partial alkoxide ($-O-K^+$) formation as described in the literature [107].

The hydration degree was also studied through water-uptake measurements. Figure 1 shows a typical water-uptake versus time graph for SEVA-C and SEVA-C treated by the previously described procedures.

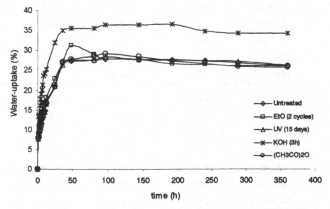

Figure 1. Water-uptake (%) versus time for treated and non-treated SEVA-C samples.

After 50 hours of immersion, the hydration degree of SEVA-C blends was about 25% due to its hydrophilic nature, mainly as a result of the starch and vinyl alcohol hydroxy groups, as it has been previously described [96]. The UV exposure treatment did not exhibit remarkable differences on the SEVA-C water-uptake behaviour. Sterilized SEVA-C samples have shown a faster water-uptake during the first hours of the experiment when a second consecutive sterilization cycle was performed, which was probably due to a loss of crystallinity of the material as found by Raman spectroscopy in a previous work [96]. Therefore, sterilized samples did not shown notable variations in the equilibrium hydration degree after 15 days of immersion, in comparison to untreated SEVA-C. Ac$_2$O treated samples hydration degree is also similar to untreated SEVA-C. The most significant differences were obtained with KOH treated SEVA-C samples, which after 50 hours of immersion have achieved its equilibrium hydration degree above 35%. The increased hydrophilicity after this treatment can be justified by the molecular hydroxy-KOH hydrogen bonding complexes, which are easily solvated, by water molecules.

Contact angle measurements, using the sessile drop technique [108], were also preformed in the surface of untreated and surface treated materials in order to investigate any changes in the wettability of the materials. Table 2 present the results concerning to the average water contact angle (θ) and average surface tension (γ_s) of the treated surfaces.

TABLE 2. Water contact angle and surface energy measurements of SEVA-C untreated and after surface treatments, as function of time.

Treatment	Time	Water contact Angle θ ($^\circ$)	γ_s (mN/m)	γ_s^d (mN/m)	γ_s^p (mN/m)
SEVA-C untreated	0	64.1	45.1	32.3	12.8
EtO sterilisation	1 cycle	70.7	44.5	29.2	15.3
	2 cycles	72.4	44.4	32	12.4
UV radiation	60 min.	62.0	44.5	29.2	15.3
	180 min.	65.4	44.4	32	12.4
(CH$_3$CO)$_2$O treatment	5 min.	66.7	45.7	34.1	11.7
	60 min.	66.7	46.5	34.4	12.0
	180 min.	64.7	46.7	34.2	12.5
	300 min.	64.6	47.0	34.3	12.6
	1440 min.	62.9	44.8	33.9	10.9
KOH treatment	5 min.	62.2	46.5	32.0	14.5
	15 min.	59.7	48.4	34.3	14.1
	30 min.	55.7	51.7	35.2	16.4
	60 min.	48.1	56.9	35.4	21.6
	180 min.	47.1	56.7	34.1	22.6
	300 min.	47.8	55.4	32.1	23.3
	1440 min.	28.6	66.7	32.8	33.8

UV treatment did not show any significant changes, neither in contact angle nor in the polar contribution, whereas dispersive and total surface energy remains approximately constant. Sterilisation treatment shows an increase in the contact angle values for 1 and 2 cycles, remaining approximately constant the total surface energy. On the other hand, a small decrease (12.8 to ≈ 11 mN/m) in the polar contribution to the surface energy was observed over all treatment times for $(CH_3CO)_2O$ treated samples, with an increase in the contact angle during short treatment times, as a consequence of partial incorporation of acetyl groups (which have a smaller polarity to the hydroxy ones) to SEVA-C blends. For longer periods of time, the total surface energy remains constant with respect to untreated samples. KOH treatment of SEVA-C samples did modify the surface parameters, such as contact angle polar and total surface energy. As the KOH treatment time increases an increment in total surface energy, as consequence of the increase of the polar contribution (12.8 to 33.8 mN/m after 1440 s) was observed. Values changed, from 45.1 mN/m for non-treated SEVA-C to 66.7 mN/m after 1440 s of KOH treatment. These significant changes in the polar contribution to the total surface energy can be attributed to molecular complexes formation as described above.

2.1.1. Apatite formation

The study of the influence of the surface pre-treatments over the induction and growing periods of a biomimetic coating was the main aim of this part work. Figure 2 presents SEM photographs focusing the first 6 and 12 hours of SBF immersion, were the effect of the surface modifications was manifestly evident. This figure clearly presents an example of effective pre-treatments such as UV radiation and KOH etched substrates.

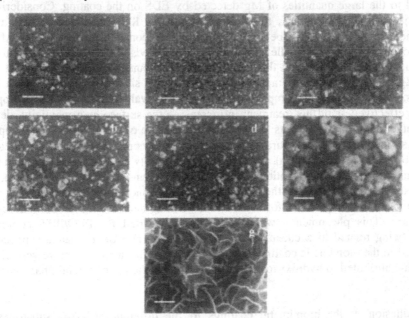

Figure 2. SEVA-C SEM photographs of a) non-treated (6 h.), b) non treated (12 h.), c) UV treated (6 h.), d) UV treated (12 h.), e) KOH treated (6 h.), f) and g) KOH treated (12 h.). ( ) corresponds to immersion time (hours) in SBF solution. The scale bar corresponds to 10 μm.

Figures 2.a and b show the surface of untreated SEVA-C after the first 6 and 12 hours of SBF immersion. The UV treated surface presented in Figure 2.c shows that after 6 hours of SBF immersion it seems to exist a greater number of apatite nuclei in the surface of SEVA-C, when comparing with the untreated substrate. After 12 hours of SBF immersion (Figure 2.d) it is already clear the difference in the quantity of formed nuclei, which are more uniformly distributed. The Ca/P ratio for the apatite formed on an UV radiation treated surface after 6 hours of nucleation is 1.23 (data not shown). This value is smaller than the stoichiometric relation found on hydroxyapatite (1.67). However, is near the characteristic value for brushite, which is a very well known apatite inductor [56]. For greater times of SBF immersion, Ca/P ratios tend to increase and then to stabilise, reaching a value of around 1.7 after 4 days SBF immersion. TF-XRD data obtained for these samples clearly match the standard XRD hydroxyapatite pattern (JCPDS 09-432 standard file). Although a more amorphous nature of the film is also evident. The inclusion on the calculation of the Ca/P ratios of Mg, Na and K elements did not affect the "Ca/P ratio", mostly in the first 24 hours ("Ca/P"=3). The composition of the apatites formed on the UV radiation modified substrates is similar to that found on an untreated substrate. After 15 days of SBF immersion it was calculated a ratio of 1.77. Again, TF-XRD data confirms that Ca-P films are mainly apatite-like.

KOH treatment has demonstrated to be the most effective in inducing the formation of an apatite layer in the SEVA-C substrate. In fact Figure 2.e shows that after only 6 hours of SBF immersion it is already possible to observe a clear apatite-like layer formed on the SEVA-C surface, which for an untreated sample only could be observed after 7 days immersion in SBF. This layer presents a singular morphology, which can be related to the large quantities of Mg detected by EDS on the coating. Considering the elements Ca, Mg, Na and K, the Ca/P ratio for the KOH treatment suffers a great increment in the first hours, being Mg the major responsible for this result. After 12 hours of immersion this "needle type" structure is still visible, and only disappears after 24 hours, which matches with the decrease in Mg amounts (Figures 2. (f, g)). It seems that Mg may play a role on apatite nucleation. This should be confirmed on future studies. The Ca/P ratio also drops to the characteristic values. This result was supported by the effectiveness of this treatment in increasing the surface hydrophylicity, detected by contact angle measurements that lead to an increase of the polar energy component of the surface which has favourable sites for salt ions complexation and for the apatite nucleation. On the order hand, the water-uptake ability have improved, allowing the material to absorb higher quantities of Ca^{2+} and PO_4^{3-} ions from the SBF solution, when immersed. As a consequence, the Ca^{2+} and PO_4^{3-} ion concentration in the surface would probably rise, leading to the formation of nucleating sites for the biomimetic coating formation. This phenomenon was already been reported for PEO/PBT copolymers [109], being related to a chelation effect. The Ca/P ratio of the calcium phosphates obtained in the biomimetic coatings, after 15 days of SBF immersion were very close to the value attributed to hydroxyapatite (1.67), which is the main mineral phase in human bone.

The adhesion of the biomimetic coatings to the different SEVA-C substrates was accessed by means of a pull-of test in a tensile test machine. The measured adhesion of the coatings was quite high, in the 37 to 47 MPa range. The presence of OH groups on the substrate seems to facilitate the connection with the apatite layer. Similar results were obtained by other authors [53] studying polyvinyl alcohol (PVA) polymers. The water-uptake ability of PVA generated a higher adhesive strength of the Ca-P when compared with other polymeric substrates. In fact, it seems that a fairly strong bond could be formed between the polar groups of the polymer and the calcium ion of the apatite layer. Figure 3 shows the preliminary results of a pull-off test, carried out in order to quantify the effect of UV radiation and KOH etching pre-treatments on the adhesive strength of the Ca-P coatings to the substrates. These coatings were formed after 15 days of SBF immersion.

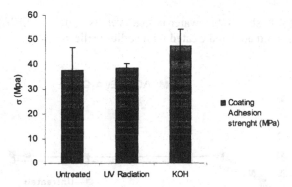

Figure 3. Adhesion strenght of the Ca/P coatings nucleating on the surface of the treated and non-treated SEVA-C substrates.

In spite of the typical scattering of these type of results, it is possible to state that the adhesion of the biomimetic coating to the substrates pre-treated with KOH was considerably enhanced. This treatment, besides diminishing the induction period for apatite nucleation, resulted in the formation of a thicker apatite coating with a higher adhesive strength.

2.2. SODIUM SILICATE TREATMENT

As mentioned before, when considering materials with complex shapes, such as porous 3D architectures, the traditional biomimetic methodology is not so effective. This can be explained by the considerable difficulty on forming an apatite coating, since the apatite nuclei will preferentially grow in the surfaces which are facing the bioactive glass particles. To overcome this obstacle a new biomimetic methodology to produce bioactive coatings on the surface of starch based or other polymeric biomaterials is being proposed by Oliveira *et al.* [110], using a sodium silicate gel as alternative nucleating agent. Sodium silicate gel can reach inside the pores of porous 3D architectures to be used on tissue replacement and in tissue engineering scaffolding. This new methodology is aimed at: (i) reducing the incubation periods; (ii) improving of the adhesion strength between the coating and substrate; (iii) being able to coat the inside of pores in porous 3D architectures to be used on tissue replacement and as tissue engineering scaffolds; (iv) producing Ca-P layers with different (tailored) Ca-P ratios.
The studied materials included injection moulded blends of starch with: (i) poly(ethylene vinyl alcohol) (SEVA-C), (ii) cellulose acetate (SCA) and (iii) polycrapolactone (SPCL). A high molecular polyethylene (HMWPE) was used as a

198

control to assess the possibility of coating polymers that do not uptake water. SEVA-C based 3D-arquitectures prepared by a previously developed microwave baking process [96], were also studied. A well-established polymeric foam - polyurethane - was used as a porous model control. To produce the bioactive coatings the materials were "impregnated" with a commercially available sodium silicate gel (Na$_2$SiO$_3$ • H$_2$O, containing ≈14% NaOH and ≈27% SiO$_2$), that acted as an alternative Ca-P nucleating agent. After the sodium silicate treatment the samples were soaked in a simulated body fluid (SBF) at 37 °C. After 7 days the ion concentration of SBF solution was raised to 1.5X in order to make the apatite nuclei to grow.

Figures 4.a and 4.b shows the water-uptake versus time for SEVA-C compact and porous structures, untreated and treated with sodium silicate.

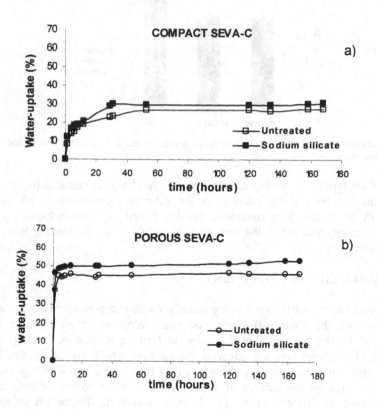

Figure 4. Water-uptake (%) versus time for untreated and sodium silicate treated a) compact and b) porous SEVA-C samples.

For compact structures, the equilibrium hydration degree of untreated samples is about 25%, after 50 hours. This hydrophilic behaviour is mainly a result of the presence of starch and vinyl alcohol hydroxyl groups, as it has been previously described [104,111]. Untreated porous structures achieve the equilibrium hydration degree, around 45%, after the first 10 hours of water uptaking. This higher hydrophilicity may result from the existence of high amounts of polar groups available in the structure after reaction of hydrogen peroxide with starch. On the other hand, porous structures have a higher specific surface then compact ones, which allows for a higher water-uptake. With sodium silicate treatment the amount of water uptaked increases around 5% in both types of structures, being this observation more significant in the case of the compact material. The materials used as controls (compact HMWPE and porous PU) did not uptake water after sodium silicate treatment.

2.2.1. Apatite formation

After immersion in SBF for several periods, the formation of an apatite layer was studied. Figures 5 and 6 show the SEM photographs of the evolution of the typical films formed on the surface of the treated materials.

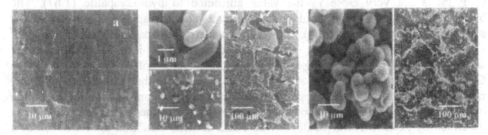

Figure 5. SEM micrographs of compact SEVA-C treated with a) sodium silicate and after b) 6 hours and c) 30 days of immersion in SBF.

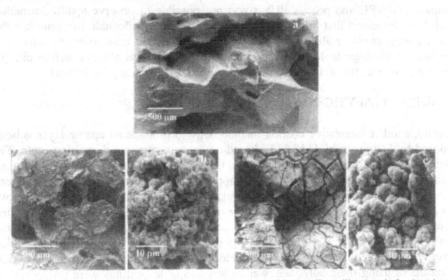

Figure 6. SEM micrographs of porous SEVA-C + H₂O₂ treated with a) sodium silicate and after b) 6 hours and c) 30 days of immersion SBF.

In sodium silicate treated materials only after 6 hours of immersion in SBF (see Figures 5 and 6) it was possible to observe the formation of very cohesive apatite-like layers that became fragmented due to the swelling of the polymers. This result was also observed for SPCL and SCA materials. For the porous materials, the apatite layer could be also observed inside the pores, clearing covering the cell walls. When comparing with the traditional biomimetic treatment, the later was not so effective on reaching the bulk of porous structures. This result is very promising for the developing of cancellous bone replacement materials and for pre-calcifying bone tissue engineering scaffolds. For the first stages of immersion in SBF, it were detected the presence of the ions Na^{2+}, K^{2+} and mostly Mg^+ in the crystalline network of the apatite. In fact, Mg^+ ions seem to play an important role in the first stages of apatite formation. This phenomenon was already been reported in Section 2.1 for KOH treated SEVA-C substrates [104] and is being presently under more detailed study. For the SEVA-C compact material the corresponding Ca/P ratios for the apatite formed in this earlier stage of nucleation is around 1.5 which is typical of tri-calcium phosphate. After 30 days the morphology of the layers tended to develop to the so-called cauliflower morphology [106], which is clearly showed in both compact and porous structures. On the other hand, the respective Ca/P ratios are very close to the value attributed to hydroxyapatite (1.67). The biomimetic methodology proposed in this case lead to an increasing on the water-uptake ability of the polymers (Figure 4), allowing the materials to absorb higher quantities of Ca^{2+} ions from the SBF solution. As a consequence, the Ca^{2+} ion concentration in the surface will be increased, leading to the formation of additional nucleating sites for the Ca-P coating formation. On the other hand, the increase of the surface hydrophilicity raising the amount of polar groups in the surface could be a contribute to the formation of silanol groups that are well known apatite inductors. This theory has been previously described by Kokubo *et al.* [55,56,58]. TF-XRD patterns exhibited the formation of a partially amorphous Ca-P film with the crystalline peaks mainly corresponding to hydroxylapatite, for the longer SBF immersion periods. On materials used as controls (compact HMWPE and porous PU) it was not possible to observe apatite formation, which have indicated that this methodology is only highly adequate for materials that have a strong swelling ability. In this case, even with the correspondent difficulties associated to biodegradable polymers raising from continuous pH and surface changes as function of time, the methodology was successful on generating Ca-P coatings.

2.3. AUTO-CATALYTIC COATING METHODOLOGY

Recently, another innovative coating methodology to produce an apatite layer is being proposed by Leonor *et al.* [112,113], based on an auto-catalytic deposition route. This original approach uses a deposition route that is totally "electroless", i.e., does not require the use of electric current for its application, being based on redox reactions. Two types of solutions are being studied, alkaline and acid baths, to produce the novel auto-catalytic Ca-P coatings. The respective compositions are generically the following: calcium chloride, sodium pyrophosphate, sodium hypophosphite, palladium chloride (alcaline bath, operated at a pH of 9.2) and calcium chloride, sodium fluoride, sodium hypophosphite, succinic acid, palladium chloride (acid bath, operated at a pH of 5.3). By means of using these solutions, well adherent apatite coatings were formed on the surface of both bionert (HMWPE) and biodegradable polymers (SEVA-C and SCA) [113]. The developed route seems to be a very promising and simple methodology for being used as a pre-implantation treatment to coat several types of materials previously to their clinical application.

Surface pre-treatments (polishing, in case of bioinert or UV activation, in case of biodegradable) were performed on the surfaces of all materials previous to immersion on the baths, which leaded to a better adhesion between the polymer and coating. For the HMWPE, with the polishing treatment, the surface roughness had increased, enhancing the coating adhesion, due to the fact that in general, a rougher surface promotes nucleation (over that of a smooth surface) as a result of the lower free surface energy [114]. The effects of UV radiation on starch based polymers may be explained due to the increasing of polar groups on the SEVA-C and SCA surface, which seems to facilitate the connection with an apatite layer, i.e., with calcium or hydroxyl ion of the apatite, as reported in others works [104, 106]. Kokubo et al. [63] demonstrated that a presence of polar groups on the surface of the polymeric substrates increase the adhesive strength since it is assumed to increase with increasing number of points at which the apatite nuclei are attached to the substrates.

2.3.1. Apatite formation

After 60 min of immersion, which was the typical coating time for both baths, the surface of the three polymers was completely covered with a calcium phosphate layer, as it is shown in the Figures 7, 8 and 9, by SEM micrographs.

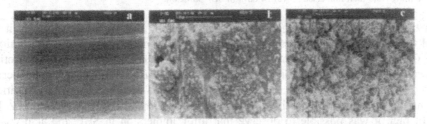

Figure 7. SEM photographs of the Ca-P coatings produced on the surfaces of HMWPE substrate. Sample before coating (a), after 60 min in acid bath (b) and then immersed for (c) 14 days in SBF solution.

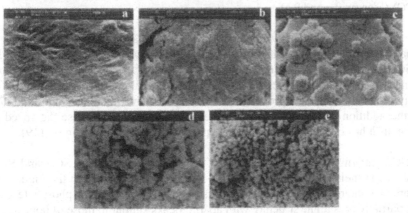

Figure 8. SEM photographs of Ca-P coatings produced on the surfaces of SCA substrate. Sample before coating (a), after (b) 60 min in alkaline bath, then immersed for (c) 14 days in SBF solution and (d) after 60 min in acid bath and then immersed for (e) 14 days in SBF solution.

202

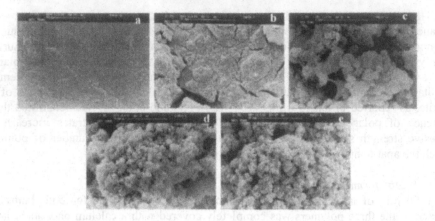

Figure 9. SEM photographs of Ca-P coatings produced on the surfaces SEVA-C substrate. Sample before coating (a), after (b) 60 min in alkaline bath, then immersed for (c) 14 days in SBF solution and after 60 min in acid bath (d) and then immersed for (e) 14 days in SBF solution.

From the characterization of the coatings generated by the two types of baths, it was found that there were considerable differences in the morphology and crystallinity of these Ca-P films. After the production of the different coatings a standard bioactivity test was performed by immersing the materials in a SBF solution for different periods up to 14 days to investigate the *in vitro* bioactivity of these coatings. Additional information on standard bioactivity tests can be found in works of Kokubo *et al.* [54-56]. In fact, it was possible to observe that after immersion in SBF, the acid coating (that typically operates at 80 °C) generates a film with a more pronounced needle like morphology (Figures 7.c, 8.e and 9.e) than the alkaline coatings (that typically operates at 60 °C) (Figures 8.c and 9.c).

These differences are probably related with the fact that operation temperature of the acid bath is usually higher than of the alkaline bath, which help to maintain a higher deposition rate [115,116]. Furthermore, the major advantages of hypophosphite containing acidic solution include its lower cost, the higher deposition rate, good stability, and better physical properties of the deposits over alkaline solution [117,118]. Also, the addition of succinic acid in the acid bath helps to increase the speed of the reaction, as it has been reported for other types of auto-catalytic coatings [119].

The XRD patterns of the coatings formed on the alkaline bath on SCA and SEVA-C substrates and their corresponding un-coated substrates are shown in the Figure 10 as a function of immersion time. It may be observed that the calcium phosphate coatings have a partially crystalline structure with apatite peaks similar to those of bone apatite.

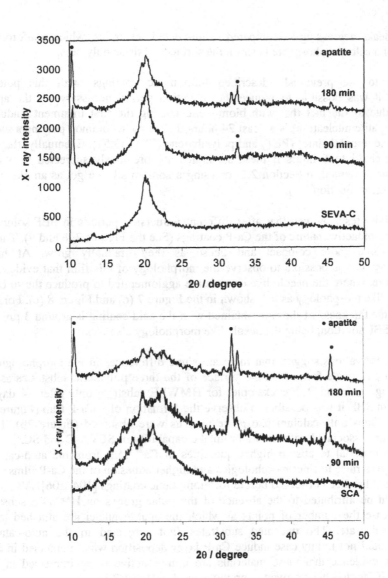

Figure 10. patterns of film formed on SEVA-C and SCA substrate after 0, 90 and 180 min, immersion in alkaline bath.

This fact was confirmed by the matching of the XRD spectra with the standard pattern of hydroxylapatite (JCPDS 9-432), although the partially amorphous nature (similar to human bone apatite) of this Ca-P film was also evident. These coatings are expected to be more reactive and to resorb faster than highly crystallinity ones. Also, for longer immersion times in the alkaline bath, it was possible to observe the gradual increase of the intensity of the apatite peaks, which corresponds to the growth of an apatite layer on the substrate. However, the intensity of the apatite peaks for SEVA-C is not so strong as for SCA due to the different water uptake capability and composition of the two materials [120,121]. It was also possible to observe the intensity of the typical SEVA-C and SCA peaks decreasing as compared to un-coated substrates, which reinforces the existence of a calcium phosphate layer on the surfaces of these polymers.

Comparing to the previously described biomimetic coatings, with this potential technology it was possible to reduce the induction period necessary for the apatite formation due to the fact that with biomimetic coating the first treatment (induction period for apatite nucleation) is at least 24 hours for the most common polymers such as poly(ethylene terephthalate) (PET) and polyethylene (PE) [53,54]. Eventually, that time can be decreased to 6 hours if these substrates are previously subjected to a KOH treatment (as presented in Section 2.1) or using a sodium silicate gel as an alternative nucleating agent (Section 2.2).

The bioactivity tests showed that after different immersion periods in SBF solution it was clear the bioactive nature of the Ca-P coatings (See the Figures 7, 8 and 9). The Ca-P film became more compact and dense as they gradually grow. At higher magnifications it was possible to observe the morphology of this film that evidenced a finer structure, where the needle like crystals are agglomerated to produce the so called cauliflower like morphology as it is shown in the Figure 7 (c) and Figure 8 (e). For SCA substrates, the thickness of the film obtained with the acid coating is around 3 μm after 14 days in SBF solution, being the needle like morphology also clear.

The SEM observations suggest that there are some differences in the morphologies of the calcium phosphate formed on the surface of the three polymeric substrates as it is shown in Fig. 7, 8 and 9. For example, for HMWPE substrate only after 14 days of immersion in SBF it was possible to observe the formation of a Ca-P film (Figure 7c). For shorter times only calcium phosphate nucleus were observed (Figure 7b). These differences are associated with the water uptake capability of SEVA-C and SCA, which allowed the material to absorb higher quantities of Ca^{2+} ions from the auto-catalytic bath. This gives raise to finer morphologies and higher adhesion of the Ca-P films to the substrate, as it has been observed before for biomimetic coatings [104, 106]. Also, these results might be attributed to the absence of the polar groups on HMWPE substrate, which decrease the number of points at which the apatite nuclei are attached to the respective substrate. The uncoated substrates (not subjected to the auto-catalytic treatment) could not in any case induce Ca-P layer deposition when immersed in SBF, which is an evidence that these materials are non-bioactive when immersed in SBF solution. This fact has been proved in previous works [104, 106].

No changes were observed on the Ca and P concentration in the solution for non pre-coated substrates (data not shown). However, for the auto-catalytic coated substrates, in the first day of immersion in SBF there was a slight increased of calcium and phosphorus concentration in the solution, which indicates some dissolution of the coating. Then as the immersion time in SBF solution increased, there was a decrease of the calcium and phosphorus concentration in the solution, indicating that these ions had been consumed during precipitation and growth of the bioactive Ca-P layer on the surface of the substrate. These results clearly indicate a bioactive character of the produced coating.

2.4. *IN-SITU* STUDIES OF APATITE FORMATION

With the invention of atomic force microscopy (AFM) [122], a new door was opened in the field of biomaterials due to its high-resolution visualization of surface morphology in different types of solutions. Therefore, the capability to image in a fluid, gives rise to the opportunity o studying biomaterial surfaces in chemical environments simulating those found *in-vivo* [123]. Since the herein reported methodologies to produce Ca-P coatings lead to significantly short induction periods for apatite formation, to understand the processes governing the formation of the first apatite nuclei during that stage can be extremely valuable. Therefore, to investigate the nucleation and growth of a Ca-P layer on the surface of the proposed starch based biodegradable materials using AFM analysis is another ongoing research line in our research group.

Just as an example of the potential demonstrated by this technique, Figure 11 shows the formation of the first apatite nuclei on the surface of a SEVA-C/HA composite, when immersed in a fluid cell at 37°C, using *in-situ* AFM analysis, performed in the contact mode.

Figure 11. AFM images showing the formation of Ca-P layer on the surface of SEVA-C + 30%HA composite in-situ.

The *in-situ* AFM observation revealed the formation of a calcium phosphate (Ca-P) on the surface of SEVA-C reinforced with HA particles after 125 hours (See Figure 11.). In the first 8 hours in-situ there was no significant change on the sample surface. But, after 24 hours, significant changes were observed: the entire surface became covered with Ca-P nuclei. The average diameter of randomly selected groups of 16 nuclei was 126.6 ± 18.6 nm. With an increase in in-situ immersion time (up to 128h), the Ca-P nuclei increased both in number and size, and coalesced. As a result, the Ca-P layer became more dense and uniform. During the immersion in the initial 8 hours, no significant change in surface roughness was observed, which corresponded to the fact that no Ca-P nuclei formed on the composite surface during this period. However, the surface roughness started to increase with an increase in immersion time after the initial 8 hours, due to the degradation of the polymeric matrix. The degradation of the polymeric matrix resulted in an increase in exposure of the HA particles to the solution, which led to the incipience of the Ca-P nucleation and further increase in roughness. The Ca-P nuclei grew in size by consuming the calcium and phosphorus ions of the SBF solution. Until the completion of the immersion for 128 h, the roughness of the surface increased continuously without reaching a plateau value, which indicated that the growth of a Ca-P layer was further proceeding on the surface of SEVA-C + 30%HA composites.

2.5. PLASMA- AND CHEMICAL- INDUCED GRAFT POLYMERIZATION

As it has been demonstrated previously in this chapter, surface modifications can be very effective on improving the adhesion of an apatite layer formed with the aid of a nucleating agent by a biomimetic process [59-63]. Moreover, surface treatments can be used to modify the surface in order to influence cell adhesion and proliferation. The recent developments of tissue engineering in the field of orthopaedic research makes it possible to envisage the association of a material with living cells that can stimulate the formation of bone tissue either ex vivo or after implantation. It is then possible to create osteoinductive materials, i.e. bioactive materials, by mainly biological routes. In this case, surface modifications have a very important role, since they can enhance the cell adhesion and proliferation at the surface of materials proposed for bone tissue engineering applications.

Surface modifications of starch blends were performed by Elvira *et al.* [124], by grafting different acrylic polar monomers with plasma induced and chemical induced polymerisation. For the plasma activation, samples were submitted to an exposure period of 30 minutes, after Ar/O_2 being introduced in the plasma reactor. Concerning to the chemical activation, conventional chemistry was used, in order to create free radicals at the surface. The samples were immersed in a $H_2O_2/(NH_4)_2S_2O_8$ solution for 30 min under a UV lamp. After surface activation treatments (plasma or chemical), samples were then immerse in aqueous monomer solution (acrylic acid, AA) for 2 hours at T=60° C. After polymerisation samples were washed overnight with distilled water to remove homopolymers and non-reacted monomers, and finally vacuum dried.

The presence of acrylic acid in the surface of the modified specimens was checked analysing the carbonyl region and the characteristic signal at about 1740 cm^{-1} correspondent to the stretching vibration of the carbonyl group existent in the polymer. This signal was observed in samples activated by both methods. Figure 12 shows the carbonyl region of SEVA-C and SEVA-C-g-pAA activated chemically, where it can be clearly observed the assignment of these signals on the spectra of grafted starch blend.

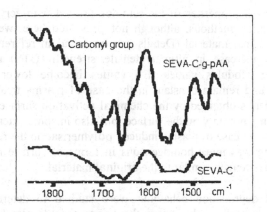

Figure 12. FTIR spectra of the carbonyl region of SEVA-C, SEVA-C-g-pAA chemically activated.

XPS analysis was also carried out to monitor the chemical composition of the grafted surfaces and to compare them to non-treated samples. Figure 13 show the C1s peak fit of SEVA-C grafted with AA by plasma induced polymerisation. It can be observed that the introduction of new –CH and CO groups increases the atom percentage of the original chemical state of the surfaces (at 285 eV) which is also revealed by the binding energy shoulders of the main peak at 285.7 eV, assigned to carboxylic –C-COOH of AA, at about 286.5 eV assigned to hydroxylic –C-OH, and at 287.7 eV assigned to carbonyl –C=O. It was also observed that when the chemical activation method was used, samples showed more intense peaks than when the plasma induced process was performed.

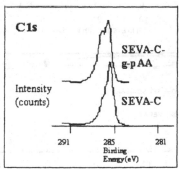

Figure 13. C1s peak fit of SEVA-C grafted with pAA by plasma induced polymerisation.

The total surface energy, as well as the corresponding polar and dispersive components were calculated (Data not shown) [124]. No significant differences could be observed in both activation methods used to create free radicals on the starch blend surfaces. In both cases the total surface energy increases when grafting AA increasing also the polar

208

component of the surface energy as a consequence of the incorporation of carboxylic group.

The tensile mechanical properties of the samples grafted with acrylic acid by plasma and chemical activation methods, although not presented here, were determined and compared to the original material (Details can be found in reference [124]). Treated samples exhibit an increase in ultimate tensile strength (UTS) for both activation treatments, but the Modulus values (E') values become lower for the chemical activation samples and remain constant in the case of plasma treated samples. Some dispersion of the results obtained by the chemical activation surface modifications was attributed to a grafting not only on the surface, but also in some extent of the bulk of the material, whereas in the case of plasma induced polymerisation the results indicated that the polymer grafting was more homogeneous in terms of surface modification as the mechanical properties remain similar to the original material.

Both swelling and degradation behaviour were studied to determine the effect of grafting acrylic hydrophilic polymer on the biodegradable starch-based material in terms of equilibrium hydration degree and weight loss percentages. As it was expected, the equilibrium hydration degree has increased. This can be observed in Figures 14.a and 14.b where are shown the swelling isotherms of SEVA-C grafted with the acrylic polymer by plasma induced polymerisation.

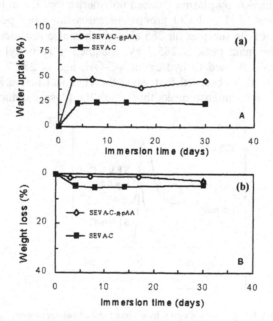

Figure 14. Swelling (a) and weight loss (b) isotherms of SEVA-C and SEVA-C-g-pAA (plasma activated) as a function of immersion time in isotonic saline solution at 37° C.

When comparing the type of grafting activation method it could be observed that the equilibrium hydration degree was higher when the grafting polymerisation was activated chemically (5-10 % higher). This is in agreement with the grafting of polymers that in some extent have penetrated into the polymers bulk, when polymerisation was chemically activated.

In terms of degradation behaviour (see Figure 14.b) a lower weight loss with respect to the original sample could be observed. This effect was explained in terms of the crosslinked nature of the grafted pAA as samples were found to be insoluble in organic solvents as dimethylsulphoxide (DMSO) whereas non-treated SEVA-C samples were completely soluble. The weight loss percentage was also found to be higher in the samples chemically activated as compared to plasma graft polymerisation, as their hydration degree was also higher.

2.5.1. Cell adhesion and proliferation
To evaluate cell adhesion and proliferation, on the surface of the materials grafted with acrylic polymers, bone marrow cells were cultured on the modified surfaces. As an example, Figures 15.a, 15.b and 5.4.c show the optical photographs of culture of bone marrow cells on tissue plates (control) after 5 days as well as on original SEVA-C and on chemically activated SEVA-C-g-pAA after one week of culture, respectively.

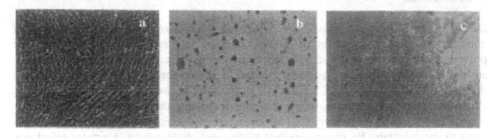

Figure 15. Optical photographs of a) control tissue culture plates, b) methylene blue staining of cultured bone marrow cells on non-modified (as processed) SEVA-C and c) methylene blue staining of cultured bone marrow cells on SEVA-C-g-pAA chemically activated.

It can be observed that the cells proliferate on the tissue culture plates (control) whereas in the case of SEVA-C only some isolated cells are observed on the surface. The most remarkable results were detected when pAA was grafted (chemically or by plasma activation) on the starch blends, as bone marrow cells showed a very good adhesion and proliferation in confluent mono-layer, which is the most desirable proliferation in bone regeneration processes.

The chemical modification of starch via grafting of vinyl monomers has demonstrated to be a very effective method to incorporate desirable surface properties into starch without sacrificing its biodegradable nature. To introduce these groups at the surfaces, several other methods are also presently being experimented in the different starch based materials. For this purpose, the active species, generated by plasma radiation of the material, are being grafted by contact with monomer solutions (acrylic acid, maleic acid and itaconic acid). The preliminary results indicate that the treatment is more successful when oxidizing plasma is utilized. In this case different oxygen-containing functional groups were introduced at the surface. The chemical composition of the surface obtained in this way is usually less defined than one would desire. Consequently we are now developing a different methodology -plasma immobilization of preadsorbed monomer. In this process a monomer with selected functionality is preadsorbed on a polymer surface and then plasma treated with an inert gas to induce covalent coupling between surface and monomer. UV modification studies are also presently being carried out in our laboratory using an UV-lamp to help placing oxygenated functional groups (CO, COOH and OH) on the surfaces. Another interesting approach to modify the surface of starch based polymers is to introduce carboxylic groups by means of an oxidation in acidic solution (HNO_3) with $KMnO_4$. The carboxylic acid groups are particularly suitable for the coupling of amine or hydroxyl containing bioactive molecules, improving cell adhesion.

For this preliminary studies *in vitro* cell adhesion and proliferation tests are being carried out on all the treated materials. Promising results are being obtained with these, being $KMnO_4$ treatment the most effective, showing clusters of cells on the surface for all materials whereas almost no cell proliferation could be observed on the control samples.

4. Conclusions

Different methodologies to produce a calcium-phosphate layer on the surface of compact or porous structures of several bioinert and starch based biodegradable polymers were successfully developed. Surface modifications performed on these substrates, before biomimetic coating, resulted in a faster formation of more Ca-P nuclei during the first stages of SBF immersion, particularly in the case of the KOH etching. With this treatment it was possible not only to reduce the induction period for the formation of a well defined apatite-like layer but also to improve the adhesion of the layer to the substrate. By using a sodium silicate gel methodology, a similar effect was observed of a reduction of the induction time for apatite formation. Furthermore, when applied to porous materials, this methodology was highly effective, since a clear apatite-like layer was observed inside the pores, clearly covering the cell walls. However, this methodology was not so effective on coating bioinert materials. On the other hand, by using the auto-catalytic coating route it was possible to produce a Ca-P film not only on the surface of starch based biodegradable polymers but also on the surface of bioinert polymers. Again very short induction periods for apatite formation were obtained. These proposed routes seem to be very promising and simple methodologies to be used as pre-implantation treatments to be applied to different types and shapes of materials, including polymers and eventually metals, previously to their clinical application. *In-situ* AFM studies were performed, contributing with valuable information for the understanding of the nucleation and growth of the apatite layers. Grafting polyacrylic acid on starch based materials by plasma and chemical induced radical polymerisation is a another approach proposed by our group in order to modify the surface properties of these materials with the different purpose of improving the cell adhesion and proliferation on of starch based materials when considering bone tissue engineering applications. Preliminary cell adhesion and proliferation tests were performed by using bone marrow cells showing a remarkable improvement with respect to original starch based samples. Therefore, the obtained results are very promising for the developing of cancellous bone replacement materials and for pre-calcifying bone tissue engineering scaffolds.

References

1. Mann, S. (1991) Flatery by Imitation, *News and Views*, **349**, 285.

2. Mann, S. (1991) Biomineralization: a Novel Approach to Crystal Engeneering, *Endeavour*, **3**, 120.

3. Addadi, L., Weiner, S. (1992) Control and Design principles in Biological Mineralization, *Angew. Chem. Int. Ed. Engl.* **31**, 153.

4. Heuer, Fink, D.J., Arias, V.J, Calvert, P.D., Kendal, K., Messing, G.L., Blackwell, J., Rieke, P.C., Thompson, D.H., Wheeler, A.P., Veis, A., Caplan, A.I. (1992) Innovative Materials Processing Strategies: A Biomimetic Approach, *Science*, **255**, 1098.

5. Hoffman, A.S. (1996) Biologically Functional Materials, in Ratner, B.D., Hoffman, A.S., Schoen, F. J., Lemnons, J. E.(eds.), *Biomaterial Science*, Academic Press, San Diego, 124.

6. Galaev, I.Y., Mattiasson, B. (1999) 'Smart' Polymers and What They Could do in Biotechnology and Medicine, *Trends in Biotechnology*, **17**, 335.

7. C. Choi, Y. Kim, A Study of the Correlation Between Organic Matrices and Nanocomposite Materials in Oyster Shell Formation, Biomaterials, 21 (2000) 213.

8. Collier, J.H., Messersmith, P.B. (2001) Phospholipid Strategies in Biomineralization and Biomaterials Research, Annu. *Rev. Mater. Res.*, **31**, 237.

9. Rodan, G.A. (1992) Introduction to bone biology, *Bone*, **13**, S3.

10. Heughebaert, J.C., Bonel, G. (1986) Composition, Structures and Properties of Calcium Phosphastes of Biological Interest, in Christel, P., Meunier, A., Lee A.J.C. (Eds.), *Biological and Biomechanical Performance of Biomaterials*, Elsevier Science Publishers, Amsterdam.

11. Simske, S.J., Ayers, R.A. and Bateman, T.A. Porous (1997) Materials for Bone Engineering, *Mater. Sci. Forum*, **250**, 151.

12. Postner, A.S. (1985) The Mineral of Bone, *Clinical Orthopedics*, **200**, 87.

13. Sikavitsas, V.I., Temenoff, J.S. and Mikos A.G. (2001) Biomaterials and bone mechanotransduction, *Biomaterials*, **22**, 2581.

14. Anderson, H.C. (1983) Normal and Abnormal Mineralization in Mamals, *Trans. ASAIO*, **27**, 702.

15. Anderson, H.C. (1983) Calcific Diseases, *Archives of Pathology and Laboratory Medicine*, **107**, 341.

16. Anderson, H.C. (1989) Biology of Disease, *Mineralization review*, **60**, 320.

17. Linde, A., Structure and Calcification of Dentin (1992) in *Calcification in Biological Sistems*, Ermano Bonucci, Roma 269.

18. Anderson, H.C., Morris, D.C. (1993) Mineralization, Physiology and Pharmacology of Bone, in Mundy G.R. and Martin T.J. (eds.), *Hand. Exptl. Pharmacol.*, Springer-Verlag, Berlin, 267.

19. Anderson, H.C. (1992) Conference Introduction and Summary, *Bone and Mineral*, **17**, 107.

20. Anderson, H.C. (1984) Mineralization by Matrix Vesicles, *Scanning Electron Microscopy*, 953.

21. Hench, L.L. (1998) Bioceramics, *J. Am. Ceramic. Soc.* **81**, 1705.

22. Thomas K.A. (1994) Hydroxyapatite coatings, *Orthopaedics*, **17** 267.

23. Greenspan, D.C. (1999) Bioactive Ceramic Implant Materials, *Current Opinion in Solid State and Materials Science*, **4**, 389.

24. Hench, L.L., Wilson, J. (1984) Surface-Active Biomaterials, *Science* **26** 630.

25. Ducheyne, P., Qiu, Q. (1999) Bioactive Ceramics: the Effect of Surface Reactivity on Bone Formation and Bone Cell Function, *Biomaterials*, **20**, 2287.

26. Ratner, B.D., Hoffman, A.S. (1996) Thin films, Grafts and Coatings, in Ratner, B.D., Hoffman, A. S., Schoen, F. J., Lemnons, J. E. (eds.), *Biomaterial Science*, Academic Press, San Diego *Biomaterials Science*, 84.

27. de Bruijn, J.D., van den Brink, I., Mendes, S., Dekker, R., Bovell, Y.P., van Blitterswijk, C.A. (1999) Bone Induction by Implants Coated with Cultures Osteogenic Bone Marrow Cells, *Advents in Dental Research*, **13**, 74.

28. Kanazawa, T. (1989) General Background on Phosphate Materials, in Kanazawa (Eds.), *Inorganic Phosphate Materials*, Elsevier, Amsterdam, 1.

29. Silver, F.H. (1994) Scope and Markets for Medical Implants, in Silver F.H. (Eds.), *Biomaterials, Medical Devices and Tissue Engineering: An Integrated Approach*, Chapman & Hall, London, 2.

30. Hench, L.L, Splinter, R.J., Allen, W. C., Greenlee, T.K. (1972) Bonding Mechanisms at the Interface of Ceramic Prosthetic Materials, *J. Biomed. Mter. Res.*, **2**, 117.

31. Puleo, D.A., Nanci, A. (1999) Understanding and Controlling the Bone-Implant Interface, *Biomaterials*, **20**, 2311.

32. Kohn, D.H. (1998) Metals in medical applications, *Current Opinion in Solid State & Materials Science*, **3**, 309.

33. Weng, J., Liu, Q., Wolke, J.G., Zhang, X., and de Groot, K. (1997) Formation and Characteristics of the Apatite Layer on Plasma-Sprayed Hydroxiapatite Coatings in Simulated Body Fluid, *Biomaterials*, **18**, 1027.

34. Zheng, X., Huang, M., and Ding, C. (2000) Bond Strength of Plasma-Sprayed Hydroxyapatite/Ti Composite Coatings, *Biomaterials*, **21** 841.

35. H.C. Gledhill, I.G. Turner, C. Doyle, In Vitro Dissolution Behaviour of Two Morphologically Different Thermally Sprayed Hydroxyapatite Coatings, Biomaterials 22 (2001) 695.

36. Fazan, F., Marquis, P.M. (2000) Dissolution Behaviour of Plasma-Sprayed Hydroxyapatite Coatings, *Journal of Materials Science: Materials in Medicine*, **11**, 787.

37. Hamdi, M., Hakamata, U.S., Ektessabi, A.M. (2000) Coating of Hydroxyapatite Thin Film By Simultaneous Vapor Deposition, *Thin Solid Films*, **13**, 484.

38. Yamashita, K., Arashi, T., Kitagaki, K., Yamada, S., Umegaki, T. (1994) Preparation of Apatite Thin Films Trough rf-Sputtering from Calcium Phosphate Glasses, *Journal of American Ceramic Society*, **77, 9**, 2401.

39. Burk, E.M., Haman, J.D., Weimer, J.J., Cheney, A.B., Rigsbee, J.M., Lucas, L.C. (2001) In fluence of Coating Strain on Calcium Phosphate Thin-Film Dissolution, *Journal of Biomedical Materials Research*, **57, 1**, 41.

40. Wang, C.X., Chen, Z.Q., Wang, M., Liu, Z.Y., Wang, P.L., Zheng, S.X. (2001) Functionally Graded Calcium Phosphate Coatings Produced by Ion Beam Sputtering/Mixing Deposition, *Biomaterials*, **22**, 1619.

41. Ong, J.L., Lucas, L.C., Lacefield, W.R., Rigney. E.D. (1992) Structure, Solubility and Bond Strength of Thin Calcium phosphate Coatings Produced by Ion Beam Sputter Deposition, *Biomaterials*, **13**, 249.

42. Kim, C.S., Ducheyne, P. (1991) Compositional Variation in the Surface and Interface of Calcium Phosphate Ceramic Coatings on Ti and Ti-6Al-4V Due to Sintering and Immersion, *Biomaterials*, **12**, 461.

43. Li, P., Ohtsuki, C., Kokubo, T., Nankanishi, K., Soga, N., de Groot., K. (1994) The Role of Hydrated Silica, Titania, and Alumina in Inducing Apatite on Implants, *Journal of Biomedical Materials Research*, **28**, 7.

44. Peltola, T., Patsi, M., Rahiala, H., Kangasniem, I., Yli-Urpo, A. (1998) Calcium Phosphate Induction by Sol-Gel-Derived Titania Coating in Titanium Substrates in Vitro, *Journal of Biomedical Materials Research*, **41**, 504.

45. Hanawa, T., Ukai, H., Murakami, K., Asaoka., K. (1995) Structure of Surface-Modified Layers of Calcium-Ion-Implanted Ti-6Al4-V And Ti-56Ni, *Material Trans JIM*, **36**, 438.

46. Mayor, B., Arias, J., Chiussi, S., Garcia, F., Pou, J., Leon Fong, B., Perez-Amor, M. (1998) Calcium Phosphate Coatings Grown at Different Substrate Temperatures by Pulsed ArF-Laser Deposition, *Thin Solid Films*, **317**, 363.

47. Fernandez-Pradas, J.M., Sardin, G., Cleries, L., Serra, P., Ferrater, C., Morenza, J.L. (1998) Deposition of Hydroxyapatite Thin Films by Excimer Laser Ablation, *Thin Solid Films*, **317**, 393.

48. Clèries, L., Fernández-Pradas, J.M., Morenza, J.L. (2000) Behavior in Simulated Body Fluid of Calcium Phosphate Coatings Obtained By Laser Ablation, *Biomaterials*, **21**, 1861.

49. Zitomirsky, I., Gol-OR, L. and Isral, I. (1997) Electrophoretic Deposition of Hydroxyapatite on Ti6Al4V, *Journal of Materials Science: Materials in Medicine*, **8**, 213.

50. Shirkhanzadeh, M. (1995) Calcium Phosphate Coatings Prepared by Electrocrystallization from Aquiod Electrolytes, *J Mater Sci Mater Med*, **6**, 90.

51. Ishizawa, H., Ogino, M. (1997) Formation and Charaterization of Anodic Titanium Oxide Containing Ca and P, *J Biomed Mater Res*, **34**, 15.

52. Rey, C. (1990) Calcium Phosphate Biomaterials and Bone Mineral. Differences in Composition, Structures and Properties, *Biomaterials*, **11**, 13.

53. Tanashi, M., Yao, T., Kokubo, T., Minoda, M., Miyamoto, T., Nakamura, T., Yamamuro, T. (1994) Apatite Coating on Organic Polymers by Biomimetic Process, *Journal of American Ceramic Society*, **77**, 2805.

54. Takadama, H., Miyaji, F., Kokubo, T., Nakamura, T. (1997) Bonelike Apatite Layer Formed on Organic Polymers by Biomimetic: TEM-EDX Observation of Initial Stage of Apatite Formation, in Sedel, L., and Rey C. (Eds.), *Bioceramics 10*, Elsevier, Amsterdam, 257.

55. Tanashi, M., Hata, K., Kokubo, T., Minoda, M., Miyamoto, T., Nakamura, T., Yamamuro, T. (1992) Effect of Substrate on Apatite Formation by a Biomimetic Process, in Yamamuro, T., Kokubo, T., Nakamura, T., *Bioceramics 5*, Kobunshi-Kankokai, Kyoto, 57.

56. Abe, Y., Kokubo, T., Yamamuro, T. (1990) Apatite Coating on Ceramics, Metals and Polymers Utilizing a Biological Process, *Journal of Materials Science: Materials in Medicine*, **1**, 233.

57. Kokubo, T., Hata, K., Nakamura, T., Yamamuro, T. (1991) Apatite Formation on Ceramics, Metals and Polymers Induced by a CaO-SiO2- Based Glass in a Simulated Body Fluid, in Bonfield, W., Hastings, G.W. and Turner, K.E. (Eds.), *Bioceramics 4*, Butterworth-Heinemann, Guilford, 120.

214

58. Kobubo, T., Kushitani, H., Sakka, S., Kitsugi, T., Yamamuro, T. (1990) Solution Able to Reproduce In Vivo Surface-Structure Changes in Bioactive Glass-Ceramic A-W, *Journal of Biomedical Materials Research*, **24**, 721.

59. Tanashi, M., Yao, T., Kokubo, T., Minoda, M., Miyamoto, T., Nakamura, T., Yamamuro, T. (1994) Apatite Coated on Organic Polymers by Biomimetic Process: Improvement in Its Adhesion to Substrate by NaOH Treatment, *Journal of Applied Biomaterials*, **5**, 339.

60. Tanashi, M., Yao, T., Kokubo, T., Minoda, M., Miyamoto, T., Nakamura, T., Yamamuro, T. (1995) Apatite Coated on Organic Polymers by Biomimetic Process: Improvement in Its Adhesion to Substrate by HCl Treatment, *Journal of Materials Science: Materials in Medicine*, **6**, 319.

61. Tanashi, M., Yao, T. Kokubo, T., Minoda, M., Miyamoto, T., Nakamura, T., Yamamuro, T. (1995) Apatite Coated on Organic Polymers by Biomimetic Process: Improvement in Its Adhesion to Substrate by Glow-discharge Treatment, *Journal of Biomedical Materials Research*, **29**, 349.

62. Liu, G.J, Takadama, H., Miyaji, F., Kokubo, T., Nakamura, T. (1996) Apatite Coated on Organic Polymer by Biomimetic Process: Improvement in Adhesion to substrate by Ultraviolet, in Kokubo, T., Nakamura, T., Miyaji, F. (Eds.), *Bioceramics 9*, Elsevier, Amsterdam, 407.

63. Liu, G.J., Miyaji, F., Kokubo, T., Takadama, H., Nakamura, T., Murakami, A. (1998) Apatite Organic Polymer Composites Prepared by a Biomimetic Process: Improvement in Adhesion of the Apatite Layer to the Substrate by Ultraviolet Irradiation, *Journal of Materials Science: Materials in Medicine*, **9**, 285.

64. Kim, H.-M., Kishimoto, K., Miyaji, F., Kokubo, T., Yao, T., Suetsugu, Y., Tanaka, J., Nakamura, T. (1999) Composition and Structure of the Apatite Formed on PET Substrates in SBF modified with Various Ionic Activity Products, *Journal of Biomedical Materials Research*, **46**, 228.

65. Kim, H.-M., Kishimoto, K., Miyaji, F., Kokubo, T., Yao, T., Suetsugu, Y., Tanaka, J., Nakamura, T. (2000) Composition and Structure of Apatite Formed on Organic Polymer in Simulated Body Fluid with a High Content of Carbonate Ion, *Journal of Materials Science: Materials in Medicine*, **11**, 421.

66. Kim, H.-M., Miyaji, F., Kokubo, T., Nakamura, T. (2000) Revised Simulated Body Fluid, in Sandro Giannini and Antonio Moroni (Eds.), *Bioceramics 13*, Trans Tech. Publications, Zurich, 47.

67. Miyaji, F., Handa, S., Kokubo, T., Nakamura, T. (1997) Apatite Formation on Polymers by Biomimetic Process Using Sodium Silicate Solution, in Sedel, L. and Rey C. (Eds.), *Bioceramics 10*, Elsevier, Amsterdam, 257.

68. Kim, H. M., Uenoyama, M., Kokubo, T., Minoda, M., Miyamoto, T., Nakamura, T. (2001) Biomimetic Apatite Formation on Polyethylene Photografted with Vinyltrimethoxysilane and Hydrolyzed, *Biomaterials*, **22**, 2489.

69. Tretinnikov, O.N., Kato, K., Ykada, Y. (1994) *In Vivo* hydroxylapatite deposition on to a film surface-grafted with organophosphate polymer, *J. Biomed: Mater. Res.*, **28**, 1365.

70. Mucalo, M.R., Yokogawa, Y, Susuki, T., Kawamoto, Y., Nagata, F., Nishizawa, K. (1995) Further Studies of Calcium Phosphate Growth on Phosphorylated Cotton Fibers, *Journal of Materials Science: Materials in Medicine*, **6**, 658.

71. Yokogawa, Y., Reyes, J.P., Mucalo, M.R., Toriyama, M., Kawamoto, Y., Susuki, T., Nishizawa, K., Nagata, F., Kamayama, T. (1997) Growth of Calcium Phosphate on Phosphorylated Chitin Fibres, *Journal of Materials Science: Materials in Medicine*, **8**, 407.

72. Miyaji, F., Kim, H., Handa, S., Kokubo, T., Nakamura, T. (1999) Bonelike Apatite coating on Organic Polymers: Novel Nucleation Process Using Sodium Silicate Solution, *Biomaterials*, **20**, 913.

73. Oyane, A., Minoda, M., Miyamoto, T., Nakanishi, K., Kim, H.-M., Miyaji, F., Kokubo, T., Nakamura, T. (1998) Surface Modification of Ethylene-Vinyl alcohol Copolymer for Inducing its Apatite-Forming Ability, in LeGeros, R.Z., LeGeros, J.P. (Eds), *Bioceramics 11*, World Scientific Publishing, 687.

74. Oyane, A., Minoda, M., Miyamoto, T., Takahashi, R., Nakanishi, K., Kim, H.-M., Miyaji, F., Kokubo, T., Nakamura, T. (1999) Apatite Formation on Ethylene-Vinyl alcohol Copolymer Modified with Silanol Groups, *Journal of Biomedical Materials Research*, **47**, 367.

75. Oyane, A., Minoda, M., Miyamoto, T., Nakanishi, K., Kawashita, M., Kokubo, T., Nakamura, T. (2000) Apatite Formation on Ethylene-vinyl Alcohol Copolymer Modified with Calcium Silicate, *Proceedings of the Sixth World Biomaterials Congress Transactions*, Hawai, USA, 1304.

76. Oyane, A., Minoda, M., Miyamoto, T., Nakanishi, K., Kawashita, M., Kokubo, T., Nakamura, T. (2000) Apatite Formation on Ethylene-vinyl Alcohol Copolymer Modified with Silane Coupling Agent and Calcium Silicate, in Giannini, S., and Moroni, A. (Eds.), *Bioceramics 13*, Trans Tech. Publications, Zurich, 75.

77. Costa, N., Maquis, P.M. (1998) Biomimetic processing of calcium phosphate coating, *Medical Engineering & Physics*, **20**, 602.

78. Taguchi, T., Kishida, A., Akashi M. (1998) Hydroxiapatite Formation on/in Hydrogels Using a Novel Alternate Soaking Process, *Chemistry Letters*, **8**, 711.

79. Furuzono, T., Taguchi, T., Kishida, A., Akashi, M., Tamada, Y. (2000) Preparation and Characterization of Apatite Deposited on Silk Fabric Using an Alternate Soaking Process, *Journal of Biomedical Materials Research*, **50**, 344.

80. Kim, H.-M., Kim, Y., Park, S.-J., Rey, C., Lee, H., Glimcher, M.J., Ko, J.S. (2000) Thin Film of Low-Crystalline Calcium Phosphate Apatite Formed at Low Temperature, *Biomaterials*, **21**, 1129.

81. Yuan, X., Mark, A., Li, J. (2001) Formation of Bone-like Apatite on Poly(L-lactic acid) Fibers by a Biomimetic Process, *Journal of Biomedical Materials Research*, **57**, 140.

82. Rhee, S.-H. and Tanaka, J. (1998) Hydroxyapatite Coating on a Collagen by a Biomimetic Method, *Journal of American Ceramic Society*, **8**, **11**, 3029.

83. Rhee, S.-H. and Tanaka, J. (1999) Effect of Citric Acid on the Nucleation of Hydroxyapatite in a Simulated Body Fluid, *Biomaterials*, **20**, 2155.

84. Rhee, S.-H. and Tanaka, J. (1999) Hydroxyapatite Formation on Cellulose Cloth Induced by Citric Acid, *Biomaterials*, **20**, 2155.

85. Galliard, T. (1987) Starch: Properties and Potential, *Critical Report on Applied Chemistry*, John Wiley & Sons, New-York.

86. Forssell, P. and Poutanen, K. (1996) Modification of Starch Properties with Plasticizers, *Biotechnology and Food Research*, **4**, 128.

87. Griffin, G.J.L. (1994) Chemistry and Technology of Biodegradable Polymers, Blackie Academic and Professional, New York.

88. Andreopoulos, A.G., Theophanides, T. (1994) Degradable Plastics: A Smart Approach to Various Aplications, *Journal of Elastomers and Plastics*, **26**, 308.

89. Gomes, M.E., Reis, R.L., Cunha, A.M. Blitterswijk, C.A., De Bruijn, J. (2001) Cytocompatibility and Response of Osteoblastic-like Cells to Starch-Based Polymers: Effect of Several Additives and Processing Conditions, *Biomaterials*, **22**, 1911.

90. Marques, A.P., Reis, R.L., Hunt, J.A. (2002) The Biocompatibility of Novel Starch-Based Polymers and Composites: In Vitro Studies, *Biomaterials*, **23**, 1471.

216

91. Mendes, S.C., Bezemer, J., Claase, M.B., Grijpma, D.W., Bellia, V, Degli-Innocenti, F., Reis, R.L., de Groot, K., Van Blitterswijk, C., De Bruijn, J. (2001) Evaluation of Two Biodegradable Polymeric Systems as Substrates for Bone Tissue Engineering, *Tissue Engineering*, in press.

92. Mendes, S.C., Reis, R.L., Bovell, Y.P., Cunha, A.M., Van Blitterswijk, C., De Bruijn, J. (2001) Biocompatibility Testing of Novel Starch Based Materials with Potential Application in Orthopaedic Surgery: a Preliminary Study, *Biomaterials*, **22**, 2057.

93. Reis, R.L. and Cunha, A.M. (1998) Reinforced Starch Based Blends: A New Alternative for Bioresorbable Load-Bearing Implants, *Antec'98 – Plastics on my Mind*, Society of Plastics Engineers, Atlanta, USA, 2733.

94. Reis, R.L. and Cunha, A.M. (1995) Characterization of Two Biodegradable Polymers of Potential Application within the Biomaterials Field, *Journal of Materials Science: Materials in Medicine*, **6**, 786.

95. Reis, R. L., Cunha, A. M., Allan, P. S. and Bevis, M. J. (1996) Mechanical Behaviour of Injection-Moulded Starch Based Polymers, *Polymers Advanced Technologies*, **7**, 1.

96. Pereira, C.S., Cunha, A.M., Reis, R.L., Vázquez, B., San Roman, J. (1998) New Starch-Based Thermoplastic Hydrogels for Use as Bone Cements or Drug-delivery Carriers, *Journal of Materials. Science: Materials in Medicine*, **9**, 825.

97. Malafaya, P.B., Elvira, C., Gallardo, A., San Román, J., Reis, R.L. (2001) Porous Starch-Based Drug Delivery Systems Processed by a Microwave Route, *Journal of Biomedical Science - Polymer Edition*, **12**, **11**, 1227.

98. Gomes, M.E., Ribeiro, A.S., Malafaya, P.B., Reis, R.L., Cunha, A.M. (2001) A New Approach Based on Injection Moulding to Produce Biodegradable starch-Based Polymeric Scaffolds: Morphology, Mechanical and Degradation Behaviour, *Biomaterials*, **22**, 883.

99. Gomes, M.E., Ribeiro, A.S., Malafaya, P.B., Reis, R.L., Cunha, A.M. (1999) A New Approach Based on Injection Moulding To Produce Biodegradable Starch Based Scaffolds, *25th Annual Meeting of the Society for Biomaterials*, Providence, USA, 224.

100. Demirgöz, D., Elvira, C., Mano, J. F., Cunha, A.M., Piskin, E., Reis, R.L. (2000) Chemical Modification of Starch Based Biodegradable Polymeric Blends: Effects on Water, Degradation Behaviour and Mechanical Properties, *Polymer Degradation and Stability*, **70**, 161.

101. Reis, R.L., Cunha, A.M., Bevis, M.J. (1998) Using Non-Conventional Processing Routes to develop Anisotropic and Biodegradable Composites of Starch Based Thermoplastics Reinforced with Bone-Like ceramics, *Journal of Applied Medical Polymers*, **2**, 49.

102. Reis, R.L., Cunha, A.M., Allan, P.S., Bevis, M.J. (1996) Improvement of the Mechanical Properties of Hydroxylapatite Reinforced Starch Based Polymers Through Processing, *Polymers in Medicine and Surgery – PIMS'96*, 195.

103. Reis, R.L., Cunha, A.M., Lacerda, S.R., Fernandes, M.H., Correia, R.N. (1996) Reinforcement of Polyethylene and Starch Based Thermoplastics with Hydroxylapatite and Bioactive Glasses, in T. Kokubo, T. Nakamura, F. Miyaji (Eds.), *Bioceramics 9*, Elsevier, Oxford, 435.

104. Oliveira, A.L., Elvira, C., Vázquez, B., San Roman, J., Reis, R. L. (1999) Surface Modifications Tailors the Characteristics of Biomimetic Coatings Nucleated on Starch Based Polymers, *Journal of Materials Science: Materials in Medicine*, **10**, 827.

105. Gomes, M.E., Ribeiro, A.S., Malafaya, P.B., Reis, R.L., Cunha, A.M. (2001) A New Approach Based on Injection Moulding to Produce Biodegradable Starch-Based Polymeric Scaffolds: Morphology, Mechnical and Degradation Behaviour, *Biomaterials*, **22**, 883.

106. Reis, R.L., Cunha, A.M., Fernandes, M.H., Correia, R.N. (1997) Treatments to Induce the Nucleation and Growth of Apatite Like Layers onto Polymeric Surfaces and Foams, *Journal of Materials Science: Materials and Medicine*, **8**, 897.

107. Liead D. (1993) in Scott, G. and Liead, D. (Eds.), *Degradable Polymers: Principles and applications*, Chapman and Hall, London.

108. Bikerman, J.J. (1970) in Physical Surfaces, Academic Press, New York.

109. van Haastert, R.M., Grote, J.J., van Blitterswijk, C.A., (1994) Osteoinduction within PEO/PBT Copolymer Implants in Cranial Defects Using Demineralized Bone Matrix, *J. Mater. Sci. Mater. Med*, **5**, 764.

110. Oliveira, A.L., Malafaya, P.B., Reis, R.L. (2000) Sodium Silicate Gel Induced Self-Mineralization of Different Compact and Porous Polymeric Structures, in Giannini, S. and Moroni, A. (Eds.), *Bioceramics 13*, Trans Tech. Publications, Zurich, 75.

111. Reis, R.L., Mendes, S.C., Cunha, A.M., Bevis, M.J. (1997) Processing and In-Vitro Degradation of Starch/EVOH Thermoplastic Blends, *Polymer International*, **43**, 347.

112. Leonor, I.B. and Reis, R.L. (2000) A Novel Auto-Catalitic Deposition Methodology to Produce Calcium-Phosphate Coatings on Polymeric Biomaterials, in Giannini, S. and Moroni, A. (Eds.), *Bioceramics 13*, Trans Tech. Publications, Zurich, 83.

113. Leonor, I.B. and Reis, R.L. (2001) A Novel Auto-Catalytic Deposition Methodology To Produce Calcium-Phosphate Coatings On Polymeric Biomaterials, *Journal of Materials Science: Materials in Medicine*, submitted.

114. Costa, N. and Maquis, P.M. (1998) Biomimetic processing of calcium phosphate coating, Medical Engineering Physics, **20**, 602.

115. Matsuoka, M., Imanishi, S., Hayashi,T. (1989) Physical properties of electroless Ni-P alloy deposits from a pyrophosphate bath, *Plating and Surface Finishing*, 54.

116. Gonzalez, O.M., White, R.E., Cocke, D.L. (1990) Autocatalytic deposition of Ni-TM-P alloys, *Plating and Surface Finishing*, 63.

117. Alami, M. Charbonnier, M. Romand, M. (1996) Plasma chemical modification of polycarbonate surfaces for electroless plating, *J. Adhesion*, **57**, 77.

118. Kim, Y-S., Sohn, H –J. (1996) Mathematical modelling of electroless nickel deposition at seady state using rotating disk electrode, *J. Electrochemical Society*, **143**, 505.

119. Fields, W.D., Duncan, R.N. Zickgraf, J.R. (1982), *Metals Handbook*, 5.

120. Vaz, C.M., Reis, R.L., Cunha, A.M. (2001) Degradation model of starch-EVOH/HA composites, *Materials Research Innovattions*, **4**, 375.

121. Araújo, M.A., Vaz, C.M. Cunha, A.M., Mota, M. (2001) In-vitro degradation behaviour of starch/EVOH biomaterials, *Polymer Degradation and Stability*, **73**, 237.

122. Bining, G., Quante, C.F., Gerber, C.H. (1986) Atomic force microscope, *Physical Review Letters*, **56**, 930.

123. Siedlecki, C.A., Marchant, R.E. (1998) Atomic force microscopy for characterization of the biomaterials interface, *Biomaterials*, **19**, 441

124. Elvira, C., Yi, F., Azevedo, C.M., Cunha, A.M., San Roman, J., Reis, R.L., Plasma- and Chemical-induced Graft Polimerisation on Starch Based Blends Aimed at Improving Cell Adhesion and Proliferation in Bone Tissue Engineering Applications, *J. Materials Science: Materials Medicine*, in press.

Tissue Engineering and Regeneration of Bone and Cartilage

BONE TISSUE ENGINEERING USING STARCH BASED SCAFFOLDS OBTAINED BY DIFFERENT METHODS

ME Gomes[1,2], A Salgado[1,2], RL Reis[1,2]

[1]*Department of Polymer Engineering, University of Minho, Campus de Azúrem, 4810-058 Guimarães, Portugal*

[2]*3B's Research Group, University of Minho, Campus de Gualtar, 4710-057 Braga, Portugal*

Abstract

The materials to be used as scaffolds in tissue engineering must fulfill a number of complex requirements, such as, biocompatibility, appropriate porous structure, mechanical properties and suitable surface chemistry. The selection of the most appropriate polymer to produce the scaffold is a very important step towards the construction of the tissue engineered product since its intrinsic properties will determine in a great extent the properties of the scaffold. However, the method of producing these scaffolds will deeply influence its final characteristics, as it can dramatically change the type and amount of porosity, the mechanical and degradation behavior, the surface properties and even the biocompatibility of the scaffold material.

Several polymers and different processing techniques are beeing studied to produce tissue engineering scaffolds, which, combined with the appropriate type of cells may constitute an effective alternative treatment to mal-functioning organs or tissues, with a dramatic impact on the health and quality of life of patients world-wide.

This paper describes our research on tissue engineering using starch based blends and several different and innovative processing methods that are used to produce tissue engineering scaffolds. Preliminary results with osteoblastic cells are also presented.

1. Introduction

1.1. Selection of scaffold material

The selection of a scaffold material is both a critical and difficult choice. There are many biocompatible materials available, metals, ceramics and polymers. However, the

221

R.L. Reis and D. Cohn (eds.),
Polymer Based Systems on Tissue Engineering, Replacement and Regeneration, 221–249.
© 2002 *Kluwer Academic Publishers. Printed in the Netherlands.*

criterion of biodegradability excludes the use of all metals and most ceramics as scaffolds materials [1,2]. Although biodegradable/bioresorbable ceramic materials, such as tri-calcium phosphate and sea coral, have been used with some success [1,2] in tissue engineering applications, they have two major limitations. First they are difficult to process into porous materials with complex shapes, and second, it is currently not possible to control their rate of degradation. Polymers, on the other hand, are easily formed into any shape. There are many biocompatible polymers available and their usefulness as biomaterials is apparent by their numerous biomedical applications [1,2].

Tissue engineering has benefited from the discovery and the clinical utilization of a wide range of biodegradable polymeric materials, both naturally derived and synthetic systems. However, in spite of the wide range of biodegradable polymers available, there is a strong tendency to choose those that have history of regulatory approval instead of letting the application guide the choice of the material [3].

Nevertheless, the increased demands placed on biomaterials for currently approved products and novel sophisticated medical implants, particularly the tissue engineering scaffolds, continue to fuel the interest in improving the performance of existing medical-grade polymers and developing new synthetic polymers. Table 1 shows some of the most important natural and synthetic polymers that have been used, or that are under consideration for use in tissue engineering applications.

Table1. Some natural and synthetic polymers that have been used, or that are under consideration for use in tissue engineering applications

	Refs
Natural polymers	
Type I colagen	3,4
Hyaluronic acid	5,6
Alginate	7
Chitosan	3,8,9
Polyhydroxybutyrate (PHB)	3,10
Synthetic polymers	
Poly(glycolic acid), poly(lactic acid) and their copolymers (PGA,PLA)	1,3,8,10-13
Poly(e-caprolactone)	3,8,10,13
Poly(dioxanone)	13
Polyethylene oxide/polybutylene teraphthalate block copolymers (PEO/PBT)	14,15
Poly(amino acids) and "pseudo"-poly(amino acids)	3,7,10,13

It is believed, that the best biodegradable polymer for biomedical applications, including those related with tissue engineering, may be found taking steps towards the development of new synthetic biomaterials that combine the most favorable properties of synthetic and natural polymers to allow, at the same time, the precise control over material properties along with the opportunity to attach biologically functional molecules that promote favorable cell-polymer interactions [16-18].

In the past few years, these biodegradable blends of corn starch with several synthetic polymers, have been originally proposed by the U.Minho researchers as potential alternatives to the commonly used biodegradable polymers, for a variety of biomedical applications [19-32], such as, for example, partially degradable bone cements [32], as hydrogels for controlled release of drugs [25], as bone substitutes in the orthopaedic field [19,21] or as scaffolds for tissue engineering, as proposed in this manuscript. These materials are based on blends of corn starch with well known synthetic polymers used in biomedical applications such as, poly(ethylene vinyl alcohol) (SEVA-C), cellulose acetate (SCA), polycaprolactone (SPCL) and poly-lactic acid (SPLA) [33-35].

Starch consists of two main polysaccharides, amylose and amylopectin, and it forms the main source of carbohydrate in the human diet [36-40]. The ability of polysaccharides to form a network structure (gel), even at very low concentrations (gelation), constitutes one of their most important functional properties as it offers an effective means of increasing the system´s mechanical and chemical stability [36,40]. A wide range of modification mechanisms of starches is known. These include self-association (induced by changes of pH, ionic strength, or physical and thermal means) and complexation with salts and covalent cross-linking.

However, starch is difficult to process and is extremely sensitive to humidity. Depending on the environment, the physical properties such as ductility, hardness and dimensional stability are affected [37]. Therefore it is common to blend starch with hydrophobic synthetic polymers to obtain better properties [41].

Crystalline starch can be used as a filler or can be transformed into thermoplastic starch which can be processed alone or in combination with specific synthetic polymers. To make starch thermoplastic, its crystalline structure has to be destroyed (gelatinization) by pressure, heat, mechanical work and plasticizers such as water, glycerin or other polyols [33,38-40].

Starch, in its native or modified form has been subjected to extensive study, initially due to its interest for the food and paper industry, textile manufacture and pharmacology and later on, due to the increased interest of biomedical and pharmaceutical research on biodegradable polymers as matrixes for controlled drug delivery systems and other applications [36].

1.2. Scaffold design and fabrication

Polymer processing is another key issue that should be addressed in the development of tissue engineering scaffolds. In fact, methods of manufacturing such scaffolds in a reproducible manner are crucial to their success, and must allow for the necessary scale-up of the developed tissue engineering technology [12,42]. The technique used to manufacture scaffolds for tissue engineering is dependent on the properties of the polymer and its intended application [2,43]. It must allow the preparation of scaffolds with complex three-dimensional geometries with controlled porosity and pore size [16,43], since these factors are associated with supplying of nutrients to transplanted and regenerated cells and thus are very important factors in tissue regeneration.

Most of the conventional processing techniques used to produce porous materials in the polymer industry are unsuitable for producing materials for medical applications because these methods usually involve the use of additives such as surfactants, stabilizers, lubricants and standard blowing agents, which can release toxic residues or by-products that are incompatible with living organisms.

Therefore various processing techniques have and are being developed to fabricate these scaffolds, such as solvent casting [12,43-45], particulate leaching [12,16,43-45], membrane lamination [44,46], fiber bonding [43,44,47], phase separation/inversion [44,48], high pressure based methods [44,49], melt based technologies [27,43,50], and micro-wave baking and expansion [22].
More recently highly porous 3-D scaffolds have been obtained using advanced textile technologies and rapid prototyping technologies such as fused deposition modelling (FDM) and 3-D printing [51]. These engineering technologies are highly controllable and reproducible and facilitate the manufacture of well-defined 3-D structures [52].
The major problem of the scaffolds produced by the methods developed so far is their lack of mechanical strenght, which does not allow for their use in hard tissue regeneration where high strength scaffolds are required [53]. Therefore, the search for better ways of producing porous scaffolds, so that physical and chemical properties can be simultaneously optimised, is currently an important issue in hard tissue engineering research [52,54]. The optimisation and customisation of these processing technologies to design and manufacture the scaffolds, require a thorough understanding of the materials and equipment that needs to be used [52,54].

1.3. Polymer scaffold processing methods

Usually, polymer scaffold processing is divided in two general groups of techniques: melt processing and solvent processing. Melt processing involves heating the polymer above the glass transition temperature (Tg) or the melting temperature (Tm) and depends on melt viscosity. Solvent processing depends on the polymer solubility in various organic solvents and on the solvent volatility. This section describes briefly some polymer scaffold processing methods that have been proposed and some examples of their application.

1.3.1. Fiber bonding

Fiber meshes consist of individual fiber either woven or knitted into three-dimensional patterns of variable pore size. The advantageous characteristic features of fiber meshes are a large surface area for cell attachment and a rapid diffusion of nutrients in favour of cell survival and growth [1,2,16,43,44]. A drawback of these scaffolds might be the lack of structural stability.

Interconnected fiber networks have been prepared by a fiber bonding technique that involves the casting of a L-PLA solution over a non-woven mesh of PGA fibers [1,2,16,43,44,47]. Solvent evaporation results in a composite material that consists of non-bonded PGA fibers embedded into a L-PLA matrix. Fiber bonding occurs during a post treatment at a temperature above the melting temperature (Tm) of PGA. Finally the L-PLA matrix is selectively dissolved in a non-solvent for PGA, and a network of bonded PGA fibers is released. Obviously, this technique is not the most appropriate for to the fine control of porosity [1,2,43,44]. Furthermore, stipulations concerning the choice of the solvent, immiscibility of the two polymers, and their relative melting temperatures restricts the general application of the technique to other polymers [1,43,44]. In addition, this method of fiber bonding does not address the problem of creating scaffolds with complex three-dimensional shapes, but it has proven successful for producing hollow tubes that have been proposed for use in intestine regeneration [1,43,44].

Fiber meshes can also be prepared using textile processing techniques [17,51,55]. For example, Freed et al. [17,56,57], developed fiber meshes using PGA fibers previously obtained by extrusion. These fibers were needled to form non-woven mesh using barbed needles to entangle the fibers and lock them together. Heat-setting further increased the dimensional stability of the mesh and smoothed the top and bottom surfaces. Finally a multihole die was used to punch the mesh into 1 cm discs [17,56]. The described scaffolds, as others obtained from other polymers by similar methods, have been widely used in tissue engineering research [17,56-62].

1.3.2. Solvent-casting and particle leaching

The solvent casting and particle leaching method consists in dispersing sieved mineral (e.g. NaCl) or organic (e.g. saccharose) particles in a polymer solution. This dispersion is then processed either by casting or by freeze-drying in order to produce porous bi- and three-dimensional supports, respectively. The porosity basically results from the selective dissolution of the particles from the polymer/salt composite, although phase separation of the polymer solution can also contribute to form the porous structure [2]. A variation of this method includes the use of vibration during dissolution of the polymer in the solvent and during solvent evaporation [63]. The porosity and pore size can be controlled independently by varying the amount and size of the salt particles, respectively. The surface area depends on both initial salt weight fraction and particle size [1,2,43,44]. The disadvantages of this method, as it has been applied so far, include the extensive use of highly toxic solvents and the limitation to produce thin wafers or membranes up to 3mm thick [1,2,12,43,44,51].

This technique was validated for PLLA and PLGA but can be applied to and any other polymer that is soluble in a solvent [1,2,12,16,43,44], such as chloroform or methylene chloride. Scaffolds produced by this method have been used in a significant number of studies concerning their application in tissue engineering [18,63-72].

1.3.3. Membrane lamination

The membrane lamination method uses membranes previously prepared by solvent casting and particle leaching. The membranes with the appropriate shape are solvent impregnated, then stacked up in a three-dimensional assembly with continuous pore structure and morphology [2]. The bulk properties of the final 3D scaffolds are identical to those of the individual membranes [1,2,12,44].

This method may allow for the construction of 3-D polymer foams with precise anatomical shapes, since it is possible to use computer-assisted modelling to design templates with the desired implant shape [2]. However this fabrication technique is time consuming, because only thin membranes can be used. Another disadvantage is that the layering of porous sheets allows only a limited number of interconnected pores [51].

1.3.4. Melt molding

to our knowledege, melt moulding has not been used as a single tecnique to produce scaffolds for tissue engineering. It is normally used in combination with porogen techniques (as in the example described bellow) or to produce a pre-shape of the final material, for example in the high pressure method that will be described further down in this section.

In a typical example, A mixture of fine PLGA powder and gelatin microspheres is loaded in a Teflon mold and then heated above the glass-transition temperature of the polymer [1,2,43,44]. The PLGA-gelatin composite is subsequently removed from the mold and gelatin microspheres are leached out by selective dissolution in distilled deionized water. In this way, porous PLGA scaffolds with a geometry identical to the shape of the mold can be produced. Polymer scaffolds of various shapes can be produced by simply changing the mold geometry. This method also offers independent control of porosity and pore size [1,2,43,44] by varying the amount and size of microspheres used, respectively.

In addition, it is possible to incorporate bioactive molecules in either polymer or gelatin microspheres for controlled drug delivery because this process does not utilises organic solvents and is carried out at relatively low temperatures for amorphous PLGA scaffolds [1,2,43,44]. Besides the choice of gelatin, other leachable components may be used. This manufacturing technique may also be applied to PLLA or PGA. However, higher temperatures are required (above the polymer melting temperatures) because these polymers are semicrystalline. This excludes the potential for protein incorporation into these systems [1,2,43,44]. Several sucessuful examples of the application of melt based techniques on the production of tissue engineering scaffolds will be given on the results section of the present chapter.

1.3.5. Freeze-drying

Phase separation of polymer solution may actually be induced by several ways [2,48]. The basic principle of the freeze-drying process relies on a thermally induced phase separation, which occurs when the temperature of a homogenous polymer solution, previously poured into a mold, is decreased. Once the phase-separated system is stabilized, the solvent-rich phase is removed by vacuum sublimation leaving behind the polymer foam. The foam morphology is of course controlled by any phase transition that occurs during the cooling step, i.e., liquid-liquid or solid-liquid demixing [2,48]. Current research shows that this method is very sensitive, i.e., the parameters have to be very well controlled [51]. Furthermore, at present, only pore sizes of 100 μm can be reproducibly obtained by this method [51].

Porous PLLA and scaffolds loaded with small hydrophobic and hydrophilic bioactive molecules have been manufactured by this method [12,43]: the polymer is dissolved in a solvent such as molten phenol or naphthalene at a low temperature, followed by dispersion of the bioactive molecule in this homogenous solution. A liquid-liquid phase separation is induced by lowering the solution temperature. The resulting bicontinuous polymer and solvent phases are then quenched to create a two-phase solid. Subsequent removal of the solidified solvent by sublimation leaves a porous polymer scaffold loaded with bioactive molecules [12,43]. Proteins retained as much as 75% of their activity after scaffold fabrication with the naphthalene system. Since phenol has a lower melting temperature than naphthalene, it might be useful for entrapment of small drugs or short peptides into polymer matrix. However, the activity of alkaline phosphatase was lost completely after fabrication even in the phenol system. Therefore, incorporating and releasing large proteins with defined conformations such as growth factors without loss of activity remains a challenge [12,43].

Polystyrene foams, produced by phase separation from a naphthalene solution [73] have been used for hepatocyte culture experiments, after their derivation with lactose and heparin.

1.3.6. Aggregation of polymer microparticles

The so-called aggregation of polymer microparticles method, consists in the aggregation, by physical or chemical means, of microparticles [2]. The porosity is nothing but the interstices between the aggregated microspheres and it is directly related to the microspheres diameter. The possible release of previous encapsulated growth factors from the microspheres is an additional advantage of this technique.

Macroporous PLA supports have been prepared by this original method consisting of the aggregation of PLA microparticles. The microsphere aggregates can be stabilized by chemical crosslinking of a poly (vinylalcohol) (PVA) precoating with glutaraldehyde

[2]. The local fusion of the aggregates PLA particles at the point of contact is also possible, particularly in the case of plasticized particles with triethylcitrate.

1.3.7. High pressure processing

In the high pressure processing technique, solid disks of the polymer, previously prepared by either compression moulding or solvent-casting, are exposed to high pressure CO_2 (5.5 MPa, 25°C) to allow saturation of CO_2 in the polymer [12,44,49]. A thermodynamic instability is then created by reducing the CO_2 gas pressure to an ambient level, which results in nucleation and expansion of dissolved CO_2, generating macropores. This process yields mostly a nonporous surface, which results from the rapid diffusion of the dissolved gas from the surface, and a closed-pore structure inside the polymer matrix, which may be problematic for cell seeding. The porosity and the pore structure are dependent on the amount of CO_2 dissolved, the rate and type of gas nucleation, and the rate of gas diffusion to the pore nuclei [12,44,49]. This method has been used to obtain poly(DL-lactic-co-glycolic acid) foams [12,44,49], with 93% porosity and a pore size of about 100 µm.

1.3.8. In-situ polymerization

All the polymer processing techniques discussed so far are methods that can be used to manufacture pre-fabricated scaffolds, which may then be used to regenerate the appropriate tissue. Pre-fabricated scaffolds are suitable for most tissue engineering applications, however, there are many instances in orthopaedic surgery when mechanical stability must be restored immediately. In such cases, a non-degradable bone cement, usually poly (methyl methacrylate) is used to fill the bone defect and provide mechanical stability [43].

Recently, a degradable poly(propylene fumarate) (PPF) based bone cement has been developed to provide the same function on a temporary basis [43,74,75]. PPF is an unsaturated linear polyester as is a viscous liquid at room temperature. It can be crosslinked at the time of surgery to form a solid degradable bone cement via addition polymerisation with N-vinyl pyrrolidone (N-VP). At the time of the crosslinking reaction, two other components are incorporated into the PPF: the first is NaCl which provides pores into which new bone can grow and the second is tricalcium phosphate which is an osteoconductive material and stimulates new bone growth. In this manner, a temporary osteoconduvtive scaffold may be formed in-situ to replace the mechanical function of bone until new bone, stimulated to form in the scaffold pores, can assume its structural role. During its liquid phase, the cement can be injected or molded into the bone defect. It is therefore well suited for this application, since many bone injuries result in defects, which are relatively inaccessible [43,74]. In addition, the use of this type of materials may represent an important step on the way to eliminate the need for the defect or injured tissue to be exposed to a conventional surgical operation, and instead may lead to a minimally invasive surgery [74]. However, most addition polymerisation reactions are exothermic and generate large quantities of heat, which is sufficient to cause some local tissue necrosis. In contrast, much less heat is generated by the crosslinking reaction between PPF and N-VP and no local tissue necrosis has been noted in *in vivo* studies [43,74,75].

1.3.9. Rapid prototyping technologies

Rapid prototyping (RP) technologies have the potential to produce 3-D constructs of complex geometries in a multi layer design within the same gross architectural structure, from a computer aided design (CAD) model of an object, without a part specific tooling or knowledge [51]. Some RP machines join together liquid, powder and sheet materials to form parts, layer by layer, directly from a computer generated model.

Rapid prototyping technologies, such as 3-D printing (3-DP) and fused deposition modeling (FDM) allow the development of manufacturing processes to create porous scaffolds that mimic the microstructure of living tissues [51]. The 3-D printing technology is based on the printing of a binder through a print head nozzle onto a powder bed, with no tooling required. The part is built sequentially in layers. The binder is delivered to the powder bed producing the first layer. The bed is then lowered to a fixed distance, powder is deposited and spread evenly across the bed, and a second layer is built. This is repeated until the entire scaffold is fabricated. The entire process is performed under room temperatures, which offer a great potential for tissue engineering because allows for the eventual incorporation of biological agents [51], such as growth factors, without inactivation.

The FDM process involves the extrusion of a polymer filament through a heated nozzle and its deposition, as thin layers, on a platform [51]. The process begins with the design of a conceptual geometric model on a CAD workstation. The design is imported into a software, which mathematically slices the conceptual model into horizontal layers. Toolpaths are generated before the data is downloaded to the FDM hardware. The FDM extrusion head operates in the x and y axis while the plataform lowers in the z-axis for each layer to be formed. In effect, the process draws the designed model (scaffold) one layer at time.

1.4. Cells for Tissue Engineering

The next step after the development of an adequate porous structure is the choice of a reliable source of cells that allows the isolation of high amounts of cells. Moreover, these late ones should be easily expandable to higher passages, non-immunogeneic and have a protein expression rate similar to the tissue to be regenerated [76].

The first, and most obvious choice because of their non-immugenecity is the isolation of cells from biopsies taken from the patients (autologous cells), followed by limited expansion *in vitro*. However this technology has several limitations, relatively few cells are available after the dissociation of the tissue, their expansion rates are relatively low, limiting in this way the number cells available to be seeded on the scaffolds. Furthermore, in certain diseases states (e.g. osteoporosis) or in elderly patients, autologous cells may not be appropriate for transplantation because their protein expression profile is under the expected values [76].

An alternative to the referred technology is the use of cells obtained from non-human donors (xenogeneic cells), which would solve the problem of low cell number yields. A lot of research work has been done the last years in order to obtain, through genetic engineering, animals that are non-immunogeneic to humans, but the possibilities of the transmission of infectious agents such as virus and the ethical/social problems related with this issue has refrained the enthusiasm for this approach [76,77].

Another promising possibility is the use of stem cell technology. Stem cells are undifferentiated cells with a high proliferation capability, being able of self-renewal, production of large number of differentiated progeny and regeneration of tissues [78]. There are different stem cell types within the human body, being the degree of commitment to a certain function their main distinguishing factor [76,79]. Stem cells present at early-stage embryo, known as embryonic stem cells (ES) are capable to differentiate in every cell type present in the human body, and because of this are called pluripotent [79]. A second type of stem cells, known as adult stem cells, have already a certain degree of commitment/differentiation, but they are still multipotent, because they are not fully differentiated, and can be found in several organs and tissues of the human body (muscle,cartilage, bone and the nervous system).

A particular emphasis in the tissue engineering field of hard tissue regeneration, has been given to a kind of adult stem cells, human mesenchymal stem cells. These cells were first described by Fridestein *et al.* [80], and are located in the bone marrow around blood vessels (as pericytes), in fat, skin, muscle and other locations [81]. They are isolated through gradient centrifugation [82] and once in culture they grow in colonies having fibroblast like appearance [83]. Although they are present in very small amount (from 0.001% to 0.01%) [84] they have high proliferation rates, can be expanded over one billion fold in culture making them ideal for the use in tissue engineering [80-85]. Moreover with the use of differentiation factors (dexamethasone, ascorbic acid and β-glycerophosphate) it is possible to differentiate these cells into osteoblasts [83,84], chondrocytes (TGF-β_3) [84,86], adipocytes (1-methyl-3-isobutylxanthine,dexametasone, insulin and indomethacin) [84], skeletal muscle cells [84,87] as well cells typical from the central nervous system (CNS) as neurons and astrocytes [88-90]. This facts show, and in contrary of what was believed, that human mesenchymal stem cells (hMSC) have a high differential plasticity so, in this sense they may be a good alternative for the use on tissue engineering in a near future. When compared with embryonic stem cells they also present the advantage of being easier to maintain in culture, there is no tumour formation risk and finally there no ethical/social problems involved. However, and in spite of all of these advantages, there are still some drawbacks when using this technology. The difficulties of obtaining an homogeneous and 100% pure culture of hMSC are high, mainly due to the inexistence of monospecific and unique molecular probes that tag these cells [81-83]. Another problem is the fact that scientist do not completely control the differentiation procedures because the majority of the differentiation pathways is still unknown, although some progress as been made in this field [91].

In the last few years, embryonic stem cells (ES) research has received a boost. ES cells are isolated from the inner cell mass (ICM) on an early stage embryo in the blastocyst stage and, when transferred back into an early embryo, they can contribute to all cells of the embryo, but not the placental tissue, and, therefore, are not able to generate a complex organism [92]. Their culture was first established in the early 1980s [93] using rat cells, and more recently the culture of human ES was achieved. Undifferentiated stem cells are characterized by two unique properties: the almost unlimited self-renewal capability and the capacity to differentiate. Other properties are the expression of stage specific embryonic antigens, as SSEA-1, the expression of germ-line transcription factor Oct-4, high telomerase activity and the regulation of ES cell self-renewal by cytokines of the IL-6 family [92]. Because of the stated characteristics these cells have to be used in tissue engineering of hard tissues. Chondrocytes and Osteoblasts were already differentiated from these cells [94,95]. However, before this cells can be used several issues still have to be solved such as the ethical problems related with the research dealing with human embryos, the discovery of the pathways involved in the differentiation pathways, the development of methods that allow the direct differentiation of ES cells and their selective differentiation and the integration and tissue specific function of the ES-cell-generated somatic cells after transplantation [92]. Two other questions that need to be solved: (i) s that ES-cell-derived somatic donor cells are not tumorogenic (it has been known that undifferentiated ES cells give rise to teratomas and teratocarcinomas) and (ii) the immunological incompatibility between ES-cell-generated donor cells [92].

2. Materials and Methods

This sections describes some of the ongoing work at the 3B's research group of the University of Minho, on the tissue engineering of bone using starch based scaffolds.

2.1. Development of starch based scaffolds

In this study, two different polymeric blends of corn starch with: i) ethylene vinyl alcohol blends (SEVA-C) and ii) cellulose acetate (SCA), both obtained from Novamont, Italy, were used.

Several blowing agents (BA) were selected for the present study. The first blowing agent selected was Hostatron System P9947, from Hoechst, Germany, which will be designated as blowing agent 1 (BA1). This blowing agent of commercial origin is mainly composed of carboxylic acid that reacts by heating, releasing CO_2 and water at about 200°C. Two other blowing agents were selected, namely Hydrocerol BIH 70 and Hydrocerol BIH 40, which will be designated by blowing agent 2 (BA2) and blowing agent 3 (BA3), respectively; both were obtained from Clariant, Germany. These blowing agents are based on citric acid and they also release CO_2 and water upon decomposition, which happens around 170°C. Finally, it was used a blowing agent of the azodicarbonamide type (trade name CELOGEN 780) which as a decomposition temperature very close to the melt temperature of the starch based polymers. The cytotoxicity of these materials has been tested [28] and the results show a non-cytotoxic behaviour.

2.1.1. Injection moulding with blowing agents

The polymeric materials were previously mixed in a rotating drum with one of the blowing agents described above, in amounts from 1% to 2,5%. Materials were then injection moulded in a Klockner Ferromatic FM-20 or a Krauss Maffei KM60-120A)

For the injection moulding of the scaffolds, it was designed and manufactured a special mould which allows for the injection moulding of the polymeric melt at a much lower pressure. It also allows for an excellent venting of the mould cavity and therefore enables to enhance significantly the expansion inside the mould when compared to a normal mould.

2.1.2. Extrusion with blowing agents:

In the extrusion process, the polymers were previously mixed with the blowing agents in a bi-axial rotating drum prior to processing in a twin-screw extruder Carvex, with a die of 12 mm of diameter. This process was optimised for mixtures of the polymer with 10% and 15% (w/w) of blowing agent 1. The weight fraction of blowing agent 2 and 3 necessary to produce the same percentage of porosity was much smaller than the one used with BA1 (between 1 to 2,5%).

2.1.3. Compression moulding - particle leaching:

The compression moulding and particle leaching method was based on blending together a starch based polymer (in the powder form) and leachable particles (in this case, salt particles) of different sizes, from 50μm to 1000μm, in sufficient amounts to provide a continuous phase of a polymer and a dispersed phase of leachable particles in the blend. The blend was then compression moulded into discs of 6 cm of diameter and approximately 1 cm of height, using a mould specially designed for this purpose. The resultant samples were then immersed in distilled water to remove the leachable particles.

2.1.4. Solvent-casting/particle-leaching:
Using the solvent casting and particle leaching method, the polymers were first dissolved in an appropriate organic solvent and mixed with salt particles of different sizes. In general, it was added a salt weight fraction of 60 to 70% (based on the total mass of polymer and salt). The size of the particles used ranged from 50 to 1000μm. The mixture of the polymeric solution with the salt particles was then poured into a mould (a glass petri-dish of 3cm diameter) and placed in an oven at 37°C in order to allow a progressive evaporation of the solvent. Finally, when the samples were completely solidified, they were immersed in distilled water, for leaching of the salt particles, and afterwards dried.

2.1.5. In-situ polymerisation
This innovative so-called in-situ polymerization process was based on a polymerization process developed in our group in order to obtain materials to be used as bone cements or hydrogels [32]. These materials were prepared by adding the liquid phase, constituted by the acrylic monomers ((AA), from Merck), and 1% (w/w) of N-dimethylaminobenzyl alcohol (DMOH), which was used as the activator of the initiation process, to the solid phase, which consisted of SEVA-C powder and 2%(wt/wt) of benzoic peroxide ((BPO), from Merck), which was used as the radical initiator, after purification by fractional recrystallization from ethanol, mp 104 °C BPO. The leachable NaCl particles were added to the liquid or to the solid phase in order o provide the porosity of the structure. The solid and the liquid phases were then mixed together with a 10% of water with respect to the total weight and poured in a dough state in poly(tetrafluoroethylene) (PTFE) moulds until complete polymerisation takes place.
After curing time, about 5 minutes, moulds are placed into the oven at 60°C overnight to ensure a complete polymerization and then vacuum dried until constant weight was attained. Finally the samples were immersed in water to leach out the salt particles, in order to better simulate the in-vivo application of these materials.

2.1.6. Materials Characterization:
The porous structure of the materials developed, namely the morphology of the pores, their size and distribution and also the interconnectivity between these pores, was characterised by scanning electron microscopy (SEM), in a Leica Cambridge S360. All the samples were previously gold coated in a Sputter Jeol JFC 1100 equipment. The porosity measurements were obtained from the photographs acquired by SEM that were processed using an image analysis software.
The mechanical properties of the developed materials were assessed on compressive experiments in an Instron 4505 universal mechanical testing machine, using a load cell

of 50 kN. The compression tests were carried out at a crosshead speed of 2 mm/min (4.7 x 10^{-5} m/s), until obtaining a maximum reduction in samples height of 60%. A minimum of six samples of each type was tested.

The degradation behaviour was assessed after several pre-fixed ageing periods (0,3, 7, 14, 30 and 60 days), in an isotonic saline solution (NaCl 0.154 M). At the end of each degradation period, the samples were removed from the solution, rinsed with distilled water and weighted, to determine the water uptake; one batch of samples was then dried up to exhaustion (6 days at 60°C) in order to determine the dry weight loss.

2.2. Cell Culture Studies

Only the two scaffolds that presented the best combination of mechanical properties, degradation behaviour and porosity, were selected for the cell culture studies. Consequently, the studied scaffolds were based on the blend of corn starch with cellulose acetate (SCA) obtained by compression moulding combined with salt leaching and by extrusion with blowing agents.

2.2.1. Cell seeding on starch based scaffolds

Human osteoblasts were obtained from calvarian bone explant cultures as described previously [96]. Cells were passaged until they reach the P3 stage, trypsinized, ressuspendend in culture medium, mixed in fibrin glue in a 3:1 ratio and aliquots of 20 µl containing $3x10^5$ were seeded on the scaffolds. Cells/scaffolds constructs were then kept in culture in a humidified atmosphere at 37°C and 5% CO_2. Humans osteoblasts seeded on regular 24 well culture trays were used as a control. During the second week of the experiment the osteogenic phenotype of the cells was stimulated by supplementing the culture medium with 10 mM β-glycerophosphate (Sigma,USA), 100 µM ascorbic acid (Sigma,USA) and 10^{-7} M dexamethasone (sigma,USA)

2.2.2. Cellular Adhesion

Cellular adhesion was assessed by fluorescence microscopy by using phalloidin conjugated with alexa fluor 488 (Molecular probes, USA). Further details on the methodology used can be found in Salgado et al. [96].

2.2.3. Cellular viability and proliferation

Cellular viability was assayed confocal laser microscopy(CLM) using fluorescein di-acetate (FDA) (Molecular Probes, USA), which stains viable cells green, and propidium iodide (PI) (Molecular Probes, USA) which stains necrotic and secondary apoptotic cells. Cellular viability and proliferation was also assayed by the MTS (3-(4,5-dimethylthiazol-2-yl)-5-(3-carboxymethoxyphenyl)-2(4-sulfophenyl)-2H tetrazolium) (Promega, USA) test is an assay in which the substrate – MTS – is bioreduced into a brown formazan by mithocondrial enzymes, which are active in living cells. In this assay intensity of the colour is directly related to the number of viable/living cells so, an increase in the colour will mean that the number of viable cells increased and hence cellular proliferation occurred. Further details on the methods here in described can be found in Salgado et al. [96].

2.2.4. Protein expression

Osteocalcin production was assessed using an ELISA kit (IBL, Germany). Supernatants were weekly collected and frozen at −20°C and then thawed prior to analysis. 25 μl of each sample were used and the assay was performed under the instructions of the manufacturer. Results are presented in nmol/ml.

3. Results and Discussion

3.1. Morphology of the porous structures

The porous structure of the samples obtained by extrusion or injection moulding of the polymers combined with blowing agents results from the gases released by decomposition of the BA during processing. Therefore, with these melt based methods it is difficult to have full control over the pore size and the interconnectivity between the pores of the materials obtained by these methods. Nevertheless, the subsequent optimisation of processing parameters and of the type and amount of blowing agent, allowed to obtain scaffolds with subsequent higher porosity, interconnectivity between pores and pore sizes that can vary from roughly between 50 up to 1000 μm. The density of the samples obtained by these processes is approximately 0.7-0.8 g/cm^3 (depending mainly on the blowing agent used), leading to a porosity of about 40-50%. Moreover, the scaffolds obtained by these melt based methods present a microporosity througout the whole structure which can play an important role in nutrients flow during the cell culturing and/or the implantation of the scaffolds. A thin layer of solid material surrounds the porous structure of the material obtained with both processes, but this can be removed easily as a final step in the processing of the scaffolds.

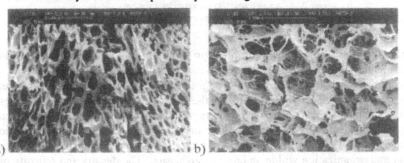

Figure 1. SCA based scaffold obtained by melt based technologies: a) injection moulding with 1% Celogen 780 b) extrusion with 2% of blowing agent 3 (Hydrocerol BIH40)

236

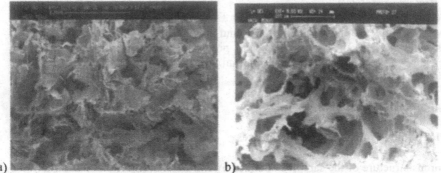

a)   b)

Figure 2. SCA based scaffold obtained by combined techniques: a) the solvent casting-particle leaching b)
compression moulding + particle leaching with 65% of salt.

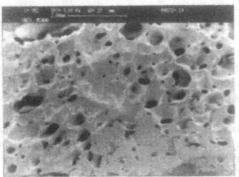

Figure 3. SCA based scaffold obtained by a method
based on *in situ* polymerization

The combined techniques such as compression moulding – particle leaching and
Solvent casting and particle leaching allowed to obtain an open-porous structure with a
good interconnectivity between the pores throughout the entire structure. Furthermore,
this method allows for the accurate control of the pore size, distribution of pore sizes
and porosity (volume of voids). These parameters of the porous structure can be tailored
by varying the size, shape and distribution of the particles and the choosen volume ratio
of polymer/particles. The control of scaffold porosity is critical for controlling cellular
colonization and organization within an engineered tissue. The density and porosity of
the scaffolds obtained by compression moulding and solvent casting combined with salt
leaching, in general, can vary from 50 to 60%, without affecting significantly the
mechanical integrity of the material.

The innovation introduced by the in-situ polymerisation method of obtaining scaffolds
for tissue engineering lies in the fact that it is possible to produce the scaffold in-situ,
i.e., it might be possible to inject the scaffold directly into the defect to treat, which can,
therefore, take immediately the shape of the defect.
These moldable polymer scaffolds fitting to the three dimensional geometry of specific
tissue defects are highly in demand for clinical applications due to the invasiveness of

the surgical implantation procedure [74]. SEM analysis of this scaffolds showed pores ranging in size roughly from 10 to 100μm in diameter (figure 3), but once again, the pore size depends on the salt particles used.

3.2. Degradation behaviour

Figures 4 and 5 show the water uptake and the weight loss as a function of the degradation period for the scaffolds obtained by the different processing methodologies developed.

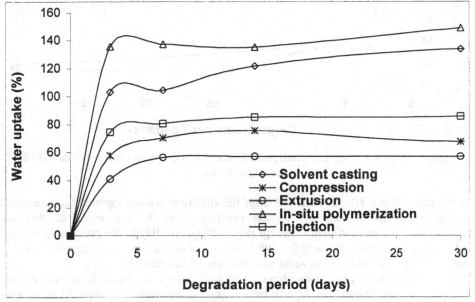

Figure 4. Water uptake versus degradation period for the SCA based scaffolds obtained by the different processing methodologies.

238

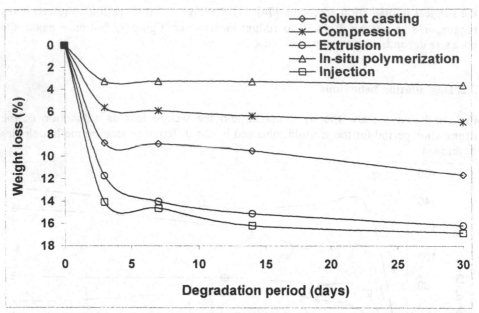

Figure 5. Weight loss versus degradation period for the SCA based scaffolds obtained by the different processing methodologies.

As expected, the scaffolds obtained from the different processing methods presented similar water uptake and degradation profiles although they exhibited different degradation rates and different water uptake ability, according to the processing method and conditions, namely those that have direct influence in the percentage of porosity, such as the amount of blowing agent and fraction of salt particles used.

The scaffolds obtained by the combination of compression moulding or solvent casting with particle leaching exhibit significantly higher water uptakes, compared to samples obtained by extrusion or injection moulding with blowing agents, which demonstrates the higher porosity and interconnectivity achieved with these methods. However, the weight loss of these scaffolds is lower than for the scaffolds obtained by extrusion, which is most probably due to thermo-mechanical degradation undergone by the materials processed by extrusion and injection moulding, which leads to an easier breakdown of the material (associated to leaching out of low molecular weight chains).

The scaffolds obtained by the in-situ polymerisation method present the highest water uptake ability, since they have water uptake properties that are typical of hydrogels. Their degradation rates are obviously lower than those presented for the above referred to scaffolds, since these type of scaffolds are composed of a blend of the starch based polymer with acrylic acid. The materials produced by this method are, in fact, not totally degradable, but they might be very useful in situations where it is necessary to ensure a high mechanical strenght and/or in situations where the defect or trauma that is necessary to treat is of difficult access, as the use of such type of scaffolds will avoid highly invasive surgery techniques.

3.3. Mechanical properties

Table 2. Mechanical properties of the scaffolds obtained by the different processing methodologies developed.

Processing Method	Compressive modulus (MPa)	Compressive strenght (MPa)
Extrusion with blowing agents	124.6 ± 27.2 to 230.8 ± 71.0	8.0 ± 0.9 to 17.6 ± 0.9
Compression molding and particle leaching.	133.7 ± 20.6 to 341.6 ± 34.3	20.56 ± 6.2 to 67.69 ± 6.2
Solvent casting and particle leaching.	170.5 ± 16.09	21.73 ± 1.1

The mechanical properties of the scaffolds obtained by extrusion with blowing agents are very dependent on the type and amount of blowing agent used, exhibiting a compressive modulus that can vary from about 124 up to 230 MPa.

Nevertheless, these scaffolds present very promising mechanical properties when compared to other scaffolds, obtained from other biodegradable polymers, and proposed for use in tissue engineering of bone. For example, PLLA/ hydroxyapatite composite foams, prepared by a process based on phase separation, presented a compression modulus bellow 12 MPa [11]. The results presented in Table 2 might be further improved by reinforcing the scaffolds with hydroxyapatite.

The compressive properties of the scaffolds obtained by compression moulding and particle leaching are also very dependent on the porosity obtained, being in some cases, superior to those obtained by the extrusion process.

As should be expected, the mechanical properties of the scaffolds obtained by the solvent casting and particle leaching method are lower when compared to the properties of the samples obtained by melt-based technologies. However, these properties may be considered very good when compared to scaffolds obtained from other materials by identical processing methods and proposed for the same type of applications. For example, a PLGA scaffold obtained by the solvent casting and particle leaching method, exhibits a modulus of 1.09 MPa [98].

Further information on the mechanical properties, morphology and degradation behaviour of starch based scaffolds may be found in references [19-32].

3.4. Cell Culture Studies

3.4.1. Cellular Adhesion

Cellular adhesion is a complex process involving initial osteoblast/materials interactions, being divided in four stages: (i) protein adsorption to the surface, (ii) contact of rounded cell, (ii) attachment of cells to the substrate, and (iv) spreading of cells [99]. Cell attachment usually occurs within 4 hours after the seeding, being this values dependent on the substrate characteristics and cell type used.

Within the present experiment cells start to spread 24 hours after cell seeding and after seven days cells showed a good confluency in the fibrin glue. One week after the experiment started it was possible to observe that cells were attached to the substrate with well extended filopodia (Figure 6), showing in this way that the materials and an adequate surface chemistry for cell attachment.

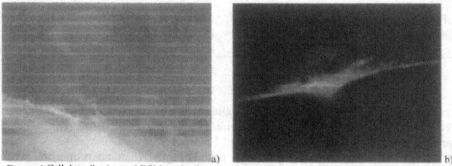
a)   b)

Figure 6. Cellular adhesion and ECM production experiments: Cells were stained with Phalloidin. From the analysis of pictures a) scaffolds obtained by extrusion and b) samples obtained by compression moulding + particle leaching, one can see clearly that cells were completely attached by focal adhesion points to the materials one week after.

3.4.2. Cellular viability and proliferation of human osteoblasts on starch based scaffolds

By observing figure 7 is clearly seen that starch based scaffolds are not harmful to the cells. After one and three weeks, in both extrusion (7a,7c) and compression moulding scaffolds (7b,7d) there are almost no necrotic cells. Furthermore, it has to be stated that for the extrusion samples the picture were taken in the central inner area of the scaffold, showing that the porosity of these scaffolds was adequate for nutrient delivery and gas exchanges.

Figure 7- Cells were stained with 2 μg/ml FDA and 100 μg/ml PI. After week 1 was possible to observe that cells proliferate mostly on the edges of both scaffolds, extrusion **(a)** and compression moulding **(c)**. However, after 3 weeks in culture the cells in the extrusion samples **(b)** occupied almost the entire area of the scaffold, while cells in the compression moulding samples showed that, in spite of being also capable of colonizing the inner part of the scaffold **(d)**, they mainly colonize the periphery of the scaffold. No necrotic cells were detected.

Another aspect of the cellular behaviour while cells were seeded on the scaffolds was that they did not behave in the same way in the two different scaffolds. While in the extrusion samples they were able to colonize the inner areas of the scaffolds (7c) in the compression moulding samples the cells they mainly colonize the periphery. The explanation for this fact is probably found within the surface chemistry of the material that in the case of compression moulding did not allow for a homogeneous cell growth through the scaffold.

MTS test confirmed the results obtained with the FDA/PI assay (Figure 8)

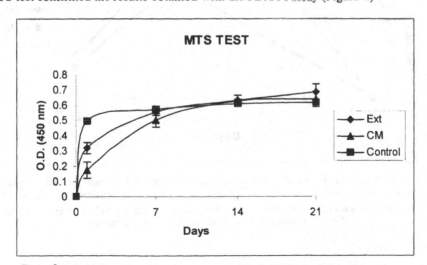

Figure 8. Cell viability and proliferation was assayed by the MTS test. Cell density used was 3×10^5 cells/scaffold. For control 3×10^5 cells were seeded on 24 well plate culture trays. Cells were kept in culture for 21 days. Between day 7 and day 14 cells were stimulated with 10 mM β-glycerophosphate (Sigma, USA), 100mM ascorbic acid (Sigma, USA) and 10^{-7} M dexamethasones (Sigma, USA).

242

The differences registered in the O.D. after only one day in culture are mainly due to different cell seeding efficacies. After week 1 the O.D. values obtained for the SCA scaffolds were almost similar to those of the control samples, proving that cells were proliferating on the starch based scaffolds. On weeks 2 and 3 cellular growth was not significant, mainly due to the stimulation with dexamethasone, which is known to decrease the cellular proliferation [100,101]. Another explanation for the fact is that once cells reached confluency on the places where they were seeded, the rate of cellular death and proliferation was equal and hence the proliferation stopped. This fact is further confirmed by the results obtained with the control samples.

3.4.3. Protein Expression

During the course of osteoblast differentiation several proteins which constitute bone extracelular matrix are expressed, being osteocalcin one of these proteins. This protein is usually involved in the deposition of new extracellular matrix [69,101] and its mineralization.

An assay, using an ELISA kit, was done in order to quantify its concentration in the supernatants [96].

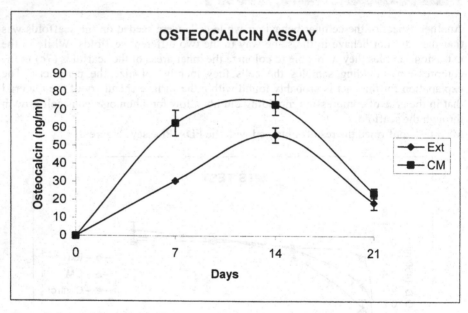

Figure 9. Osteocalcin assay: Results are shown in osteocalcin (nmol/ml) in function of days. On day 7 cells were stimulated with 10 mM β-glycerophosphate (Sigma, USA), 100mM ascorbic acid (Sigma, USA) and 10^{-7} M dexamethasones (Sigma, USA). 7 days after stimulation osteocalcin levels increase (14 days in culture), reaching the lowest values after 3 weeks in culture.

The peak observed during the second week of the experiment was due to the stimulation with ascorbic acid, which is known to stimulate the deposition of bone extracellular proteins [101] such as osteocalcin and indicates that bone matrix was being deposed. Moreover, the values throughout the experiment were similar to those found in the human serum, showing in this way that the cells had maintained their osteogenic phenotype.

Again, additional information on the behaviour of starch based materials when in contact with cells can be found in different works. The biocompatibility has been evaluated in works done by Mendes SC *et al.*, *Gomes ME et al.*, Marques AP *et al.* [28,29,102], direct contact in Marques AP *et al.*[102], inflamatory response in Marques AP *et al.*[103], protein adsorption in Alves CM *et al.*[104]. The behaviour of the cells when seeded in starch based scaffolds is described in works by Salgado AJ *et al.*[96].

4. Conclusions

The explosion in tissue engineering research has accentuated the need for new classes of biodegradable polymers which either possess particularly desirable tissue-specific properties, or which have wider applicability and can be tailored to several tissue systems, since the range of potential tissue engineered systems is broad.

In this study, several innovative processing routes, based on conventional and non-conventional technologies, were developed towards the production of several starch based scaffolds with different architectures and properties. These scaffolds present adequate porous structure, that can be tailored according to the intended application, with pore sizes in the range of those which are believed to be the most appropriate for bone cell culturing and/or bone tissue ingrowth. The interconnectivity and the porosity obtained is dependent on the processing method. The resultant scaffolds from the above described methods presented a similar degradation profile although they exhibited different degradation rates, that depend on the used processing method and processing conditions Furthermore, although only few data on mechanical properties of scaffolds for tissue engineering is found in the literature, it is possible to conclude that mechanical properties of all the resultant scaffold are very promising, when compared to scaffolds obtained from other biodegradable polymers. This is specially true when the properties are compared with those of the gold standard of biodegradable scaffolds, the PLA/PGA systems.

It is then possible to conclude that by selecting and optimising a methodology from the processing routes described in this study, it is possible to tailor the porous morphologies and properties of the resulting scaffolds in order to meet the requirements for different sites of application.

Furthermore, cell culturing experiments using osteoblasts showed that osteoblastic cells remain viable after 3 weeks of culturing and the biochemical assays performed

244

demonstrate that cells maintained the osteogenic phenotype throught the experiments and that deposition of mineralised extracellular matrix could be detected.

Therefore scaffolds obtained from these materials using one of the described methodologies may constitute an important alternative to the materials currently used in tissue engineering, namely on bone tissue engineering.

References

1. Thomson RC, Wake MC, Yaszemski M, Mikos AG. Biodegradable polymer scaffolds to regenerate organs. Adv Polym Sci 1995; 122: 247-274.

2. Maquet V, Jerome R. Design of Macroporous Biodegradable Polymer Scaffolds for Cell Transplantation, Mater Sci Forum 1997; 250: 15-42.

3. Pachence JM, Kohn J. Biodegradable polymers for tissue engineering. In Lanza R, Langer R, Chick W, editors. Principles of Tissue Engineering, Academic Press, San Diego, 1997; 273-293.

4. Hayashi T. Biodegradable polymers for biomedical uses. Prog Polym Sci, 1994; 19: 663-702.

5. Solchaga L, Dennis J, Goldberg V, Caplan A. Hyaluronic acid-based polymers as cell carriers for tissue-engineered repair of bone and cartilage. J Ortho Res, 1999; 17:205-213.

6. Solchaga L, Goldberg V, Caplan A. Hyaluronic acid-based polymers in tissue-engineered cartilage repair. In New frontiers in Medical Sciences: Redefining Hyaluronan. Abatangelo G, Weigel P (eds), Oxford, 2000, Elsevier Science, 233-245.

7. Kim BS, Mooney D. Development of biocompatible synthetic extracellular matrices for tissue engineering. TIB TECH, 1998; 16:224-230

8. Ratner B. Biomedical applications of synthetic polymers. In Allen G, Bevington JC, Aggarwal SL,editors. Comprehensive Polymer Science. Pergamon Press,Oxford, 1989 ;7 : 201-249

9. Madihally S, Matthew H. Porous chitosan scaffolds for tissue engineering. Biomaterials, 1999; 20:1133-1142.

10. Kohn J, Langer R. Bioresorbable and bioerodible materials. In Biomaterials Science, Ratner B, Hoffman A, Schoen F, Lemons J, editors. Academic Press, San Diego, 1996. 64-72.

11. Zhang R, Ma PX. Poly(α-hydroxyl acids)/hydroxyapatite porous composites for bone-tissue engineering.i.preparation and morphology. J Biomed Mater Res 1999; 44, 446-455.

12. Agrawal CM, Athanasiou KA, Heckman JD. Biodegradable PLA-PGA polymers for tissue engineering in orthopaedics. Mater Sci Forum 1997; 250: 115-228.

13. Middleton JC, Tipton AJ. Synthetic biodegradable polymers as orthopaedic devices. Biomaterials, 2000; 21:2335-2346.

14. Radder A, Leenders H, van Blitterswijk C. Application of PEO/PBT copolymers for bone replacement. J Biomed Mater Res, 1996; 30: 341-351.

15. Xiao Y, Riesle J, van Blitterswijk. Static and dinamic fibroblast seeding and cultivation in porous PEO/PBT scaffolds. J Mater Sci: Mater Med, 1999; 10: 773-777.

16. Langer R. Selected advances in drug delivery and tissue engineering. J Contr Rel 1999; 62:7-11

17. Langer R, Vacanti J. Tissue engineering: the challenges ahead. Scientific American, 1999; 280: 62-65.

18. Chen G, Ushida T, Tateishi T. A biodegradable hybrid sponge nested with collagen microsponges. J Biomed Mater Res, 2000; 51: 273-279.

19. Reis RL, Mendes SC, Cunha AM, Bevis MJ. Processing and in-vitro degradation of starch/EVOH thermoplastic blends. Polym Int 1997; 43: 347- 353

20. Reis RL, Cunha AM, Fernandes MH, Correia, RN. Treatments to induce the nucleation and growth of apatite-like layers on polymeric surfaces and foams. J Mater Sci: Mater in Med 1997; 8: 897-905.

21. Reis RL, Cunha AM, Bevis MJ. Using non-conventional processing routes to develop anisotropic and biodegradable composites of starch based thermoplastics reinforced with bone-like ceramics. J App Med Pol, 1998; 2: 49-53

22. Ribeiro A, Malafaya PB, Reis RL. Microwave based methodologies for the production of polymeric and ceramic porous arquitechtures to be used in bone replacement, tissue engineering and drug delivery. 25th Annual Meeting of the Society for Biomaterials, Providence, USA 1999; 555

23. Oliveira AL, Elvira C, Vázquez B, San Roman J, Reis RL. Surface modifications tailors the characteristics of biomimetic coatings nucleated on starch based polymers. J Mater Sci: Mater in Med 1999; 10:827-835

24. Mano JF, Vaz CM, Mendes SC, Reis RL, Cunha AM. Dynamic mechanical properties of hydroxylapatite reinforcement and porous starch based degraded biomaterials. J Mater Sci: Mater in Med 1999; 10:857-862

25. Pereira CS, Cunha AM, Reis, RL, Vázquez B, San Roman J. New starch-based thermoplastic hydrogels for use as bone cements or drug-delivery carriers. J Mater Sci: Mater in Med 1998; 9: 825-833.

26. Pereira CS, Gomes ME, Reis RL, Cunha AM. Hard cellular materials in the human body: properties and production of foamed polymers for bone replacement. NATO/ASI Series,1998. Ed. N. Rivier, J. F. Sadoc, Kluwer Press, Drodercht; 193-204

27. Gomes M.E., Ribeiro A.S., Malafaya P.B., Reis R. L., Cunha A.M., A new approach based on injection moulding to produce biodegradable starch-based polymeric scaffolds: morphology, mechanical and degradation behaviour, Biomaterials 2001, 22:883-889

28. Gomes M.E., Reis R.I., Cunha A.M., Blitterswijk C. A., Bruijn J.D., Cytocompatibility and response of osteoblastic-like cells to starch-based polymers: effect of several additives and processing conditions, Biomaterials, 2001, 22:1911-1017.

29. Mendes S, Bovell Y, Reis R, van Blitterswijk C, de Bruijn J. Biocompatibility testing of novel starch based polymers and composites with potential application in orthopaedic surgery. Biomaterials, 2001, 22:2057-2064.

30. Reis RL. , (1999) Starch and Polyethylene Based Bone-Analogue Composite Biomaterials. PhD Thesis. Univ. Minho. Portugal.

31. Vaz CM, Cunha AM, Reis RL. Degradation model of starch-EVOH/HA composites.Mat Res Innov, 2001, 4:375-380

32. Espigares I, Elvira C, Mano JF, San Román J, Reis RL. New biodegradable and bioactive acrylic bone cements based on starch blends and ceramic fillers", Biomaterials, 2002, in press.

33. Bastiolli C. Starch-polymer composites. In Degradable Polymers, Scott G, Gilead D (eds), 1995 Chapman & Hall, London. p 112-137

34. Bastioli C, Bellotti C, Del Giudici L, Gilli G. Mater-Bi: Properties and biodegradability. J Env Poly Deg, 1993; 1: 181-191

35. Bastioli C. The starch based thermoplastics. Recycle'94, Switzerland, 1994

36. Shefer A, Shefer S, Kost J, Langer R. Structural characterization of starch networks in the solid state by cross-polarization magic-angle-spinning 13C NMR spectroscopy and wide angle X-ray diffraction. Macromolecules, 1992; 25:6756-6760.

37. Bhattacharya M, Vaidya U, Zhang D, Narayan R. Properties of Blends of starch and synthetic polymers containing anhydride groups.II.effect of amylopectin to amylose ratio in starch. J App Pol Sci, 1995; 57: 539-554

38. Willet J, Jasberg B, Swanson C. Rheology of thermoplastic starch: effects of temperature, moisture content and additives on melt viscosity. Pol Eng and Sci, 1995; 35:202-210

39. Van Soest J, Borger D. Structure and properties of compression-molded thermoplastic starch materials from normal and high-amylose maize starch. J Appl Polym Sci, 1997; 64: 631-644

40. Van Soest J, Vliegenthart J. Cristallinity in starch plastics: consequences for material properties. TIBTECH, 1997; 15: 208-213

41. Ramkumar D, Bhattacharya M, Zhang D. High shear and transient viscosity of starch and maleated ethylene vinyl acetate copolymer blends. Polym Networks Blends, 1997; 7: 51-59

42. Langer R, Vacanti, JP. Tissue engineering. Science 1993; 260: 920-925.

43. Thomson R, Yaszemski M, Mikos A. Polymer Scaffold processing. In Lanza R, Langer R, Chick W, editors. Principles of Tissue Engineering. Academic Press, 1997. p 263-272.

44. Lu L, Mikos A. The importance of new processing techniques in tissue engineering. MRS Bulletin 1996; 21: 28-32.

45. Mikos AG, Thorsen AJ, Czerwonka LA, Bao Y, Langer RB. Preparation and characterization of poly(l-lactid acid) foams. Polymer 1994;1068 -1077

46. Mikos AG, Sarakinos G, Leite SM, Vacanti JP, Langer R. Laminated three-dimensional biodegradable foams for use in tissue engineering. Biomaterials 1993; 14: 323-330

47. Mikos AG, Bao Y, Cima LG, Ingeber DE, Vacanti JP, Langer RB. Preparation of poly(glycolic acid) bonded fiber structures for cell attachment and transplantation. J Biomed Mater Res 1993; 27: 183-189

48. Hinrichs W. Porous polymer structures for tissue regeneration. PhD Thesis. Univ. Twente. The Netherlands 1992.

49. Mooney DJ, Baldwin DF, Suh NP, Vacanti JP. Novel approach to fabricate porous sponges of poly(d,l-lactid-co-glycolic acid) without the use of organic solvents. Biomaterials 1996; 17: 1417- 1422

50. Thompson RC, Yaszemski MJ, Powders JM. Fabrication of biodegradable polymer scaffolds to engineer trabecular bone. J Biomat Sci-Polym Edn 1995; 7: 23-28

51. Hutmacher DW. Scaffolds in tissue engineering bone and cartilage. Biomaterials, 2000; 21:2529 – 2543.

52. Hutmacher DW, Teoh SH, Zein I, Renawake M, Lau S. Tissue engineering Research: the engineer's role. Med Dev Tech, 2000; 1:33-39.

53. Rotter N, Aigner J, Nauman A, Planck H, Hammer C, Burmester G, Sittinger M. Cartilage reconstruction in head and neck surgery: comparasion of resorbable polymer scaffolds for tissue engineering of human septal cartilage. J Biomed Mater Res 1998; 42:347-356.

54. Jiang G, Shi D. Coating of hidroxylapatite on highly porous Al_2O_3 substrate for bone substitutes. J Biomed Mater Res 1997; 43: 77-88

55. Guidoin M, Marois Y, Bejui J, Poddevin N, King M, Guidoin R. Analysis of retrived polymer fiber based replacements for the ACL. Biomaterials, 2000; 2461-2474

56. Vunjak-Novakovic G, Obradovic B, Martin I, Bursac P, Langer R, Freed LE. Biotchnology Progress, 1998;14: 193-202

57. Freed LE, Hollander A, Martin I, Barry J, Langer R, Vunjak-Novakovic G. Chondrogenesis in a cell-polymer-bioreactor system. Exp Cel Res, 1998; 240: 58-65

58. Holder W, Gruber H, Moore A, Culberson C, Anderson W, Burg K, Mooney D. Cellular ingrowth and thickness changes in poly-L-lactide and polyglycolide matrices implanted subcutaneously in the rat. J Biomed Mater Res, 1998; 41: 412-421

59. Gao J, Niklason L, Langer R. Surface hydrolysis of poly(glycolic acid) meshes increases the seeding density of vascular smooth muscle cells. J Biomed Mater Res, 1998; 42:417-424

60. Aigner J, Tegeler J, Hutzler P, Campoccia D, Pavesio A, Hammer C, Kastenbauer E, Naumann A. Cartilage tissue engineering with novel nonwoven structured biomaterial based on hyaluronic acid benzyl ester. J Biomed Mater Res, 1998; 42: 172-181

61. Rotter N, Aigner J, Naumann A, Planck H, Hammer c, Burmester G, Sittinger M. Cartilage reconstruction of resorbable polymer scaffolds for tissue engineering of human septal cartilage. J Biomed Mater Res, 1998; 42: 347-356

62. Sittinger M, Reitzel D, Dauner M, Hierlemann H, Hammer C, Kastenbauer E, Plank H, Burmester G, Bujia J. J Biomed Mater Res, 1996; 33:57-63

63. Agrawal C, McKinney J, Lanctot D, Athanasiou. Effects of fluid flow on the in vitro degradation kinetics of biodegradable scaffolds for tissue engineering. Biomaterials, 2000; 21:2443-2452

64. Lu L, Peter S, Lyman M, Lai HL, Leite S, Tamada J, Uyama S, Vacanti J, Langer R, Mikos A. In vitro and in vivo degradation of porous poly(DL-lactic-co-glygolic acid) foams. Biomaterials, 2000; 21: 1837-1845

65. Murphy W, Kohn D, Mooney D. Growth of continuos bonelike mineral within porous poly(lactid-co-glycolide) scaffolds in vitro. J Biomed Mater Res, 2000; 50: 50-58

66. Thompson R, Mikos A, Beahm E, Lemon J, Satterfiels W, Aufdemorte T, Miller M. Guided tissue fabrication from periosteum using preformed biodegradable polymer scaffolds. Biomaterials, 1999; 20:2007-2018

67. Patrick C, Chauvin P, Hobley J, Reece G. Preadipocyte seeded PLGA scaffolds for adipose tissue engineering. Tissue Eng, 1999; 5: 139-151

68. Laurencin C, El-Amin S, Ibim S, Willoughby D, Attawia M, Allcock H, Ambrosio A. A ighly porous 3-dimensional polyphosphazene polymer matrix for skeletal tissue regeneration. J Biomed Mater Res, 1996; 30:133-138

69. Holy CE, Shoichet M, Davies J. engineering three-dimensional bone tissues in vitro using biodegradable scaffolds: investigating initial cell-seeding density and culture period. J Biomed Mater Res, 2000; 51:376-382

248

70. Kim H, Smith J, Valentini R. Bone morphogenetic protein 2-coated porous poly-L-lactic acid scaffolds: release kinetics and induction of pluripotent C3H10T1/2 cells. Tissue Eng, 1998; 4: 35-51

71. Mooney D, Park S, Kaufmann P, McNamara K, Vacanti J, Langer R. biodegradable sponges for hepatocyte transplantation. J Biomed Mater Res, 1995; 29: 959-965

72. Rivard C, Chaput C, DesRosiers E, Yahia L, Selmani A. Fibroblast seeding and culture in biodegradable porous substrates. J Biomed Mater Res, 1995; 6:65-68

73. Gutsche A, Lo H, Zurlo J, Yager J, Leong K. Engineering of a sugar-derivatized porous network for hepatocyte culture. Biomaterials, 1996; 17:387-393

74. Temenhoff JS, Mikos AG. Injectable materials for orthopaedic tissue engineering. Biomaterials, 2000; 21:2405-2412

75. He S, Yaszemski M, Yasko A, Engel P, Mikos A. Injectable biodegradable polymer composites based on poly(propylene fumarate) crosslinked with poly(ethylene glycol)-dimethacrylate. Biomaterials, 2000; 20:2389-2394

76. Heath CA. Cells for tissue engineering. TIBTECH 2000; 18: 17-19.

77. Platt JL. The immunological barriers to xenotransplantation. Critc Rev Immunol 1996; 16:331-358.

78. Blau HM, Brazelton TR, Weimann JM. The evolving concept of a stem cell: Entity or Function. Cell 2001; 105: 829-841.

79. Roche PA, Grodin MA. The ethical challenge of stem cell research. Women's Health Issues 2000; 10 (3): 139.

80. Friedenstain AJ. Determined and inducible osteogenic precursor cells. In: Hard Tissue Growth, Repair and Remineralization. Amsterdam, Elsevier 1973; 11: 169-185.

81. Caplan AI, Bruder SP. Mesenchymal stem cells: Building blocks for molecular medicine in the 21st century. Trends in Mol Med, 2001; 7(6): 259-264.

82. Haynesworth SE, Goshima J, Goldberg VM, Caplan AI, Characterization of cells with osteogenic potential from human marrow. Bone 1992; 13:81-88.

83. Jaiswall N, Haynesworth SE, Caplan AI, Bruder SP. Osteogenic differentiation of purified, culture-expanded human mesenchymal stem cell in vitro. J Cell Biochem 1997; 64:295-312.

84. Pittinger MF, Mackay AM, Beck SC, Jaiswal RK, Douglas R, Mosxca JD, Moorman MA, Simonetti DW, Craig S, Marshak DR. Multilineage potential of adult mesenchymal stem cells. Science 1999; 284: 143-147.

85. Bruder SP, Jaiwal N, Haynesworth SE. Growth Kinetics self rewal, and the osteogenic potential of purified human mesenchymal stem cells during extensive subcultivation and following cryopreservation. J Cell Biochem 1997; 64(2); 278-294.

86. Mackay AM, Beck SC, Murphy JM, Barry FP, Chichester CO, Pittenger MF. Chondrogenic differentiation of cultured human mesenchymal stem cells from marrow. Tissue Eng 1998; 4(4): 415-428.

87. Liechty KW, MacKenzie TC, Shaaban AF, Radu A., Moseley AM, Deans R, Marshak R., Flake AW. Human mesenchymal stem cells engraft and demonstrate site-specific differentiation after in utero transplantation in sheep. Nat Med 2000; 6: 1282-1286.

88. Kopen GC, Prockop DJ, Phinney DG. Marrow stromal cells migrate throughout forebrain and cerebellum, and they differentiate into astrocytes after injection into neonatal mouse brains. Proc. Nat. Acad. Sci. 1999; 96:10711-10716.

89. Sanchez-Ramos J, Song S, Cardozo-Pelaez F, Hazzi C, Stedeford T, willing A, Freeman TB, Saporta S, Janssen W, Patel N. Adult bone marrow stromal cells differentiate into neural cells *in vitro*. Exp Neur 2000; 164: 247-256.

90. Woodbury D, Schwarz EJ, Prockop DJ, Black IB. Adult rat and human bone marrow stromall cells differentiate into neurons. J Neurosci Res; 151: 55-69.

91. Jaiswal RK, Jaiswal N, Bruder Sp, Mbalaviele G, Marshak DR, Pittenger MF. Adult mesenchymal stem cell differentiation to the osteogenic and adipogenic lineage is regulated by mytogen-activated protein Kinase. J Biol Chem (2000); 275: 9645-9652.

92. Wobus AM. Potential of embryonic stem cells. Mol Asp Med (2001); 22:149-164.

93. Evans MJ, Kaufmann MH. Establishment in culture of puripotent cells from mouse embryos. Nature (1981); 292: 154-156

94. Buttery LD, Bourne S, Xynos JD, Wood H, Hughes FJ, Hughes Sp, Episkopou V, Polak JM. Differentiation of osteoblasts and *in vitro* bone formation from murine embryonic stem cells. Tissue Eng (2001); 7(1): 89-99.

95. Kramer J, Hegert C, Guan K, Wobus AN, Muller PK, Rohwedel J. Embryonic stem cell-derived chondrogenic differentiation in vitro: activation by BMP-2 and BMP-4. Mech of Devel (2000); 92: 193-205.

96. Salgado AJ ,. Gomes ME, Chou A, Coutinho OP, Reis RL, D.W. Hutmacher. Preliminary Study on the adhesion and proliferation of human osteoblasts on stach based scaffolds. Mat Sci Eng: C, 2001, in press.

97. Burkoth A, Anseth K. a review of photocrosslinked polyanhydrides: in situ forming degradable networks. Biomaterials. 21 (2000) 2395-2404.

98. Nam YS, Yoon JJ, Park TG. A novel fabrication method of Macroporous biodegradable polymer scaffolds using gas foaming salt as a porogen additive. J Biomed Mater Res:Appl Biomater. 53 (2000) 1-7

99. Rizzi SC, Health D.J., Coombes A.G.A, Bock N, Textor M, Downes S, Biodegradable polymer/hydroxyapatite composites: surface analysis and initial attachment of human osteoblasts, J Biomed Mater Res (2001) 55 475-486.

100. Gundle R, Stewart K, Screen J, Beresford JN, Isolation and culture of human bone-derived cells in Marrow Stromal Cell Culture, Cambridge University Press , Cambridge, 1998.

101. Aubin, JE Turksen K, Heersche JNM, Osteoblastic cell lineage in Cellular and molecular biology of bone *in* Cellular Molecular Biology edited by M. Noda, Academic Press, New York, 1994.

102. Marques AP, Reis RL,. Hunt JA, The biocompatibility of novel starch-based polymers and composites. Biomaterials (2002); 23: 1471-1478.

103. Marques AP, Reis RL, Hunt JA, Evaluation of the Potential of Starch-Based Biodegradable Polymers in the Activation of Human Inflammatory Cells, J. Materials Science: Materials in Medicine, (2002), *submitted*.

104. C. M. Alves, R. L. Reis, J. A. Hunt, "Preliminary Study on Human Protein Adsorption and Blood Cells Adhesion to Starch-Based Biomaterials", J. Materials Science: Materials in Medicine, (2002), *submitted*.

BONE TISSUE ENGINEERING BY CELL TRANSPLANTATION*

G.N. BANCROFT AND A.G. MIKOS
Department of Bioengineering, Rice University
6100 Main, Houston, TX, 77005-1892, U.S.A.

Abstract

Bone defects that require skeletal reconstruction are a common problem facing physicians today. These defects often arise as a result of trauma, tumor resection, infection, or skeletal abnormalities. Traditional therapies to overcome these bone deficits include bone grafts, bone cement, and synthetic bone substitutes including plastics, ceramics, and metals. Each of these has limitations that preclude it from being an ideal bone replacement, indicating a need for bone tissue engineering.

Tissue engineering seeks to regenerate this lost bone tissue by favorably utilizing the interactions between cells, growth factors, and scaffolding in novel bone tissue engineering constructs and therapies. Although different tissue engineering approaches may place emphasis on the cellular component, the involved growth factors, or implanted matrices, this review focuses on the use of cell transplantation to regenerate bone and enhance bone healing. By transplanting different cell populations including bone marrow cells, mesenchymal stem cells, osteoblasts, or genetically modified cells, researchers have shown the potential of these cell-based tissue engineering approaches to engineer new bone tissue. Because of this demonstrated ability, cell transplantation will undoubtedly play an important role in eventual tissue engineering therapies for treating bone defects.

1. Introduction

It is estimated that over one million surgeries to restore lost bone function are performed each year in the United States alone (1). Such surgeries become necessary whenever sufficient volume of bone has either been destroyed through injury or necessitated its removal due to infection, neoplastic growth, or other disease. Such defects often impart a cosmetic as well as functional deficit. In addition, non-union of long bones after fracture is a serious complication of impaired fracture healing, which can result in long term disability. Although non-union develops in only one to five percent of all diaphyseal fractures, it is estimated that 100,000 non-unions of long bones occur in the United States annually (2). In addition, because a significant number of

---

R.L. Reis and D. Cohn (eds.),
Polymer Based Systems on Tissue Engineering, Replacement and Regeneration, 251–263.
© 2002 *Kluwer Academic Publishers. Printed in the Netherlands.*

fractures occur in the young, the degree of disability is substantial. There is also a significant need for bone regeneration therapy in spinal arthrodesis, fixation of prosthetic implants, and restoration of maxillofacial structures (3). This has led to an estimated annual market for bone grafts of $500 million dollars in the United States alone (4).

Conventional therapeutic replacements for bony tissues involve bone autografts and allografts, poly(methyl methacrylate) (PMMA) bone cement, and preformed orthopedic hardware implants of various materials including metals, ceramics, and plastics (5). While each of these therapies offers certain benefits and advantages, they all have limitations and disadvantages that make them less than ideal. Autogenous bone provides osteoprogenitor cells, an osteoinductive matrix and mechanical support (6), but is scarce and available in limited amounts. It is difficult to use in irregular defects and can be resorbed by the body before osteogenesis is complete (7,8). In addition, morbidity rates at the donor site including pain, paresthesia, and infection can be as high as 8-10% (9). Allograft bone is more readily available, but is associated with inflammation and infection (10). In addition, the processing may cause a lessening or complete loss of osteoinductive factors (11). PMMA bone cement can be molded to fill defects of any shape, but particle formation as well as stress shielding, which can result in resorption or weakening of bone adjacent to the bone cement, can occur (12). If this occurs, surgical revision may be required, which is complicated by the fact the PMMA bone cement is nondegradable. Orthopedic hardware also presents problems with stress shielding in addition to being prone to eventual mechanical failure. And as most implants are not biodegradable, removal or revision of the implant requires removal of healthy bone tissue. Due to these limitations in current therapies, bone repair and replacement is an area that can benefit from applications of tissue engineering.

2. Overview of Bone

Bone is a complex specialized tissue characterized by its rigidity and hardness, properties vital to its main functions of providing mechanical support and protection and serving as a frame for the locomotion of the musculoskeletal system. In addition, it also acts as a reservoir of mineral salts and contains the marrow responsible for the production of hematopoietic cells.

On a microscopic level, bone is composed of support cells termed osteoblasts and osteocytes, a non-mineral matrix of collagen and glycosaminoglycans called osteoid, inorganic mineral salts deposited within this matrix, and remodeling cells called osteoclasts. Osteoblasts are responsible for secreting and nourishing the osteoid into which the mineral salts are deposited. These deposited mineral salts are responsible for the rigidity of bone. All osteocytes are derived from osteoblasts. When an osteoblast, after secreting osteoid and mineral salts, becomes isolated in a cavity of bony matrix, it becomes an osteocyte. Osteoclasts are constantly eroding deposited bone (the mineralized osteoid) while osteoblasts are constantly laying down osteoid and minerals to replace it. In this way bone is a dynamic tissue, being formed and broken down in a continuous cycle in response to physical and hormonal factors. It is through this continual turnover that bone can undergo remodeling to modify the underlying structure of the bone to meet physical stresses including fracture repair.

On a macroscopic level, bone exists in one of two forms: cortical bone and trabecular bone (also known as cancellous bone). Cortical bone is the dense bony tissue that makes up the outer tubular shell of the long bones and outer surface of the small and flat bones of the skeleton. Trabecular bone is of lesser density and is found near the ends of long bones, in the interior of small bones, and between the surfaces of flat bones. Cortical bone consists of parallel cylindrical units called osteons oriented along the long axis of the long bones. The orientation of the osteons determines the directionality of their mechanical properties. Trabecular bone is much less dense than cortical bone and can be considered to consist of an array of plates and rods of bony tissue that form an open-celled foam (13). The marrow resides in the space between this meshwork of internal bone struts in trabecular bone.

When bone is injured, a programmed cascade of events begins with an immediate acute inflammatory response (14). Disruption of blood vessels creates an initial hematoma at the site of injury. Neutrophils and macrophages migrate into this hematoma, phagocytosing necrotic debris and, along with other cell types, secreting cytokines and growth factors to create a complicated milieu of growth factors and cytokines responsible for inducing the chemotaxis and mitotic expansion of mesenchymal precursor cells in the surrounding tissues (15). In addition, ingrowth of capillaries and fibroblasts from surrounding tissues produce a fibrovascular granulation tissue. Additional mesenchymal osteoblastic and osteoprogenitor cells are recruited from several sources including the bone itself, the bone marrow, the periosteum (the outer lining of bone), and the pericytic cells of the nearby blood vessels. As these progenitor cells differentiate and aggregate at the site of injury, a repair blastema is formed. The differentiation of these precursor cells along an osteoblastic pathway proceeds, and osteoblasts derived from these mesenchymal precursor cells migrate into this fibrovascular tissue and begin the process of laying down new osteoid forming a bony callus. This initial rapid production of osteoid by the osteoblasts results in a haphazard organization of collagen fibers termed woven bone. As this mechanically weak woven bone is remodeled by the degradative effects of osteoclasts and reforming actions of osteoblasts, it is reshaped into regular parallel arrangements of collagen fibers in sheets characterizing the appearance of normal, non-pathologic lamellar bone. Successful bone tissue engineering should recapitulate these events and mimic this natural bone repair process. At times the body's own cells are insufficient in number or activity to effectively complete the repair process, and a fracture non-union occurs. In other instances, the defect size is too large for the body's own repair response to initiate healing on its own. It is in these situations that bone tissue engineering strategies can be used to generate bone at the defect site.

3. Bone Tissue Engineering Approaches

Almost all tissue engineering strategies revolve around the interactions of three components: cells, scaffold, and growth factors. Bone tissue engineering is no different, requiring a successful and productive interaction between osteoprogenitor or osteoblastic cells, osteoinductive growth factors, and an osteoconductive matrix. The osteoprogenitor or osteoblastic cells can be derived from the host tissues through recruitment or induction or can be transplanted to the defect site from an outside source. These cells, whatever the source, are directly responsible for helping to form the newly engineered bone tissue. The osteoinductive growth factors can also be delivered from an outside source or secreted by transplanted cells. These growth factors induce the proper recruitment, differentiation, and proliferation of the appropriate cell types at the defect site. The osteoconductive matrix can be used to provide mechanical support and to serve as a scaffold for attachment and growth of the cellular component guiding the regenerating bone tissue as well as serving as a delivery vehicle for transplanted cells or growth factors. The interactions between these three components in the defect site are substantial. Growth factors can alter the cellular activity of the involved cells such that other growth factors are secreted. These secreted growth factors can be sequestered on the implanted matrix affecting their orientation and presentation to other cells. Proliferating cells can modify the scaffold by producing extracellular matrix proteins on it. The degradation products of the scaffold can alter the cellular environment, and so on. These interactions continuously go back and forth as the engineered tissue is formed. Because of the importance of these three components, tissue engineering approaches to bone tissue engineering have developed around the same three factors: matrix-based strategies, growth factor based strategies, and cell-based strategies. However, there is a blending of approaches even within these categories as transplanted cells in cell-based strategies are often carried on a matrix of some sort or altered to secrete a growth factor. Consequently, more advanced bone tissue engineering constructs combine these approaches.

In matrix-based strategies, a space-filling scaffold of some sort is implanted in the defect. A wide variety of materials have been investigated and proposed for use in bone defects including, but not limited to, metals such as titanium fiber, ceramics such as hydroxy apapitite or tricalcium phosphate, polymers, polymer/ceramic composites, and even marine coral (16-22). Once implanted, such matrices are dependent on the infiltration and recruitment of appropriate osteoprogenitors from the body secreting their own growth factors to facilitate a proper repair. While such matrices offer some advantages as they are readily available and usable "off-the-shelf", their lack of osteoinductivity remains a real hindrance to their use by themselves. However, such scaffolds can be utilized as carriers for growth factors or matrices for transplanted cells.

In growth factor-based strategies, one or more purified or extracted growth factors are delivered to the defect site to induce bone tissue growth and repair. Such therapies use the osteoinductivity of the chosen factor to enhance the proliferation and differentiation of cells present in the defect site to cause bone tissue formation. Numerous growth factors have been employed including Bone Morphogenetic Proteins (BMPs), Transforming Growth Factor-ß (TGF-ß), and Platelet-Derived Growth Factor (PDGF) (23,24). The success of this approach has been shown in many animal models and pre-clinical trials, but questions about dosing and scale-up to humans remain valid concerns. In addition, the high cost needed for such large doses and limited availability of some of the proteins, as well as the lack of an optimal carrier for delivery of these growth factors, represent hindrances to their widespread implementation.

The remainder of this review will focus on the last approach, cell-based strategies that use cell transplantation for bone tissue engineering. In this approach, cells with osteogenic potential and/or osteoinductive ability are transplanted into a bone defect site typically on a scaffold or with some carrier. By their own actions and/or by recruiting other cells, the transplanted cells contribute to healing at the defect site by forming new bone. Cell transplantation could be especially beneficial for patients in whom the bone wound bed is lacking either in appropriate cellular activity, such as patients with diabetes, osteoporosis, and aging, or cell number, such as patients with vascular disease or irradiation after tumor resection (25,26). Studies investigating the ability of transplanted cells to engineer new bone have typically focused on the transplantation of four cell populations: fresh bone marrow, mesenchymal stem cells (MSCs) that have been selectively expanded in culture, committed differentiated osteoblasts, and cells of various lineages that have been genetically modified to promote bone formation.

3.1 TRANSPLANTS OF FRESH BONE MARROW

The osteogenic capabilities of bone marrow have been know for some time (27). This is due to the presence of osteoprogenitor cells within fresh bone marrow (28). Because of this, transplant of fresh bone marrow cells offered the first non-bone source of a cellular component capable of affecting bone regeneration. This capability has been demonstrated in several studies in different animal models (29-32) with Werntz et al showing that autologous marrow has the ability to heal a segmental bone defect in a rat model (33). This study found that bone marrow, when combined with collagen, could give a rate of union comparable to that of autogenous bone grafting. This new bone formed by the transplanted marrow also had similar mechanical properties to those of cancellous bone grafts.

Connolly et al demonstrated the further clinical osteogenic ability of bone marrow to successfully heal fracture non-unions (34, 35). In this approach, marrow was harvested from the iliac crest of a patient with a fracture non-union. This marrow, with or without a carrier, was injected through a fluoroscope-positioned needle into the site of the non-union. As the morbidity of bone marrow harvesting is very low, bone marrow can be used to augment all currently used synthetic grafts, allografts, and other biomaterials in clinical use. Although it has only seen limited clinical use, bone marrow could conceivably add to the osteogenic capabilities of implants and help to establish a more normal fracture hematoma, especially after extensive irrigation (23).

More recent studies that utilized a combination of fresh marrow transplants with an osteoinductive growth factor, recombinant human BMP-2 (rhBMP-2), have further extended the osteogenic abilities of fresh bone marrow. Lane examined the ability of bone marrow alone, rhBMP-2 alone, and a combination of the two on an inert biodegradable polymer scaffold to heal a critical-sized segmental defect in a rat model (36). The results of this study showed the beneficial synergy that resulted when the marrow and rhBMP-2 were combined together rather than each used alone. When used together, marrow and rhBMP-2 showed enhanced performance over each one used alone and even over autogenous cancellous bone in terms of bone formation, union, and biomechanical strength. Tomin et al combined marrow and rhBMP-2 in a vascularized silicone mold, using the bone marrow as a cellular source to grow bone units at an ectopic site for transplantation (37). The osteoprogenitor cells in the marrow under the influence of the rhBMP-2 formed bony ossicles that were then used in critical sized femoral defects in rats. Samples from the ectopic mold sites were harvested at two and four weeks and transplanted to the femoral defect site. At two weeks, the ossicles consisted primarily of woven bone while ossicles at four weeks had already progressed to a lamellar structure. Interestingly, ossicles harvested at two weeks proved better at healing the segmental defect, and performed on par with syngeneic bone graft in terms of bone formation, union, and biomechanical strength. Ossicles harvested at four weeks gave an unsatisfactory repair with lesioned cystic bone forming in the femoral shaft. Such studies shed insight on potential future therapies and demonstrate not just the osteoinductivity of rhBMP-2, but the added value of cell transplantation to bone tissue engineering strategies, reinforcing the idea that eventual therapies will incorporate refined aspects of matrix-, growth factor-, and cell-based approaches.

3.2 TRANSPLANTATION OF MESENCHYMAL STEM CELLS

Although fresh bone marrow contains osteoprogenitor cells, a key limitation is the relatively low concentration of these cells in the marrow being at most 0.001% of the cellular component in a marrow sample (38,39). In addition, the amount of these cells in a given person declines with age such that progenitor cells represent about 1 in 10,000 marrow cells in a newborn to 1 in 2,000,000 marrow cells in an elderly patient. It is believed that this decrease in progenitor cells accounts for the decreased rate of bone repair in elderly patients (40).

Because of this limitation of fresh bone marrow, techniques have been developed to select and expand the population of progenitor cells from bone marrow from humans and animals (39,41-43). Termed mesenchymal stem cells (MSCs), these multipotent cells are capable of differentiating into several mesenchymal cell types including osteoblasts, chondrocytes, and adipodcytes (44). Thought to be the progenitor cells in marrow that proliferate and differentiate to form new tissue in response to injury or exogenous factors like rhBMP-2, such cells hold great promise for regeneration of other tissues besides bone. These MSCs, once isolated from a marrow sample, can be expanded in culture until a sufficient number desirable for transplantation has been reached. Cultured MSCs have been passaged through 30 population doublings *in vitro* representing over a billion-fold expansion without loss of osteogenic potential (45). When seeded onto porous ceramic cubes and implanted in nude mice to examine the MSCs *in vivo* osteogenic potential, constructs loaded with MSCs formed bone much earlier than cubes loaded in a comparable manner with fresh bone marrow cells, demonstrating the benefits of concentrating the progenitor population present in bone marrow (46).

Culture-expanded MSCs from different animals and even humans have demonstrated their ability to satisfactorily heal osseous defects when transplanted into bone defect sites. Kadiyala et al transplanted MSCs seeded on a ceramic scaffold to treat segmental femur defects in rats (47). Syngeneic rat culture-expanded MSCs were loaded into a porous hydroxyapatite/ß-tricalcium phosphate (HA/ ß-TCP) cylinder and transplanted into a critical-sized 8 mm defect in a rat femur. After 8 weeks, significant amounts of bone were present within the pores of the MSC-seeded constructs while cell-free constructs demonstrated only fibrovascular tissue with negligible new bone formation. Importantly, new bone growth at the interface between the MSC-seeded constructs and the defect edges resulted in a continuous length of bone bridging the defect. In addition, only MSC-seeded constructs succeeded in forming a periosteal callus thereby demonstrating that transplanted progenitor cells can recapitulate the normal bone healing process.
A later study showed this capability in a larger animal model (48). Bruder et al transplanted autologous culture-expanded MSCs seeded on a porous HA/ß-TCP scaffold into a 21 mm osteoperiosteal defect in a canine segmental femur defect model. After 16 weeks, untreated defects resulted in non-unions with minimal bone formed at the severed ends of the defect. However, in animals that received transplanted MSCs, union was demonstrated radiographically. And as in the previously mentioned study, a periosteal callus continuous with the normal adjacent periosteum formed around the implant, eventually spanning the defect and remodeling to a collar of bone integrating with both ends of the defect. This was not seen in animals that received cell-free HA/ß-TCP implants. This again demonstrates the ability of transplanted cells to affect a repair process mimicking the natural repair process.

In a different approach, Richards et al. used cultured marrow progenitor cells to enhance bone formation in rats during distraction osteogenesis (49). Rats received a single injection of undifferentiated marrow stromal cells in a collagen gel carrier in the distraction gap in a limb undergoing distraction osteogenesis anywhere from 6 to 18 days after surgery. Limbs receiving single injections of the gel/cell solution formed on average over 20% more bone volume by than limbs receiving injections of the gel alone by 36 days after surgery. This study demonstrates that marrow progenitor cells without a scaffold can contribute to bone formation and regeneration.

The ability of culture-expanded MSCs isolated from humans to heal a bone defect has also been demonstrated (50). In a manner similar to that mentioned above, human

MSCs were seeded onto porous ceramic cylinders and implanted into a segmental defect in the femurs of athymic rats. By 12 weeks after implantation, bone union had occurred in the animals receiving MSC-seeded constructs. In addition, mechanical testing revealed the superior mechanical properties of the resulting bone in the MSC-seeded constructs. Experiments like this demonstrate the future potential of stem cell transplantation in human therapies for bone tissue engineering.

3.3 TRANSPLANTATION OF OSTEOBLASTS

As previously mentioned, MSCs have the innate ability to differentiate into different mesenchymal cell types, including chondrocytes, adipocytes, and osteoblasts. Under certain specific culture conditions using defined media supplements, MSCs can not only be selected and expanded but also differentiated into these different cell types during *in vitro* culture. It has been shown that culturing MSCs in media supplemented with dexamethasone, ascorbic acid, and ß-glycerophosphate initiates a development process that directs the cells down an osteoblastic differentiation pathway (38, 44). It thus becomes possible to harvest a small sample of marrow from a patient, select and expand the MSC population, differentiate these cells into osteoblasts, and have a specific osteogenic cell collection available for transplantation and use in bone tissue engineering applications (38).

Predifferentiating MSCs along the osteoblastic pathway prior to transplantation has been shown to accelerate the formation of *in vivo, de novo* bone when implanted. When predifferentiated stromal osteoblasts are seeded onto porous ceramic cubes and implanted subcutaneously, they began forming bone 1 week after implantation. This is much earlier than was found with comparable constructs seeded with fresh bone marrow or undifferentiated MSCs (51). Similar results were found when differentiated rat bone marrow-derived osteoblasts were seeded onto poly(DL-lactic-*co*-glycolic acid) (PLGA) scaffolds and implanted into the mesentery of syngeneic rats (52). Within a week, ectopic bone tissue formed within the pores of the scaffold. Such work is especially exciting because PLGA is a true biodegradable material unlike ceramics like hydroxyapatite that can take many years to resorb. Experiments like this point the way to eventual therapies involving cell transplantation where the scaffolding material degrades leaving behind only the regenerated bone tissue.

Because of such strong osteogenic potential, several investigators have been investigating different methods to utilize culture-derived osteoblasts in therapeutic applications. One approach is to use predifferentiated osteoblasts in a manner analogous to those of undifferentiated MSCs, that is, loading a scaffold with culture-derived osteoblasts and implanting it in a defect site. The ability of predifferentiated osteoblasts to generate new bone when transplanted was demonstrated by Breitbart et al, who showed that predifferentiated osteoblasts derived from rabbit periosteum transplanted on a resorbable polymer scaffold into a full thickness cranial defect could generate significant amounts of new bone (53). A novel approach towards regenerating new bone using transplanted predifferentiated osteoblasts has been proposed by Yoshikawa et al (54). In their investigation, rat bone marrow stromal cells were cultured on hydroxyapatite cubes in osteogenic media to differentiate the cells into osteoblasts. These constructs were then implanted subcutaneously in syngeneic rats. Implants harvested at 52 weeks showed continuous evidence of osteogenesis with bony tissue filling the pores undergoing active remodeling by both osteoblasts and osteoclasts and even having active hematopoietic marrow tissue inside the constructs. This morphology is similar to cancellous bone. Based on these results, the investigators propose using similar methods to generate ectopic sites of tissue engineered cancellous bone for use in bone grafting therapies.

Other researchers have proposed harnessing the *in vitro* capabilities of cultured osteoblasts (55). Osteoblasts, as previously described, secrete and mineralize osteoid. This resulting mineralized extracellular matrix is specific to bone and contains biological components (like BMPs, TGF-ß, and other growth factors and cytokines) to facilitate osteoinduction. Ohgushi and Caplan have proposed a process which they term "osteogenic matrix coating" to enhance the integration of orthopedic implants (55). Harvested MSCs from a bone marrow sample from a patient could be differentiated into osteoblasts *in vitro* and cultured on the surface of orthopedic prostheses prior to implantation. During this culture period, the osteoblasts would deposit an osteogenic extracellular matrix on the implant, which they propose would then enhance integration and prevent loosening once the prosthesis is implanted into the patient. This would offer the superior osteoinductive ability of transplanting osteoblasts and their matrix to defect sites. Recent work on advanced culturing techniques is extending this principle not just to the outer surfaces of implants but also to entire porous constructs. Goldstein et al, using a perfusion bioreactor, cultured stromal osteoblasts in porous PLGA scaffolds, obtaining not only a potential enhancement of the osteoblastic differentiation of the cells, but also a more uniform distribution of cells throughout the scaffold than was obtained with conventional tissue culture techniques (56). Experiments like this offer the potential of generating an "osteogenic matrix coating" of osteoblasts and osteoinductive extracellular matrix not just on non-degradable prostheses but also on and throughout porous biodegradable polymer tissue engineering constructs. This could add an osteoinductive ability to the already beneficial physical and degradative properties of synthetic polymer scaffolds used for cell transplantation (57). Different and novel approaches like these mentioned above demonstrate the future potential utility of osteoblast transplantation in bone tissue engineering.

3.4 TRANSPLANTATION OF GENETICALLY MODIFIED CELLS

The last cell population to be discussed represents a blurring of the lines between cell transplantation-based therapies and growth factor-based therapies for bone tissue engineering. Genetically modified cells are engineered to produce osteoinductive factors that may recruit, proliferate, differentiate, and/or enhance the osteogenic activity of the patient's own cells in a paracrine fashion or even act upon the originally transplanted cells in an autocrine fashion causing themselves to form new bone. Due to the importance of the latter, most investigators have used modified stromal cells in their experiments.

This approach was taken by Lieberman et al, who used a mice stromal cell line transduced to secrete rhBMP-2 (58). When transplanted to a femur defect in mice, these cells induced the formation of coarse trabecular bone complete with new cortical formation in the defect site. This was superior to the resulting delicate lace-like trabecular bone lacking real cortical formation obtained with implantation of the protein itself, rhBMP-2, in a devitalized bone matrix carrier. These results demonstrate not just the advantage of transplanting genetically modified cells, but also the importance of delivery kinetics in growth-factor-based therapies. Other researchers performing similar studies have demonstrated the osteogenic abilities of cells genetically modified to secrete rhBMP-2 (59).

Using another one of the BMPs, Krebsbach et al was able to use cells genetically modified to express BMP-7 to successfully treat a critical size defect in rat calvaria (60). Especially unique to this work was the selection of dermal fibroblasts as the cell transduced. Krebsbach et al found that the BMP-7 secreting cells not only aided in osteogenesis, but also differentiated into osteoblast-like cells themselves. Such work highlights the power of both BMPs and genetic engineering and expands the cells usable for transplantation in bone tissue engineering to cells not usually considered osteogenic.

In a different approach, Boden et al used rat marrow cells transfected with cDNA for LIM mineralization protein-1 (LMP-1) to induce lumbar spine fusion (61). LMP-1 is not a secreted protein, rather it is an intracellular signaling molecule implicated as being very early in the order of proteins involved in the osteoblastic differentiation cascade, and it acts intracellularly to affect osteoblastic differentiation. Its osteogenic ability is thought to stem not only from its positive effects on cells transfected, but also because transfected cells express and secrete a variety of osteoinductive factors and cytokines (61,62). Work like this expands the field of genetic engineering of cells for transplantation in bone tissue engineering from moving from secretion of one specific growth factor to invoking the osteoinductive power of multiple proteins involved in osteogenesis.

4. Conclusions and Future

This discussion has focused on bone tissue engineering approaches that used cell transplantation as an integral aspect. Undoubtedly, some of the approaches discussed above and their subsequent modifications will be used in actual therapies that enter clinical use. Although this review was focused, finalized bone tissue engineering strategies will probably not be confined to solely matrix-, growth factor-, or purely cell-based therapies, but incorporate optimized aspects of each. Refinements in scaffold designs and materials, enhanced understanding and selection of bone growth factors, and advancements in culturing techniques and methodologies will be combined to create new tissue engineered constructs for regenerating bone for patients who are unable to currently receive satisfactory treatment. Cell transplantation will certainly play an important role in these future therapies.

References

1. Langer R. and Vacanti J.P. (1993) Tissue Engineering, *Science* **260**, 920.
2. Calandruccio R. (1981) *Musculoskeletal System Research: Current and Future Research Needs*, American Academy of Orthopaedic Surgeons, Chicago.
3. Bruder S.P. and Fox B.S. (1999) Tissue Engineering of Bone: Cell Based Strategies, *Clin. Orthop.* 376S, S68-S83.
4. Tihanksy C.E. and Fayemi W.M. (1996) *Going for Broke: Orthopaedic Biomaterials for the Twenty-First Century*, Medical Device Industry Report. Genesis Merchant Group Securities, San Fransisco.
5. Peter S.J., Miller S.T., Zhu G., Yasko A.W. and Mikos A.G. (1998) In vivo degradation of a poly(propylene fumarate)/beta-tricalcium phosphate injectable composite scaffold, *J. Biomed. Mater. Res.* **41**, 1-7.
6. Burwell R.G. (1994) History of bone grafting and bone substitutes with special reference to osteogenic induction, in M.R. Urist and R.G. Burwell (eds.), *Bone Grafts, Derivatives and Substitutes*, Butterworth-Heinemann Ltd., Oxford, pp. 3-102.
7. Brown L.K.B. and Cruess R.L. (1982) Bone and Cartilage transplantation surgery, *J. Bone Jt. Surg. Am.* **64A**, 270-279.
8. Enneking W.F., Eady J.L. and Burchardt H. (1980) Autogenous cortical bone grafts in the reconstruction of segmental skeletal defects, *J. Bone Jt. Surg. Am.* **62A**, 1039-58.
9. Younger E.M. and Chapman M.W. (1989) Morbidity at bone graft donor sites, *J. Orthop. Trauma* **3**, 192-195.
10. Strong D.M., Friedlaender G.E., Tomford W.W., Springfield D.S., Shives T.C., Burchardt H., Enneking W.F. and Mankin H.J. (1996) Immunologic responses in human recipients of osseous and osteochondral allografts, *Clin. Orthop.* **326**, 107-114.
11. Bostrom R.D. and Mikos A.G. (1997) Tissue Engineering of Bone, in A. Atala, D. Mooney, J.P. Vacanti and R. Langer (eds.), *Synthetic biodegradable polymer scaffolds*, Birkhauser, Boston, pp. 215-234.
12. Spector M. (1992) Biomaterial failure, *Orth. Clin. North Am.* **23**, 211-217.
13. Gibson L.J. (1985) The mechanical behavior of cancellous bone, *J. Biomech.* **18**, 317-328.
14. Black J. (1992) The inflammatory process, in J. Black (ed.) *Biological performance of materials*, Marcel Dekker, New York, pp. 125-145.
15. Joyce M.E., Jingushi S., Scully S.P. and Bolander M.E. (1991) Role of growth factors in fracture healing, *Prog. Clin. Biol. Res.* **365**, 391-416.
16. Wolff D., Goldberg V.M. and Stevenson S. (1994) Histomorphometric analysis of the repair of a segmental diaphyseal defect with ceramic and titanium fibermetal implants: Effects of bone marrow, *J. Orthop. Res.* **12** 439-446.
17. Holmes R.E., Bucholz R.W. and Mooney V. (1987) Porous hydroxyapatite as a bone graft substitute in diaphyseal defects: A histometric study, J. Orthop. Res. **5**, 114-121.
18. Johnson K.D., Frierson K.E., Keller T.S., Cook C., Scheinberg R., Zerwekh J., Meyers L. and Sciadini M.F. (1996) Porous ceramics as bone graft substitutes in long bone defects: A biomechancal, histological, and radiographic analysis, *J. Orthop. Res.* **14**, 351-369.

262

19. Hutmacher D.W. (2000) Scaffolds in tissue engineering bone and cartilage, *Biomaterials* **21**, 2529-2543.
20. Marra K.G., Szem J.W., Kumta P.N., DiMilla P.A. and Weiss L.E. (1999) In vitro analysis of biodegradable polymer blend/hydroxyapatite composites for bone tissue engineering, *J. Biomed. Mat. Res.* **47**(3), 324-335.
21. Zhang R. and Ma P.X. (1999) Porous poly(L-lactic acid)/apatite composites created by biomimetic process, *J. Biomed. Mater. Res.* **45**(4), 285-293.
22. Guilelmin G., Patat J.L., Fournie J. and Chetail M. (1987) The use of coral as a bone graft substitute, *J. Biomed. Mater. Res.* **21**, 557-567.
23. Lane J.M., Tomin E. and Bostrum M.P.G. (1999) Biosynthetic bone grafting, *Clin. Orthop.* **367S**, S107-S117.
24. Boden S.D. (1999) Bioactive factors for bone tissue engineering, *Clin. Orthop.* **367S**, S84-S94.
25. Inoue K., Ohgushi H., Yoshikawa T., Okumura M., Sempuku T., Tamai S. and Dohi Y. (1997) The effect of aging on bone formation in porous hydroxyapatite: Biochemical and histological analysis, *J. Bone Min. Res.* **12**, 989-994.
26. Quarto R., Thomas D. and Liang T. (1995) Bone progenitor cell deficits and the age-associated decline in bone repair capacity, *Calcif. Tissue Int.* **56**, 123-129.
27. Burwell R. (1964) Studies in the transplantation of bone VII. The fresh composite homograft-autograft of cancellous bone: An analysis of factors leading to osteogenesis in marrow transplants and in marrow-containing bone grafts, *J. Bone Joint Surg.* **46B**, 110-140.
28. Owen M. (1985) Lineage of osteogenic cells and their relationship to the stromal system, in W.J. Peck (ed.) *Bone and Mineral* (3rd edition), Elsevier, Amsterdam, pp. 1-25.
29. Grundel R.E., Chapman M.W., Yee T. and Moore D.C. (1991) Autogenic bone marrow and porous biphasic calcium phosphate ceramic for segmental bone defects in the canine ulna, *Clin. Orthop.* **266**, 244-258.
30. Ohgushi H., Goldberg V.M. and Caplan A. (1989) Repair of bone defects with marrow cells and porous ceramics, *Acta Orthop. Scand.* **60**, 334-339.
31. Paley D., Young M.C., Wiley A.M., Fornasier V.L. and Jackson R.W. (1986) Percutaneous bone marrow grafting of fractures and bony defects: An experimental study in rabbits, *Clin. Orthop.* **208**, 300-312.
32. Tiedeman J.J., Connolly J.F., Strates B.S. and Lippiello L. (1991) Treatment of nonunion by percutaneous injection of bone marrow and demineralized bone matrix. An experimental study in dogs, *Clin. Orthop.* **268**, 294-302.
33. Werntz J.R. (1996) Qualitative and quantitative analysis of orthotopic bone regeneration by marrow, *J. Orthop. Res.* **14**, 85-93.
34. Connolly J., Guse R., Lippiello L. and Dehne R. (1989) Development of an osteogenic bone-marrow preparation, *J. Bone Joint Surg.* **71A**, 684-691.
35. Connolly J., Guse R., Tiedeman J. and Dehne R. (1991) Autologous marrow injection as a substitute for operative grafting of tibial nonunions, *Clin. Orthop.* **266**, 259-270.
36. Lane J.M. (1999) Bone marrow and recombinant human bone morphogenetic protein-2 in osseous repair, *Clin. Orthop.* **361**, 216-227.
37. Tomin E., Lane J., Nakamichi K., Hsu J., Schneider K. and Weiland A. (1999) Performance of molded vascularized bone grafts induced by BMP-2 and marrow in a rat segmental defect model, *Trans. Orthop. Res. Soc.* **24**, 621.
38. Bruder S.P., Jaiswal N. and Haynesworth S.E. (1997) Growth kinetics, self-renewal and the osteogenic potential of purified human mesenchymal stem cells during extensive subcultivation and following cryopreservation, *J. Cell. Biochem.* **64**, 278-294.
39. Haynesworth S.E., Goshima J., Goldberg V.M. and Caplan A.I. (1992) Characterization of cells with osteogenic potentiall from human marrow, *Bone* **13**, 69-80.
40. Haynesworth S.E., Goldberg V.M. and Caplan A.I. (1994) Diminution of the number of mesenchymal stem cells as a cause for skeletal aging, in J.A. Buckwater, V.M. Goldberg and S.L.Y. Woo (eds.) *Musculoskeletal soft-tissue aging: Impact on mobility*, Am Acad Ortho Surg, Chicago, pp. 79-87.
41. Kadiyala S., Young R.G., Thiede M.A. and Bruder S.P. (1997) Culture-expanded canine mesenchymal stem cells possess osteochondrogenic potential in vivo and in vitro, *Cell Transplant.* **6**, 125-134.
42. Lennon D.P., Haynesworth S.E., Bruder S.P., Jaiswal N. and Caplan A.I. (1996) Development of a serum screen for mesenchymal progenitor cells from bone marrow, *In Vitro Cell Dev. Biol.* **32**, 602-611.
43. Maniatopoulos C., Sodek J. and Melcher A.H. (1988) Bone formation in vitro by stromal cells obtained from marrow of young adult rats, *Cell Tissue Res.* **254**, 317-330.

44. Pittenger M.F., Mackay A.M., Beck S.C., Jaiswal R.K., Douglas R., Mosca J.D., Moorman M.A., Simonetti D.W., Craig S. and Marshak D.R. (1999) Multi-lineage potential of adult human mesenchymal stem cells, *Science* **284**, 143-147.
45. Jaiswal N., Haynesworth S.E., Caplan A.I. and Bruder S.P. (1997) Osteogenic differentiation of purified, culture-expanded human mesenchymal stem cells in vitro, *J. Cell. Biochem.* **64**, 295-312.
46. Goshima J., Goldberg V.M. and Caplan A.I. (1991) The osteogenic potential of culture-expanded rat marrow mesenchymal cells assayed in vivo in calcium phosphate ceramic blocks, *Clin. Orthop.* **262**, 298-311.
47. Kadiyala S., Jaiswal N. and Bruder S.P. (1997) Culture-expanded, bone marrow-derived mesenchymal stem cells can regenerate a critical-sized segmental bone defect, *Tissue Eng.* **3**, 173-185.
48. Bruder S.P., Kraus K.H., Goldberg V.M. and Kadiyala S. (1998) The effect of implants loaded with autologous mesenchymal stem cells on the healing of canine segmental bone defects, *J. Bone Joint Surg.* **80A**, 985-996.
49. Richards M., Huibregtse B.A., Caplan A.I., Goulet J.A. and Goldstein S.A. (1999) Marrow-derived progenitor cell injections enhance new bone formation during distraction, *J. Clin. Orthop. Res.* **17**, 900-908.
50. Bruder S.P., Kurth A.A., Shea M., Hayes W.C., Jaiswal N. and Kadiyala S. (1998) Bone regeneration by implantation of purified, culture-expanded human mesenchymal stem cells, *J. Orthop. Res.* **16**, 155-162.
51. Yoshikawa T., Ohgushi H. and Tamai S. (1996) Immediate bone forming capability of prefabricated osteogenic hydroxyapatite, *J. Biomed. Mater. Res.* **32**, 481-492.
52. Ishaug-Riley S.L., Crane G.M. , Gurlek A., Miller M.J., Yasko A.W., Yaszemski M.J. and Mikos A.G. (1997) Ectopic bone formation by marrow stromal osteoblast transplantation using poly(DL-lactic-co-glycolic acid) foams implanted into the rat mesentery, *J. Biomed. Mater. Res.* **36**, 1-8.
53. Breitbart A.S., Grande D.A., Kessler R., Ryaby J.T., Fitzsimmons R.J. and Grant R.T. (1998) Tissue engineered bone repair of calvarial defects using cultured periosteal cells, *Plastic and Reconstructive Surgery* **101**, 567-574.
54. Yoshikawa T., Ohgushi H., Nakajima H., Yamada E., Ichijima K., Tamai S. and Ohta T. (2000) In vivo osteogenic durability of cultured bone in porous ceramics, *Transplantation* **69**, 128-134.
55. Ohgushi H. and Caplan A.I. (1999) Stem cell technology and bioceramics: From cell to gene therapy, *J. Biomed. Mater. Res. (Appl. Biomaterials)* **48**, 913-927.
56. Goldstein A.S., Juarez T.M., Helmke C.D., Gustin M.C. and Mikos A.G. (2001) Effect of Convection on Osteoblastic Cell Growth and Function in Biodegradable Polymer Foam Scaffolds, *Biomaterials* **22**, 1279-1288.
57. Behravesh E., Engel P.S., Yasko A.W. and Mikos A.G. (1999) Synthetic Biodegradable Polymers for Orthopaedic Applications, *Clin. Orthop.* **367S**, S118-S125.
58. Lieberman J.R., Le L.Q., Wu L., Finerman G.A., Berk A., Witte O.N. and Stevenson S. (1998) Regional gene therapy with a BMP-2 producing murine stromal cell line induces heterotopic and orthotopic bone formation in rodents., *J. Orthop. Res.* **16**, 330-339.
59. Gazit D., Turgeman G., Kelley P., Wang E., Jalenak M., Zilberman Y. and Moutsatsos I. Engineered pluripotent mesenchymal cells integrate and differentiate regenerating bone: a novel cell-mediated gene therapy, *J. Gene Med.* **1**(2), 121-133.
60. Krebsbach P.H., Gu K., Franceschi R.T. and Rutherford R.B. (2000) Gene therapy-directed osteogenesis: BMP-7-transduced human fibroblasts from bone in vivo. *Human Gene Therapy* **11**, 1201-1210.
61. Boden S.D., Titus L., Hair G., Liu Y., Viggeswarapu M., Nanes M.S. and Baranowski C. (1998) Lumbar spine fusion by local gene therapy with a cDNA encoding a novel osteoconductive protein (LMP-1), *Spine* **23**, 2486-2492.
62. Boden S.D., Liu Y., Hair G.A., Helms J.A., Hu D., Racine M., Nanes M.S. and Titus L. (1998) LMP-1, a LIM-domain protein, mediates BMP-6 effects on bone formation, *Endocrinology* **139**, 5125-5134.

CULTURED BONE ON BIOMATERIAL SUBSTRATES

A Tissue Engineering Approach to Treat Bone Defects

S.C. MENDES J.D. DE BRUIJN AND C.A. VAN BLITTERSWIJK
IsoTis NV
Professor Bronkhorstlaan 10
3723 MB Bilthoven
The Netherlands

1. Abstract

In the present work, a tissue engineering approach to treat bone defects was investigated. Such strategy was based on the use of patient own cultured bone marrow stromal cells (BMSCs) in association with biomaterials to produce autologous living bone equivalents. When engineering such implants, three main factors had to be taken into account: (i) the cells, (ii) the culture technology and (iii) the biomaterial scaffolds. The capacity of BMSCs to proliferate, differentiate along the osteogenic lineage and form a bone like tissue was demonstrated in various in vitro assays making use of biochemical, immunological, microscopic and gene expression techniques. The ability of the cells to produce bone in vivo was established using an ectopic (extra osseous) implantation model. Results indicated that BMSC cultures were composed of a heterogeneous population containing a subpopulation of cells with high proliferative capacity and with potential to differentiate into bone forming cells. Both the growth and the differentiation pattern of these cells could be manipulated, to a certain degree, through the use of bioactive factors during culture. After implantation, the bone forming capacity of the cultures proved to be related to the amount of early osteoprogenitors and precursors cells that could be induced into starting the osteogenic differentiation process. In bone marrow aspirates, this subpopulation appeared to decrease with donor age and to be strongly dependent on the donor, indicating that the aspiration procedure plays an important role in the obtained bone marrow cell population. In order to evaluate the in vivo bone formation capacity of BMSC cultures prior to implantation, an experimental method was developed in which the amount of early osteoprogenitors and precursors cells could be quantified. With regard to the technology design, data indicated that the culture of cells on the biomaterial scaffolds prior to implantation resulted in implants with faster in vivo bone forming ability as compared to scaffolds implanted shortly after cell seeding. In addition, two biodegradable polymeric systems were proposed as scaffolds to be used in the described bone engineering approach after evaluating their ability to support bone

265

R.L. Reis and D. Cohn (eds.),
Polymer Based Systems on Tissue Engineering, Replacement and Regeneration, 265–298.
© 2002 *Kluwer Academic Publishers. Printed in the Netherlands.*

marrow cell growth, differentiation and in vivo bone formation. In summary, although the complete knowledge of the factors controlling BMSC growth and osteogenic differentiation still needs to be further expanded, the obtained results suggest that the bone tissue engineering approach described in this work presents a great potential for the repair of bone defects and will become an advantageous alternative to the traditional autologous bone grafting.

2. Introduction

2.1. BONE

Bone is a complex dynamic tissue that is constantly being remodelled throughout adult life (resorbed and re-deposited). It is a natural composite material, mainly composed of mineral (60% in weight), an organic matrix (30% in weight) and water (10% in weight) [1]. The mineral part of bone confers stiffness to the tissue and consists of calcium phosphates, from which the major component is hydroxyapatite [2]. The organic matrix of bone confers tensile strength and is composed of a well organised network of proteins, from which collagen type I is the main constituent. The non collagenous proteins include osteonectin, osteopontin, bone sialoprotein, osteocalcin, decorin and biglycan [2-3].

Bone has mainly three functions: (i) It is a major organ for calcium homeostasis and it stores phosphate, magnesium, potassium and bicarbonate; (ii) it is the most abundant site of hematopoiesis in the human adult and (iii) it provides mechanical support for soft tissue and attachment sites for the muscles [4-5]. In adult life, physiological remodelling consists of bone resorption followed by bone deposition in approximately the same location. Bone resorption is accomplished by multinucleated giant cells of hematopoietic origin, named osteoclasts, while bone deposition occurs via osteoblasts, which are from stromal origin [2].

Bone exists in two forms, cortical and trabecular. The cortical bone, also called compact bone, is rigid, dense, anisotropic and plays a major role in mechanical support. It comprises the outer shell of the long bones, as well as the outer surface of small and flat bones. Trabecular or cancellous bone is less dense than cortical bone but it is metabolically more active. It occurs near the ends of long bones, in the interior of small bones and between the surfaces of flat bones [4-6].

Bone formation occurs by either of two processes, intramembranous or endochondral. In the intramembranous process, mesenchymal progenitors condense and differentiate directly into osteoblasts, while in the endochondral ossification process the same progenitors first form a cartilage template that is later replaced by bone. Intramembranous ossification is mainly responsible for the development of flat bones from the skull and for the addition of bone on the periosteal surfaces of long bones. Endochondral ossification occurs in the formation of long bones, vertebrae and fracture repair [5-6].

2.2. MATERIALS FOR OSSEOUS RECONSTRUCTION

Bone tissue regeneration remains an important challenge in the field of orthopaedic and oral-maxillofacial surgery. Spinal fusion, augmentation of fracture healing and reconstruction of bone defects resulting from trauma, tumour, infections, biochemical disorders and abnormal skeletal development are some of the clinical situations in which surgical intervention is required. The type of graft materials available to treat such problems essentially include autologous and allogeneic bone, as well as a wide range of synthetic biomaterials such as metals, ceramics, polymers and composites.

2.2.1. Autologous Bone

Currently the use of autologous (host) bone grafts is broadly considered as the golden standard therapy for bone repair and regeneration [5, 7-10]. Besides lacking immunogenicity, autologous bone possesses a range of intrinsic properties that make it an optimal implant material to achieve bone healing. These grafts are osteogenic, osteoinductive and osteoconductive. The osteogenic potential of autologous grafts is provided by bone forming cells present in the bone marrow, which are directly delivered at the implant site [11-12]. The grafts are also osteoinductive, that is, they are able to recruit mesenchymal cells located near the implant or from blood vessels and induce them to differentiate into osteogenic cells, through the exposure of osteoinductive growth factors of which the bone morphogenetic proteins (BMP's) are the most commonly studied [7, 9, 12-13]. Finally, the three-dimensional structure of the bone matrix, mainly composed of hydroxyapatite and collagen, allows for the infiltration of osteogenic cells that establish direct contact with the material (osteoconductivity) [9, 12, 14]. Although autologous bone grafting has the requirements for optimal bone regeneration, its use is also associated with serious drawbacks. The harvest of the graft implies an extra and invasive surgical procedure and the removal of bone often causes morbidity at the donor site [7, 9-10, 11-12]. Post-operative continuous pain [9, 15-17], hypersensitivity [9], pelvic instability [15-16, 18], infection [12, 17] and paresthesia [9, 12] are other possible complications associated with autologous bone grafting which affect 10 to 30% of the patients [11, 17]. The limited amount of bone that can be collected constitutes another disadvantage of these grafts.

2.2.2. Allogeneic Bone

The use of allogeneic (donor) bone for osseous reconstruction can solve some of the problems associated with autologous grafts since the harvest procedure is eliminated and the quantity of available tissue is no longer an issue. Nevertheless, these type of grafts present a poor degree of cellularity, less revascularisation and a higher resorption rate as compared to autologous grafts [9, 12], which may be responsible for the slower rate of new bone tissue formation observed in several studies [15, 19-21]. In addition, the immunogenic potential of these grafts and the risks of virus transmission to the recipient constitute serious disadvantages [10, 20, 22]. Processing techniques such as demineralisation, freeze-drying and irradiation have shown to reduce the patient's immune response, however, processing also alters the structure of the graft and reduces its potential to induce bone healing, while the possibility of disease transmission still remains [9].

2.2.3. Synthetic Biomaterials
Due to the limitations associated with bone derived grafts, several synthetic biomaterials are currently available, or under investigation, to be used as bone replacements. Four main classes biomaterials can be distinguished: metals, ceramics, polymers and composites. For many years, metal implants, mainly titanium and titanium alloys, have been used in orthopaedic and dental surgery for load bearing bone replacement. In joint replacement surgery, particularly total hip arthroplasty, these type of implants have achieved good clinical results, restoring patient mobility and providing pain relief [23]. These implants have high mechanical performance and do not evoke major adverse tissue responses. Nevertheless, they also present low bonding strength with bone which can result in osteolysis if micro movements occur [23-26].
Ceramic materials have been widely studied as bone grafts substitutes. Among them, hydroxyapatite (HA) and tricalcium phosphate (TCP) have received the most attention due to their similarity to the inorganic component of bone and teeth [9-10, 27-30]. TCP is reported to possess greater biodegradation rate as compared to HA but its mechanical properties are, however, inferior [9, 11, 27]. Blends of the two components allow to obtain biphasic calcium phosphates with a wide range of mechanical properties and resorbable rates, that can be tailored according to the specific application [11, 27, 31]. Extensive studies demonstrated that these materials are non toxic and do not evoke immunologic responses [30-32]. In addition, they promote bone ingrowth and form a strong intimate bond with bone [23, 29-32]. Due to those advantageous properties calcium phosphates have found applications in orthopaedic, dental and cranio-maxillofacial fields [9, 33-34]. Nevertheless, their relatively poor mechanical performance restricts their use to non load bearing applications [35]. Calcium phosphates are also used as coatings on metallic and polymeric substrates to promote a direct bond between bone and the implant, which results in improved osseointegration and firm implant fixation [25]. Additionally, HA powder is commonly used as a polymeric filler aiming to obtain composites with higher mechanical performance [36-37].
To date several polymeric materials have been suggested as bone graft substitutes. Among the non biodegradable polymers, ultra high molecular weight polyethylene (UHMWPE) and poly(methyl methacrylate) (PMMA) have been extensively used. The main application of UHMWPE consists on the manufacture of acetabular cups, while PMMA has been used as bone cement and dental prosthesis [23]. Synthetic biodegradable polymers have also been proposed as bone grafts substitutes. These materials are "easily" processed into highly porous and complex three dimensional shapes. In addition, their degradation and mechanical properties can be tailored by adjusting the composition and molecular weight of the polymers. To date the polymeric systems that have been investigated for bone repair include poly(α-hydroxy esters) [10, 38-41], poly(dioxanone) [42], poly(propylene fumarate) [23, 43], poly(ethylene glycol) [44], poly(urethanes) [45], starch based systems [37] and copolymers of poly(ethylene glycol)-terephthalate and poly(butylene terephthalate) [46-47].

2.2.4. Biomaterials with Intrinsic Osteoinductivity
Although successful results have been achieved when using biomaterial approaches, none of the materials in the four above mentioned classes (metals, ceramics, polymers, composites) possess osteogenic properties. Additionally, it is generally agreed that they

lack intrinsic osteoinductivity. As a consequence, their clinical application is restricted to relatively small osseous defects and their performance is inferior as compared to autologous bone grafts. Nevertheless, during the last decade, increasing evidence pointed out that specific calcium phosphate ceramics induced bone formation after implantation in soft tissues. In 1969 Winter and Simpson [48] reported bone induction by macroporous sponges of polyhydroxyethyl methacrylate after subcutaneous implantation in pigs and, a few decades later, Ripamonti [49] found bone in hydroxyapatite ceramics after intramuscular implantation in baboons. Since then, several studies demonstrated that a number of porous calcium phosphate ceramics and cements, as well as glass ceramics, were capable of inducing osteogenesis when implanted in ectopic (non bony) sites [50-55]. Results suggested that osteoinduction was material related and the specific chemical and structural characteristics of the materials, including their microstructure, were very important factors playing a determinant role on their osteoinductive capacity. Additionally, both Yuan et al. [56] and Ripamonti [57] reported the osteoinductivity of porous calcium phosphate ceramics to be strongly dependent on the animal species. In summary, materials with intrinsic osteoinductivity do exist and are excellent candidates as grafts for bone reconstruction. However, the biological mechanisms of osteoinduction, as well as the required biomaterial characteristics still need further investigation. In addition, factors related to the animal species variability observed in bone induction are not yet understood and the time required for bone formation, often 2 to 3 months, is also a limiting factor.

2.3. NOVEL STRATEGIES FOR BONE REPAIR AND REGENERATION

In 1993 Langer and Vacanti [58] defined tissue engineering as an 'Interdisciplinary field that applies the principles of engineering and life sciences toward the development of biological substitutes that restore, maintain, or improve tissue function'. With regard to bone tissue engineering, mainly two strategies have been implemented to generate new tissue: (I) Chemical stimulation of bone formation through the use of bone inducing substances and (II) The construction of hybrid implants composed of osteogenic cells/tissue and a biomaterial scaffold.

2.3.1. Chemical Stimulation of Bone Healing

Bone tissue contains peptide regulator molecules generally named growth factors that are capable of modulating bone cell activity. Bone growth factors are mainly produced by osteoblasts and are incorporated into the extracellular matrix during the process of bone formation. These factors are known to stimulate neighbouring cells to proliferate and increase protein synthesis (paracrine effect) and also act on the osteoblasts themselves inducing higher metabolic activity (autocrine effect) [59]. Numerous in vitro studies have reported that bone growth factors have several regulating effects on cells from the osteoblastic lineage and in vivo studies have demonstrated that some factors can induce bone formation and/or stimulate healing. Therefore, these agents became an area of intensive investigation. To date numerous growth factors have been identified and produced by recombinant gene technology, among those are bone morphogenetic proteins (BMP's), transforming growth factors β (TGF's β) and fibroblast growth factors (FGF's).

In 1965 Urist [13] demonstrated that demineralised bone matrix free of viable cells could induce bone formation when implanted subcutaneously. Bone induction was attributed to a substance which had the property of inducing undifferentiated mesenchymal cells to differentiate towards osteoprogenitors. Later on this substance was identified as a protein, which Urist et al. [60] named bone morphogenetic protein. Since then 12 different bone morphogenetic proteins have been identified (BMP 1-12). The BMP's belong to the transforming growth factor β superfamily and are so far the only growth factors that can stimulate the differentiation of mesenchymal stem cells into the chondro and osteoblastic direction [12, 59-60]. In vitro studies demonstrate that these proteins stimulate the differentiation of pluripotent cell lines and bone marrow stromal cells, from human and animal origin, into the osteogenic lineage in a dose dependant manner [61-65]. In vivo these proteins were found to induce ectopic bone formation in several animal models [66-67]. In addition, numerous studies reported the capability of BMP's to heal bone defects and /or induce orthotopic (osseous location) bone formation in a wide range of animal species including rats [68], rabbits [69], dogs [70] and baboons [67]. These proteins have also been successfully used for spinal fusions [65] and augmentation of alveolar bone [70]. Nevertheless, the dosage required for such treatment is strongly dependent on the animal model and a direct relation is observed between the amount of BMP's required and the size of the animal [67-68]. Additionally, when implanted alone these proteins diffuse too rapidly for bone induction to occur successfully, therefore, the success of BMP's in bone reconstruction is dependent on the existence of an appropriate carrier to maintain their activity at minimal dosage, preferably allowing a controlled release. Possible carriers tested for BMP's delivery include demineralised bone matrix [70], collagen [63, 70], calcium phosphate ceramics [61, 67-69], hyaluroran [71] and various synthetic polymers [66, 70-71]. With regard to the clinical use of BMP's in humans, few studies have been performed and in those reports very high physiological doses of protein, ranging from 1.8 to 3.4 mg, were used [72-73]. As a consequence, important safety questions were raised, especially because these agents are capable of inducing ectopic bone formation in regions neighbouring but external to their carrier [74]. Moreover, these proteins are not specific modulators of hard tissue, for example, the central nervous system is reported to contain BMP receptors [75]. In summary, prior to the clinical use of BMP's for bone reconstruction, the establishment of the proper dosage has to be further investigated, as well as the possible secondary effects that may result from their use.

Transforming growth factors β (TGF's β) are cytokines with a wide range of activities in bone, connective tissue and the immunological system [59]. In general, they stimulate cells of the mesenchymal origin having profound effects on osteogenic cell proliferation, differentiation and matrix synthesis [59]. Although TGF's β are reported as potent mitogens for bone marrow stromal cells [59, 65, 76] their effects on bone cell differentiation are controversial. Collagen type I synthesis is stimulated by TGFβ [65], while alkaline phosphatase activity and expression, as well as matrix mineralisation, are inhibited [65]. The effects of these factors in bony sites are contradictory and appear to vary with the set up of the specific study. Sumner et al. [77] demonstrated that TGF β enhanced bone ingrowth of implants inserted in trabecular bone in dogs. On the contrary, in a study by Aspenberg et al. [68], using a bone conduction chamber with porous hydroxyapatite in rats tibiae, it was shown that the bone ingrowth distance had a trend towards inhibitions in implants treated with TGF β, as compared to controls.

Additionally, a negative correlation between the TGF β dosage and bone ingrowth distance was found.

To date, two fibroblast growth factors (FGF's) were identified, acidic (aFGF) and basic (bFGF) [12, 59]. bFGF increases mitogenesis on fibroblastic, chondrogenic and osteogenic cells [12, 59]. In bone marrow stromal cell cultures, it enhances cell growth, while maintaining the cells in an immature state [78]. In vivo studies also suggest bFGF exerts a stimulatory effect on proliferation of osteoblastic cells, however, excess dosage my also result in reduced cell growth [79]. In addition, FGF's are also angiogenic factors stimulating revascularisation during bone healing [12, 59].

2.3.2. Cell Therapy Approaches for Bone Reconstruction

The solution to the problems associated with bone replacement may lie in the creation of a vital autologous bone substitute using patient own osteogenic cells in association with a biomaterial. The biomaterial besides of providing volume, will function both as a carrier for the transplanted cells/tissue and as a scaffold for the formation of new bone tissue. The goal is, therefore, to develop an alternative to the traditional autologous bone graft that achieves similar success in bone regeneration, but without the limitations inherent to autologous grafting. Although an extra surgical procedure will still be needed to harvest the osteogenic cells, this will be much less invasive as compared to the collection of bone and it will not bear the post-operative complications associated with autologous bone grafting. Additionally, large quantities of osteogenic cells/tissue can be obtained from small biopsies after culture expansion. In this approach, factors such as cell source, cell proliferation and osteogenic differentiation, as well as the material scaffold are of extreme importance to successfully engineer bone tissue. With regard to cell source, various cell types from several tissues and locations have been investigated. These include calvarial [80-81] and periosteal cells [82-83], osteoblasts of trabecular bone from various locations [84-85], chondrocytes [86] and even vascular pericytes [87] and cells from extramedullary adipose tissue [88]. Nevertheless, the most widely used source of osteogenic cells is bone marrow and the rationale for its choice is both scientific and practical. Bone marrow has long been recognised as a source of osteoprogenitor cells that can differentiate towards bone forming cells when cultured under adequate conditions [89-92]. In addition, bone marrow has been claimed to be the most abundant source of osteoprogenitors, which possess high proliferative ability and great capacity for differentiation [93-94]. From a practical point of view, bone marrow is the most accessible source of osteogenic cells since it can be collected using a relatively simple aspiration procedure, which is much less invasive than collecting bone, cartilage or another type of tissue.

Bone Marrow Stromal Cells (BMSCs). Bone marrow is a complex tissue composed of two main cellular systems: hematopoietic and stromal. The stromal tissue consists of a network of cells with very little extracellular matrix that provides mechanical support for hematopoietic cells. The bone marrow stroma also expresses cell signalling factors that participate in the development of blood cells, while hematopoietic cells are also known to influence the activity of the stromal compartment [95-96]. The cell types comprising the stromal system include reticular cells, smooth muscle cells, endothelial cells, adipocytes and cells from the osteogenic lineage [97].

Friedenstein et al. [89] and Owen [91] performed pioneering studies in the characterisation of BMSCs using both in vitro culture systems and in vivo models. In these studies, when bone marrow stromal cells were plated in culture at low densities they readily adhered and formed fibroblastic colonies, each derived from a single precursor cell, the colony forming unit fibroblast (CFU-F) [98]. When marrow cells were plated at high densities, the colonies merge and the cells grew as monolayers. It has been demonstrated that CFU-F are heterogeneous in size (reflecting various growth rates), morphology and potential for differentiation, suggesting that they originate from progenitors at various stages of differentiation [91, 98]. The high proliferative ability of some of the CFU-F together with the known regenerative capacity of BMSCs led Friedenstein [89] to propose the existence of stromal stem cells that give rise to committed progenitors for different cell types. Stem cells were then defined as able to self-renew, multipotential and capable of regenerating tissue after injury [98]. This hypothesis was consistent with results from a study in which single colony derived mouse BMSCs were implanted on ectopic sites in syngeneic hosts. Approximately 15% of the implanted colonies produced bone adipose and marrow reticular tissue with the establishment of hematopoiesis by host cells. Another 15% of the transplanted colonies formed bone without associated marrow and the rest either gave rise to fibrous tissue formation or did not form any tissue [97, for review]. This experiment suggested the existence, among the CFU-F population, of both multipotential cells and precursors with a more limited potential. Since the early studies from Friedenstein and Owen, numerous reports have provided evidence that bone marrow tissue contains progenitor cells that after extended culture, are capable of giving rise to several phenotypes, including adipocytic, chondrogenic and osteogenic lineages [99-102].

Osteogenic Cell Differentiation. With regard to osteogenic cell differentiation, the existence of a lineage hierarchy in which a multipotential precursor cell gives rise to cells with a more restricted potential and these ultimately originate monopotential progenitors has been proposed [100]. The osteogenic differentiation process may be characterised by the sequential acquisition and/or loss of specific extracellular matrix molecules and cell surface markers (fig. 1). Four maturational stages in osteoblast development have been identified in bone in situ: the preosteoblast, osteoblast, osteocyte and bone lining cell [6]. The preosteoblast is the immediate precursor of the osteoblast and it is localised in the adjacent cell layers from the bone producing osteoblasts. These cells possess alkaline phosphatase (ALP) activity and limited capacity for proliferation [6]. Osteoblasts are postproliferative cells with cuboidal morphology and strong ALP activity. These cells synthesise bone matrix proteins, some hormone receptors, cytokines and growth factors. Osteoblasts produce bone tissue and line the matrix at sites of active matrix production [4, 6]. Bone lining cells present a flat, thin and elongated morphology and are thought to be inactive osteoblasts [4, 6]. When osteoblasts become incorporated in the newly formed bone matrix they are termed osteocytes. These cells are considered the most mature stage of the osteoblastic lineage and present a decreased ALP activity as compared to osteoblasts [4, 6].

Expression of the kidney/bone/liver isoform of ALP is directly related with bone formation, and it is widely accepted that an increase in ALP activity in a population of osteogenic cells corresponds to a shift to a more differentiated state [6, 78, 88, 90, 103-105]. ALP is present in both preosteoblasts and osteoblasts and studies suggest that its expression is detected in differentiating osteoblastic cells preceding the expression of the non collagenous proteins [106]. Although the exact role of ALP is unknown, studies suggest that it is involved in the mineralisation process since an inhibition of ALP activity inhibits bone matrix mineralisation [6]. Collagen type I (coll-I) constitutes approximately 90% of the total organic matrix in bone and although synthesized by many cell types it is intensively produced by osteoblasts being, therefore, considered as a characteristic marker of the osteoblast phenotype [6, 90, 107-109]. This protein is also expressed in preosteoblasts [6]. Osteopontin (OPN) is synthesised by osteoblastic cells, however, it is also produced by many cells of non skeletal tissues [6]. On the contrary, bone sialoprotein (BSP) is almost exclusively produced by hypertrophic chondrocytes, preosteoblasts, osteoblasts and osteocytes [6]. In a recent study by Cooper et al. [109], it was suggested that the expression of BSP but not osteocalcin in human bone marrow stromal cell cultures preceded histological evidence of in vivo bone formation. Osteocalcin (OCN) or bone gla protein is undetectable in preosteoblasts but highly expressed in mature osteoblasts. This protein is considered the latest of the expression markers along the process of osteogenic differentiation [6]. In addition to the bone matrix proteins mentioned, osteoblasts also secrete other proteins such as osteonectin, decorin and CD44 [6]. A wide list of hormones, growth and transcriptions factors have also been reported to regulate osteogenic activity and/or differentiation. Among those, osteogenic cells are known to possess receptors for parathyroid hormone (PTHrP, PTH-R1) and basic fibroblastic growth factor (FGFR-1) [6]. Additionally, the transcription factor cbfa1 is known to play an important role in osteoblast development [110].

To better characterise and identify the osteogenic cell differentiation process in the bone marrow stromal cell system, the isolation of a subset of cells with the highest proliferative ability and great capacity for osteogenic differentiation would be of utmost importance. Although several monoclonal antibodies are reported to bind with BMSCs at early stages of differentiation, including SH2, SH3, SH4 [102, 111] and HOP-26 [112], the IgM monoclonal antibody Stro-1 is the most widely used [113-118]. It recognises a specific population of human BMSCs, in which osteoprogenitors appear to reside [113-114, 116]. Although within the stromal compartment there are cells with the Stro-1 epitope which are not CFU-F's, all detectable CFU-F's are exclusively present on the Stro-1 positive population [113]. Using this antibody in combination with an antibody against the kidney/bone/liver isoform of ALP it has been possible to identify osteogenic cells at three different stages of differentiation, supposedly stromal precursors, osteoprogenitors and mature osteoblasts [117]. In addition, the expression of the transcription factor cbfa1 was found to be restricted to fractions expressing Stro-1 and/or ALP [117].

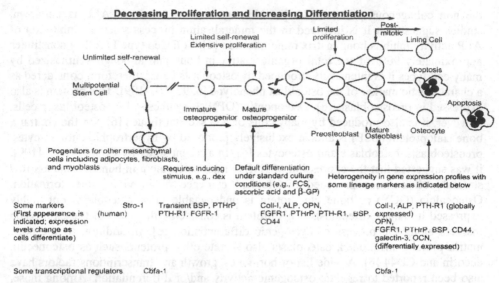

Figure 1. Proposed steps in the osteoblastic lineage, implying recognisable stages of proliferation and differentiation. Markers of the osteoblast phenotype and their expression during differentiation. Adapted from Aubin [100].

Plasticity of bone marrow stromal cells. As described above, bone marrow stromal cells can form various tissues, including bone, cartilage and fat. Another, extremely interesting characteristic of these cells is that they present a certain degree of plasticity with regard to lineage commitment. In 1991, Bennett et al. [119] showed that differentiated marrow adipocytes could differentiate in vitro back to a more proliferative stage and them form osteogenic tissue in vivo. Another example of cell commitment plasticity was reported by Galotto et al. [120], in a study in which fully differentiated chondrocytes have shown to dedifferentiate during culture and then, express the osteoblastic phenotype. These studies clearly reveal that, during culture, the lineage commitment of bone marrow stromal cells is reversible, whether this plasticity also occurs in an in vivo situation is still unclear.

Osteogenic Character of BMSCs in Vitro. In 1988 Maniatopulos et al. [90] cultured BMSCs from the femora of adult rats and reported that these cells differentiated along the osteogenic lineage, as revealed by their ability to form mineralised nodules in which the extracellular matrix was mainly composed of collagen type I and also contained osteonectin and osteocalcin. In addition, cells associated with the nodules exhibited high ALP activity. Since then, numerous studies have described the osteogenic character of BMSCs both from animal and human origin using similar criteria in defining osteogenic potential, that is expression and/or synthesis of bone matrix proteins, ALP and capacity to form a mineralised tissue [78, 104-105, 107-109, 121-124]. Nevertheless, the osteogenic character of the cultured cells and tissue has shown to be dependent on the culture conditions. The mostly widely known bioactive factors that have an influence the proliferation and differentiation of cultured bone marrow stromal cells are: serum, ascorbic acid, inorganic phosphate and glucocorticoids. The selected batch of serum added to the culture medium was shown to be extremely important for both the growth and osteogenic differentiation of BMSCs [125]. Ascorbic acid (vitamin C) was found essential for collagen synthesis and secretion. It also increases the levels of procollagen mRNA during culture [90, 104]. For mineralisation to occur, the culture medium must contain an inorganic source of phosphate which is normally obtained by the addition of sodium β-glycerophosphate to the culture medium [90, 104]. Glucocorticoids when administrated in vivo, especially at high dosage, are known to suppress bone formation and stimulate bone resorption, inducing osteoporosis [126]. Nevertheless, they exert a powerful influence on BMSCs osteogenic differentiation during culture. Dexamethasone (dex) has been extensively reported to stimulate osteogenic differentiation in cultures of BMSC's from animal and human origin [78, 104, 107-108, 122, 127-131]. Signs of differentiation induced by dexamethasone include morphological changes from an elongated to a more cuboidal cell shape [78, 104, 127, 129] and an increase in the expression and/or activity of ALP [104, 107-108, 122, 127-128, 130-131]. Additionally, this bioactive factor has been reported as essential for the mineralisation of BMSCs cultures [104, 107, 122, 127]. With regard to the effect of dexamethasone on the expression and/or synthesis of osteocalcin and osteopontin, both stimulatory [130-131] and inhibitory [107-108, 127] effects have been reported. These discrepancies may be a result of different culture conditions and experimental set-ups.

Several other biologically active factors such as BMP's, TGF's β and FGF's are also known to affect the proliferation and/or osteogenic character of BMSCs *(see above: chemical stimulation of bone healing)*.

Osteogenic Character of BMSCs in Vivo. Although the in vitro phenotype of BMSCs cultures provides valuable information on their osteogenic character, the behaviour of these cells after implantation gives the ultimate answer on whether these cells can form bone tissue. However, several factors may affect the outcome of the studies, such as species origin of the cells, culture conditions prior to implantation and implantation model. With regard to ectopic implantation models, both diffusion chambers and open systems have been used to test the osteogenic potential of BMSCs. Diffusion chambers allow for the diffusion of nutrients from the host but isolate the implanted cell population from invasion by recipient cells. As a consequence, vascularisation does not occur in the transplanted cells and the tissues formed are from donor origin [95]. In studies using cultured human BMSCs cultured without ascorbic acid and dex, both Haynesworth et al. [132] and Gundle et al. [85] reported the absence of bone or cartilage tissue after implantation in diffusion chambers in nude mice. Additionally, both types of tissue were detected when cells were cultured in the presence of ascorbic acid and dex prior to implantation [85]. Moreover, in the above mentioned study by Haynesworth et al. [132] in vivo bone formation was obtained, using the same cell preparations, when implantation was performed in an open system, using a porous calcium phosphate as a scaffold material. These results suggest that open systems are more sensitive in identifying the in vivo osteogenic potential of cells, which may be related to the lack of vascularisation in diffusion chambers.

Bone formation by rat BMSCs was widely investigated by subcutaneous implantation in nude or syngeneic hosts, using several porous calcium phosphate ceramics as biomaterial scaffolds [92, 133-137]. In this type of implants, bone formation was shown to start on the surface of the ceramic, advancing towards the centre of the pores.

Ohgushi et al. [133] reported bone formation in both HA and TCP ceramics combined with fresh bone marrow, after 4 weeks of implantation. At the end of 8 weeks survival, the extent of bone in the implants significantly increased and in some pores regeneration of bone marrow was detected. Yoshikawa et al. [134] also showed bone formation by cultured rat BMSCs on porous HA but only in samples where cells had been treated with dexamethasone. In this study, during the entire implantation period (1, 2, 3, 4 and 8 weeks) cartilage formation was not detected and therefore the process of bone formation was considered to be intramembranous. On the contrary, de Bruijn et al. [136] reported the formation of both bone and cartilaginous tissue by rat cells continuously cultured in the presence of dex, after 4 weeks of implantation. Nevertheless, cartilage like tissue was only found in samples with high cell seeding densities. Additionally, in a study by Dennis et al. [135], the culture of rat BMSCs in the presence of dex was not required to obtain in vivo formation of bone. Riley et al. [138] did report ectopic bone formation by rat BMSC cultured on poly(DL-lactic-co-glycolic acid) foams. Bone was formed as early as one week post implantation. Nevertheless, the maximum penetration of bone into the sponges was approximately 250µm after 4 weeks of implantation.

Mouse BMSC were also found to form bone and bone marrow when subcutaneously implanted in combination with a wide range of material scaffolds, such as collagen sponges and matrices, polyvinyl sponges and HA/TCP blocks and powder [139].

Rabbit BMSCs have demonstrated the capacity to produce bone tissue in ectopic sites when seeded both on calcium phosphate ceramics [140] and hyaluronic acid-based polymers [141].

Finally, the in vivo osteogenic potential of goat BMSCs cultured on porous HA has also been proven after subcutaneous implantation in immunodeficient mice. Results demonstrated that the ability of these cell populations to produce bone in vitro was not dependent on the presence of factors such as ascorbic acid, sodium β-glycerophosphate or dex in the culture medium [136].

With regard human BMSCs, several investigators have demonstrated to the ability of these populations to form bone in ectopic sites [92, 132, 136, 139, 142-144]. Nevertheless, bone formation by human BMSCs did not consistently occur with all tested cultures. Ohgushi et al. [92] reported bone formation by human BMSC cultures loaded in porous HA from 2 of the 6 donors tested, after an implantation period of 4 weeks at subcutaneous sites in immunodeficient mice. In the same study, fresh human bone marrow from 5 of the 7 assessed donors exhibited in vivo osteogenic potential. In a similar study using cultured BMSCs from 11 donors, Haynesworth et al. [132] reported bone formation in most of the biphasic calcium phosphate ceramics subcutaouslly implanted with cells in nude mice. However, cultured cells from one of the donors did not form bone during the implantation periods tested (3 and 6 weeks). In studies by Krebsbach et al. [139] and Kuznetsov et al. [142], the in vivo osteogenic potential of cultured human BMSCs seeded on to various scaffold materials was tested. Cells seeded on calcium phosphate materials (hydroxyapatite/ tricalcium phosphate powder and blocks) consistently formed bone, while cells seeded on collagen sponges or gelatin produced bone sporadically but only when cultured with dex. In addition, bone formation was never observed in polyvinyl sponges and poly (L-lactic acid). The capacity of human BMSCs to induce the formation of bone marrow like tissue was also established in some of the above mentioned studies [136-137, 139, 142].

In vivo ectopic osteogenesis, although providing valuable information on the osteogenic potential of the cells, does not simulate the microenvironment of an osseous site, which they will encounter if used in bone reconstruction. Few studies have used orthotopic (bone site) models for the implantation of culture expanded BMSCs. Porous HA/TCP scaffolds seeded with cultured BMSCs, from both rat [145] and human origin [146], were found to heal clinically relevant segmental bone defects in rat femora while defects filled with the scaffold alone did not heal. In those studies the extent of bone present on the implants was significantly increased by the presence of the cultured cells. Accordingly, in critical size segmental bone defects in dogs [147] union did not occur when the defects were left empty, while it was established in both defects filled with HA/TCP cylinders and HA/TCP cylinders loaded with cultured BMSCs. Nevertheless, the amount of bone present on the samples loaded with cells was significantly greater as compared to cell free implants. The use of cells seeded on calcium phosphate ceramics has also been reported to improve healing of critical size segmental bone defects in sheep [148-149].

Cell Therapies for Bone Reconstruction: Different Strategies. At present, in the bone tissue engineering field three different strategies make use of patient own bone marrow cells to engineer autologous osteogenic grafts. One of these strategies consists in BMSC harvest, followed by cell seeding on a biomaterial scaffold and immediate implantation into the defect site (fig. 2, I); In other approach, the harvested cells are first culture expanded and then seeded on a suitable scaffold shortly before implantation (fig. 2, II). In the third strategy, after harvesting, the cell numbers are expanded in culture and when a sufficient number of cells is obtained they are seeded on a biomaterial scaffold, in which cells are further cultured to promote the formation of a bone-like tissue layer on the implant prior to implantation (fig. 2, III).

The first above mentioned approach (implantation of the total bone marrow cell population) has clearly logistic advantages since it is possible to collect a bone marrow aspirate shortly before the reconstructive procedure takes place. The bone marrow sample is then seeded on the biomaterial scaffold that can be immediately implanted into the patient defect site. Nevertheless, with this strategy BMSC numbers will be limited and higher quantities of aspirate will be required, which besides of may posing a problem to the patient, it is known to increase contamination by peripheral blood and decrease the final concentration of osteoprogenitors in the sample [94]. Results from animal studies using this strategy are somewhat contradictory. For example, in an above mentioned study by Kadiyala et al. [145], using critical size segmental bone defects in rats femora, the addition of fresh bone marrow to the biomaterial implants did not induce differences in the rate and extent of bone formation as compared to the cell free implants, while being significantly lower than on implants seeded with culture expanded cells. Accordingly, Boden et al. [69] using a rabbit model reported that HA seeded with fresh bone marrow was not an acceptable bone graft substitute for posterolateral spine fusion. However, Louisia et al. [150] have reported that HA combined with fresh bone marrow was able to bridge osteoperiosteal gaps in rabbits after two months, while HA alone could not produce union. With regard to the second strategy, several investigators have reported the ability of culture expanded BMSCs to form bone in ectopic sites when seeded in a biomaterial shortly before implantation [85, 92, 132-133, 135, 139-142]. In this approach, BMSCs are seeded on the biomaterials either in the presence or absence of fetal bovine serum. When serum free cell

278

suspensions are used, investigators utilise proteins such as fibronectin and fibrin to stimulate cell adhesion to the biomaterial substrate [144, 148]. In vivobone formation by hybrid constructs composed of biomaterial covered with a layer of in vitro formed bone-like tissue was also demonstrated in several studies [134, 136-138, 151]. This last approach appears to present some significant advantages since the cells have already started to produce bone matrix in vitro, which is expected to accelerate in vivo bone formation. In addition, the in vitro formed bone matrix may contain several proteins and growth factors that can enhance bone formation. To our knowledge, a study comparing the in vivo osteogenic potential of these two strategies has not yet been reported.

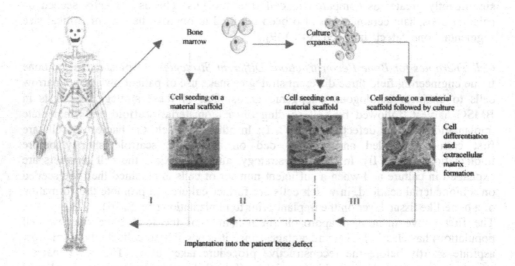

Figure 2. Different strategies for bone tissue engineering.

2.4. AIMS OF THE STUDIES

The main goal of the investigations described in this work was to identify and optimise parameters that affect the osteogenic character of BMSCs, aiming at the application of these cells in the treatment of large bone defects. In such an approach the growth and differentiation characteristics of the cells, which are affected by external stimuli during culture, as well as the model design used for the construction of engineered tissue are of utmost importance. Additionally, the choice of the biomaterial scaffold that will support cell growth, differentiation and the formation of bone will affect the final osteogenic potential of the implants. Therefore, several studies were performed with the following objectives:

- To identify and test bioactive factors that affect the proliferation characteristics and osteogenic potential of human BMSCs, aiming to optimise in vitro culture conditions;

- To evaluate whether human BMSCs characteristics are dependent on the donor and, if so, to determine which donor related parameters influence the cultures both at a proliferation and differentiation level;
- To characterise the development of the osteogenic lineage during human BMSCs in vitro culture and
- To identify which features are displayed by human BMSCs during culture and which subset of cells would be determinant for bone formation after implantation;
- To characterise the role of the extracellular matrix formed by the cells during in vitro culture on the osteogenic capacity of the implants;
- To evaluate different biomaterials as scaffolds for bone tissue engineering.

3. Results and Discussion

3.1. HETEROGENEITY OF HUMAN BONE MARROW STROMAL CELL CULTURES

The heterogeneous character of bone marrow is known to be reduced during culture due to the progressive lost of non adherent cells, macrophages, endothelial cells and cells with low proliferative capacity [89, 95, 103, 139, 142]. Nevertheless, results from our studies have shown that, even after extensive culture, the human bone marrow stromal cell (HBMSC) population still remains heterogeneous. Cultures were found to react with the monoclonal antibody Stro-1, although, this reactivity was restricted to a subset of cells and was not displayed by the entire population. During culture, the temporal pattern of stro-1 expression showed an increase during the preconfluent period followed by a progressive decline (fig. 3).

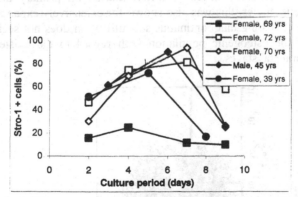

Figure 3. Development of Stro-1 expression in HBMSCs from five donors cultured in the presence of dex.

The expression of this epitope on cultured HBMSCs was also found to decrease with the degree of subcultivation. Since it is generally agreed that the osteoprogenitor cells reside exclusively in the Stro-1 reactive population [113-114, 116], these findings demonstrate that other cell types than stromal precursors were present in the cultures. In addition, subsets of cells were found to possess epitopes for antibodies known to react

280

with many different cell types (such as 1B10, CD34, CD146 and CD166 [113, 152-153]), which further demonstrated the heterogeneity of the cultured HBMSC population. Moreover, within the osteogenic population, heterogeneity was also observed with regard to cell differentiation stage. Data on gene expression confirmed that HBMSC cultures contained mRNA for both early (eg. alkaline phosphatase) and late osteogenic markers (osteocalcin) (fig. 4).

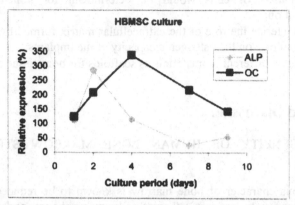

Figure 4. Semi-quantification of ALP and OC mRNA levels in HBMSCs cultured up to nine days in the presence of dex. Results obtained by RT-PCR and expressed as a percentage of the house keeping gene expression (ß- actine).

These findings are consistent with results from several other studies [107-108, 130, 154], in which HBMSC cultures were reported to coexpress bone cell related markers associated to different developmental stages. In addition, Stewart et al. [117] have demonstrated that dual labelling for Stro-1 and alkaline phosphatase allowed to identify osteogenic cells at different stages of differentiation on primary and first passage cultures and, in our studies, this was also shown for extensively expanded (4th passage) cultures (fig. 5), indicating that continuous subcultivation does not seem to reduce the heterogeneity of the osteogenic population with regard to the existence of cells in different developmental stages.

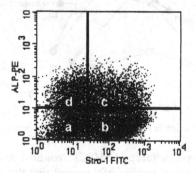

Figure 5. Representative dot plot illustrating Stro-1 and ALP dual expression by 4th passage HBMSCs. Stro-1 is detected on the x-axis (FITC label) and ALP is detected on the y-axis (PE label). Four different populations can be defined: Stro-1⁻/ALP⁻ (a), Stro-1⁺/ALP⁻ (b), Stro-1⁺/ALP⁺ (c) and Stro-1⁻/ALP⁺ (d).

Contradictory to these results are those from Pittinger et al. [102] who have claimed that a homogeneous population of stem cells can be isolated from bone marrow using standard density gradient procedures followed by in vitro culture. However, the characterisation of the cultures homogeneous character was based on their uniform reactivity with SH2 and SH3 antibodies, previously reported to recognise antigens for primitive cells of the osteoblastic phenotype [111], as well as on a lack of reactivity with antigens common on cells of the hematopoietic lineage. In our point of view, the homogeneous character attributed to the HBMSC cultures is at least controversial. Firstly, the characterisation of the so called 'pure' population did not include the Stro-1 antibody, which is widely reported to react with a distinct population of HBMSCs that contains all detectable colony forming units fibroblasts (CFU-Fs) and, therefore, all osteoprogenitor cells [113-118]. Moreover, results from the same study showed that individual colonies displayed varying degrees of multipotentiality, which further supports the existence of heterogeneity among cells. In fact, in our studies, HBMSC cultures from several donors were also found to uniformly (>93%) express SH2 antigen, independently of donor or culture period, while reactivity with Stro-1 was restricted to a subpopulation of cells, which was dependent on the specific cell donor (fig. 3). These findings are indicative that SH2 binds to a broader cell population and not exclusively to stromal stem cells. Furthermore, our results demonstrate that the differentiation stage of the cultured stromal cells is in constant evolution since Stro-1 and ALP expression was shown to vary along the culture period. In summary, although immunoselection with specific antibodies has shown to reduce the heterogeneous character of HBMSC cultures, true homogeneous populations of stromal stem cells have not yet been identified and the required culture conditions to maintain either the stem cell character or a certain differentiation stage still need to be determined. Nevertheless, bone marrow stromal cell populations represent one of the most accessible sources of stem and/or progenitor cells, which makes them excellent candidates for therapeutical use. The possibility to expand and direct the differentiation of these cells provides the opportunity to study events associated with osteogenic cell commitment and differentiation.

3.2. IN VITRO OSTEOGENIC POTENTIAL OF HBMSCs

The development of the osteoblastic lineage from bone marrow stromal precursors is characterised by a sequence of events involving cell proliferation, expression of bone related markers (cell differentiation) and synthesis and deposition of a collagenous extracellular matrix [90]. In the construction of bone tissue engineered implants, the optimisation of the culture conditions to better control these events is essential for the success of the technique. With respect to the cell proliferation step, we tested several growth factors in an attempt to optimise cell proliferation rate, which will reduce the waiting period for the patient. Our results suggested that, although basic fibroblastic growth factor (bFGF), epidermal growth factor (EGF) and transforming growth factor β1 (TGF-β1) actually participated in the proliferation mechanisms of these cells, bFGF and EGF were the most active in promoting cell growth and in maintaining the fibroblastic like morphology. These findings are in agreement with a report by Martin et al. [78], which demonstrated that bFGF and EGF are potent mitogens for HBMSCs. Additionally, in the same study, bFGF was reported to maintain cells in a more immature state during proliferation, inhibiting morphological changes from a fibroblastic morphology to a more flattened phenotype. Regarding the use of β mercaptoethanol (βME) to promote cell growth, our data indicated no stimulatory effect, contrary to the reported by Triffit et al. [155].

With respect to cell differentiation, in our studies the osteogenic potential of HBMSC cultures was characterised by the expression of bone matrix proteins, alkaline phosphatase and capacity to form a collagenous extracellular matrix. Several immunoreactivity and gene expression assays were used in these studies and results demonstrated that the cultures were immunoreactive and expressed mRNA for a wide range of markers associated with the osteoblast phenotype (fig. 4 and 6).

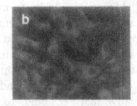

Figure 6. Immunoreactivity of HBMSCs for procollagen I (a) osteonectin (b) and osteopontin (c).

Moreover, the ability of HBMSCs to synthesise a collagenous extracellular matrix was established. Both scanning electron microscopy observations and immunostaining results revealed that the tissue engineered implants consisted of material covered with multilayers of cells embedded in an extracellular matrix rich in collagen type I (fig. 7).

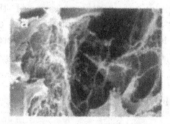

Figure 7. (a) Scanning electron micrograph illustrating the presence of collagen fibers on the tissue engineered constructs (1000x), (b) Light microscopy micrograph of HBMSCs cultured on porous calcium phosphate particles and stained with collagen type I antibody.

The synthetic glucocorticoid dexamethasone (dex), has been extensively reported to stimulate osteogenic differentiation of HBMSC cultures [104, 117, 127-129]. In our studies, signs of differentiation induced by dex included morphological changes from an elongated to a more polygonal cell shape and an increase in the relative amount of cells expressing alkaline phosphatase (fig. 8).

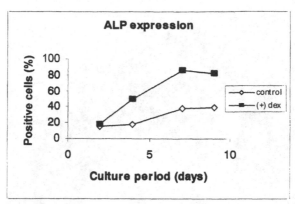

Figure 8. Temporal expression of ALP in HBMSC cultures: Effect of dexamethasone treatment.

Additionally, in sub and near confluent cultures stimulation by dex increased the fraction of cells positive for Stro-1. These effects are in accordance with a model in which dex promotes the recruitment of cells into the osteogenic lineage and further stimulates their maturation [127].

With regard to the fraction of cells expressing bone cell markers such as Stro-1, alkaline phosphatase, pro-collagen I and osteopontin, a strong donor dependency was observed. These findings are in agreement with several other studies, in which a large variability in the expression of osteogenic makers by HBMSC derived from different donors was reported [104, 117, 156]. Differences on the physiological status of the donor, as well as the aspiration site and procedure can account for these variations. With regard to the aspiration site, Phinney and coworkers [156] detected a large variation in the activity of alkaline phosphatase enzyme in HBMSC cultures from different donors despite the fact that all aspirates were obtained from the iliac crest. Furthermore, they observed clear differences in alkaline phosphatase activity of cultures established from the same donor over a 6 month period, which indicated that the method of bone marrow harvest plays a major role in producing cellular heterogeneity, pointing out the importance of developing standardised and optimised aspiration procedures. In fact, in order to produce an autologous artificial bone tissue, it is crucial that an appropriate bone marrow aspirate is collected from the patient. The cell content of the aspirate, as well as the proliferation and differentiation capacity of the cells are essential factors to be considered and will determine the final outcome of the technology.

3.3. IN VIVO OSTEOGENIC POTENTIAL OF HBMSCs

With regard to the effect of dex on the bone forming capacity of HBMSCs, our results demonstrated that, in the majority of the assessed cultures, stimulation by dex was not required to obtain in vivo bone formation by HBMSCs (fig. 9).

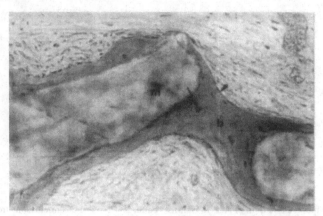

Figure 9. Representative light micrograph illustrating in vivo formed bone by HBMSCs, after subcutaneous implantation in nude mice for 6 weeks. De novo formed bone tissue (b) was deposited against the material (m) surfaces. This tissue consisted of a mineralised matrix with embedded osteocytes (arrow) and layers of osteoblasts (arrow head), (200x).

These findings are in agreement with those reported by Martin et al. [78] and suggest, as proposed by Kuznetsov et al. [142], that the HBMSC population contains subpopulations of both committed osteoprogenitors and undifferentiated cells. In the committed population, stimulation by dex, although not necessary, may stimulate further differentiation leading to an earlier start of bone formation. On the undifferentiated population, dex appears to recruit cells into the osteogenic lineage. The use of dex during the differentiation stage is, therefore, advisable to ensure that a sufficient number of HBMSCs will differentiate towards the osteoblastic lineage. In addition, our results indicated that dex contributed to a higher reproducibility in the degree of bone formation from donor to donor, increasing the extent of osteogenesis in samples with low bone forming ability.

The effect of donor age on both growth rate and in vivo osteogenic potential of HBMSC cultures was assessed in this work. With regard to the growth characteristics, an age related decrease was observed in the proliferation rate of cultures from donors older than 50 years as compared to cultures from younger donors (fig. 10).

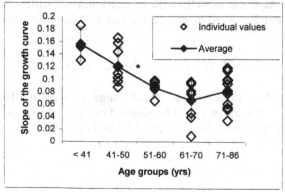

Figure 10. Growth characteristics of HBMSCs as a function of age. *Statistical decrease in the proliferation rate of HBMSC from donors older than 50 years (p=0.003).

These findings agree with a recent study by Muschler and co-workers [157], in which an age related decrease in the number of nucleated cells per ml of bone marrow aspirate was observed. In a report by Phinney et al. [156], no age related effect could be detected on the growth rate of HBMSCs, nevertheless, the results from both studies do not conflict since the age range investigated by Phinney and coworkers ranged from 19 to 45 years, where we also did not detect differences in cell growth. In our studies, the decrease observed in the growth rate of HBMSC cultures from older donors is probably due to a reduction in the number of proliferative precursors (osteoprogenitors) present in bone marrow as age increases. This hypothesis is in conformity with findings reported by Bab et al. [158], in which the number of colony forming unit fibroblasts (CFU-F) from human bone marrow also exhibited an age related decrease. With regard to the effect of donor age on the in vivo bone forming capacity of HBMSCs, our results indicated that cultures from several donors in all age groups possessed in vivo osteogenic potential. However, the increase of age above 50 years resulted in a decrease in the frequency of cases in which in vivo bone formation was observed (fig. 11).

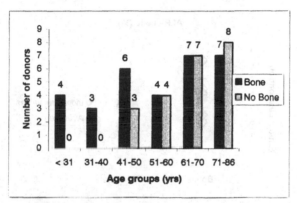

Figure 11. Effect of donor age on the in vivo osteogenic character of HBMSC. Results obtained after six weeks of subcutaneous implantation of the tissue engineered samples in nude mice.

These findings also point out a reduction in the amount of osteoprogenitor cells in bone marrow as age increases, and agree with reports from animal [159] and human [160-161] studies, in which the number of BMSCs colonies expressing alkaline phosphatase decreased during aging. Nevertheless, and as previously mentioned, the bone marrow aspiration procedure may strongly affect the obtained cell population, therefore, in older patients, an optimisation of the aspiration procedure may increase the success rate of the approach. Another crucial factor to take into account when evaluating the in vivo osteogenic capacity of bone tissue engineered constructs (material with osteogenic cells and tissue) is vascularisation. After implantation, vascular supply must be rapidly established into the implantation region in order to bring nutrients and bioactive factors essential for cell survival and function.

3.4. IDENTIFICATION AND QUANTIFICATION OF THE SUBPOPULATION OF CELLS IMPORTANT FOR IN VIVO BONE FORMATION

As previously stated, in our studies, in vivo bone formation by HBMSCs was not consistently observed in all cases. Therefore, a method was developed to identify and quantify the subpopulation of cells important for in vivo osteogenesis. Since both preosteoblasts and osteoblasts possess a limited proliferative capacity and in our studies HBMSCs were extensively expanded prior to implantation, it seemed reasonable to assume that the highly proliferative cells (that is early progenitors) would be the most important population for the production of bone. Due to the lack of procedures to isolate these cells, we proposed an indirect quantification method based on the hypothesis that after dex stimulation, the increase on the proportion of cells expressing early osteogenic markers would provide a measurement for the amount of early (and therefore inducible) osteoprogenitor cells in culture. After calculating the degree of stimulation by dex displayed by each culture, with regard to ALP expression, the results were compared to their ability to form bone in an in vivo situation (fig. 12).

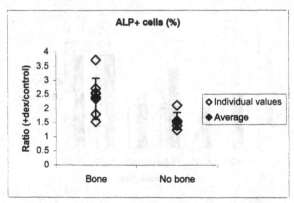

Figure 12. Relative increase in the fraction of ALP⁺ cells in bone forming and non bone forming cultures, after dex treatment. (◊) Individual values of 14 donors; (♦) Average of each population; (*) Statistical significance was observed: p = 0.021 in t test and p = 0.029 in Mann-Whitney U test.

The observations indicated that the degree of culture stimulation by dex was indeed related to the ability of the cultures to form bone tissue in vivo, showing that the ratio between the proportion of cells positive for ALP in the (+) dex and control conditions provides a simple method to assess the early osteoprogenitor cell content (that is, inducible osteogenic cells) of a given population. In summary, the method developed can be extremely relevant for the use of HBMSCs in bone reconstruction, since it allows the detection of cultures with low osteogenic potential pointing out the need for a second bone marrow aspiration procedure or for the use of e.g. bone growth factors in the culture medium to enhance their osteogenicity.

3.5. THE ROLE OF A BONE-LIKE EXTRACELLULAR MATRIX ON THE TISSUE ENGINEERED IMPLANTS

In 1991, Davies et al. [162] described the process of in vitro bone formation by cultured rat BMSCs. Morphology, biochemical and gene expression analysis indicated that the in vitro formed bone closely resembled the natural bone in the early stages of in vivo bone formation [162-164]. In our studies, the osteogenic potential of implants containing a layer of cultured autologous bone-like tissue was compared to the osteogeneicity of constructs that were implanted shortly after cell seeding and before extracellular matrix formation had started. Results demonstrated that bone-like matrix containing implants clearly induced faster bone formation as compared to the cell seeded scaffolds. The faster in vivo bone formation observed on the implants containing a bone-like tissue layer can be attributed to a combination of two factors. Firstly, the cultured cells were in a further stage in the process of osteogenic differentiation, since they had been in the presence of the differentiation factor dexamethasone for a longer period. Secondly, and as suggested by Yoshikawa et al. [143], the in vivo osteogenic potential of these implants can also be related to bone proteins and growth factors that are present in the formed extracellular matrix and contribute to enhance their osteogenicity. In fact, our studies revealed that such constructs were composed of material covered with cells embedded within a collagenous extracellular matrix (rich in collagen type I)and the cells in question expressed mRNA for alkaline phosphatase, osteopontin, osteocalcin and receptor human bone morphogenetic protein 2. Furthermore, when cells were cultured on the scaffolds prior to implantation they formed a bone-like tissue layer not only in the inner but also on the outer surface of the implants. As a result, after implantation, a bone layer delineated the implants outer surfaces and encapsulated the constructs in some areas (fig. 13).

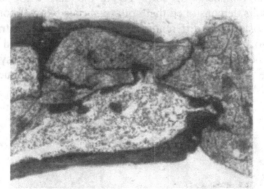

Figure 13. Light micrographs illustrating areas in which bone was formed outside the pore area of HA. (100x).

These observations are contradictory to those by Ohgushi et al. [133], who reported that, in calcium phosphate scaffolds, in vivo bone formation was always restricted to the implant inner pores. In our view, this discrepancy of results may be due to the fact that in the above mentioned work cells were seeded on the material scaffolds and directly implanted. Therefore, fibrous tissue from the host could invade not only the implant pores but also directly contact the implants outer surface.

In summary, if the results obtained using a ectopic implantation model in a small animal were extrapolated to a clinical situation, it is reasonable to assume that implant stability will be achieved earlier if bone-like tissue is present on the grafts at the time of implantation.

3.6. EVALUATION OF DIFFERENT MATERIALS AS SCAFFOLDS FOR BONE TISSUE ENGINEERING

In the development of cell therapy approaches for bone reconstruction, there is a need to engineer adequate materials that will serve as substrates for cell growth, differentiation and bone tissue formation, as well as delivery vehicles for cells and/or tissue at the implantation site. In addition, the scaffold also provides volume, reducing the amount of tissue required to fill the defects. The scaffold should, therefore, allow attachment, growth and differentiation of osteoprogenitor cells. It should also have high porosity to facilitate the ingrowth of vascular tissue that will ensure the survival of the transplanted cells and/or tissue. The selection of the specific material will depend on the site to be reconstructed. In load bearing sites high mechanical support will be required, while in non load bearing defects the mechanical requirements will be much lower. Ideally, the scaffold would also be easily processed into the desired three dimensional shape and it would degrade after bone tissue formation, allowing to obtain a totally natural regenerated tissue. In this report, two biodegradable polymeric systems were evaluated as scaffolds for bone tissue engineering, aiming at non load bearing applications. One of the systems has already been approved for human clinical use [165] and it consists of a block copolymer composed by poly(ethylene glycol)-terephthalate and poly(butylene terephthalate) (PEGT/PBT). The second polymeric system evaluated is composed of corn starch blended with poly(e-caprolactone) (SPCL). In vitro results demonstrated that both materials allowed for bone marrow cell attachment, growth, osteogenic differentiation and extracellular matrix formation. With regard to the in vivo osteogenic potential of the tissue engineered constructs, results have shown that bone marrow cells cultured on both polymeric systems induced the formation of large quantities of self maintained bone tissue, that supported hematopoiesis (fig. 14).

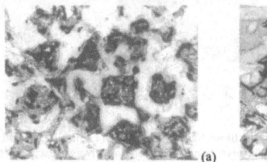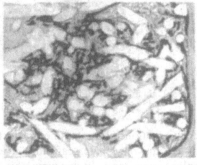

(a)   (b)

Figure 14. Light micrographs illustrating representative sections of osteogenesis in (a) PEGT/PBT and (b) SPCL scaffolds in which rat bone marrow cells were cultured for 7 days prior to implantation under the skin of nude mice for 4 weeks. (40x).

In addition, histomorphometric measurements indicated that the extent of de novo formed bone on both types of polymeric scaffolds was similar to that found in hydroxyapatite. Although direct comparisons between these studies and others using different biodegradable polymeric systems [138, 166-168] are difficult due to the diverse study set ups and material characteristics, it should be noted that to our knowledge, such high degree of bone tissue formation after 4 weeks of implantation has never been reported by others.

3.7. FUTURE APPLICATIONS AND GENERAL CONSIDERATIONS

The tissue engineering approach described in this report is a very powerful technology and the obtained results indicate that such approach would solve most of the drawbacks associated with the traditional bone replacement therapies. This technology can be applied to a wide variety of clinical situations such as spinal fusions, augmentation of bone in the jaw region, reconstruction of bone defects due to the excision of tumours and deformities and replacement of low quality bone in hip arthroplasty revisions. Nevertheless, the period of time required to produce the tissue engineered bone, 4 to 5 weeks depending on the defect size, is a limiting factor since it excludes its application in acute trauma situations.

Results from current clinical trials do not envision problems with regard to the schedule of operations with weeks in advance and the time required to produce the tissue engineered bone. Nevertheless, the present technology can become more flexible if the entire procedure is divided into two steps. Cells can be expanded in culture and then cryopreserved prior to seeding and final culture on the biomaterial scaffold. This will provide the health care institution with more freedom to schedule operations and to later adjust this schedule. Another approach could be the storage of the tissue engineered bone prior to implantation, which then could be used in an off the shelf manner. Nevertheless, the optimal storage conditions, as well as the maximum period of storage without loss of cell viability and osteogenic potential needs to be investigated in future.

Current research is already directed in reducing handling during the period of in vitro culture in order to prevent any kind of contamination. The design of bioreactors in which cells can be expanded directly in the biomaterial scaffold from the beginning to the end of the procedure, will not only reduce risks of contamination but also make the approach more cost effective. Another field of interest that is currently under investigation in our group is the development of biomaterial particulates with very small diameter, which allow producing injectable bone fillers. The biomaterial with the cells can be injected into the jaws or vertebrae to fill defects in which low mechanical performance is required. This kind of approach possesses a major advantage for both patients and clinicians since it only requires a minimal invasive surgery procedure to reconstruct the defect. First results in this area indicated the feasibility of the technique in an ectopic implantation model.

The studies described in this report mainly concern the investigation of a tissue engineering approach in the treatment of isolated bone defects. However, in cases in which all bones are affected, such as osteoporosis and osteogenesis imperfecta, it is not feasible to consider the treatment of all bones by replacement with tissue engineered bone. In the case of age related bone loss (osteoporosis), it can be envisioned that expansion of early osteoprogenitor cells in culture, followed by their systemic administration into the patient, may cure and/or diminish the severity of the disease. With regard to the treatment of diseases involving genetic mutations, molecular engineering of cells is an area that may lead to the development of techniques, which will allow correcting several bone deficiencies. During in vitro expansion, HBMSCs may be genetically manipulated to produce a desired cellular product and then systemically distributed to establish a normal bone marrow microenvironment.

4. Concluding Remarks

The studies described in this work contributed to further characterise the osteogenic potential of cultured bone marrow stromal cells, as well to identify the cell subpopulation mainly responsible for this osteogenic character. In addition, some of the in vitro manipulations required for their extensive subcultivation and in vivo bone formation were defined. Finally, new and adequate scaffold materials were presented. The obtained results demonstrate the potential of the bone tissue engineering technology and indicate that the use of such cell therapy approach to treat bone defects may improve the quality of life for many patients. Nevertheless, to successfully and reproducibly regenerate bone using a tissue engineering strategy in a wide variety of clinical situations, a great deal of knowledge and protocols still need to be further developed. Standardised and optimised bone marrow aspiration procedures have to be defined in order to obtain cell populations with optimal progenitor cell content. In addition, the complete understanding and control of the cells differentiation mechanisms will only be achieved through the development of antibodies that will allow to isolate homogeneous populations of undifferentiated cells and through the definition of the culture conditions required to maintain either an undifferentiated cell character or a certain differentiation stage.

References

1. Athanasiou, K.A., Zhu, C-F., Lanctot, D.R., Agrawal, C.M. and Wang, X. (2000) Fundamentals of biomechanics in tissue engineering of bone, *Tissue Eng.* **6**, 361-381.
2. Buckwater, J.A. and Cooper, R.R. (1987) Bone structure and function. *Instr. Course Lect.* **16**, 27-48.
3. Derkx, P., Nigg, A.L., Bosman, F.T., Birkenhäger-Frenkel, D.H., Houtsmuller, A.B., Pols, H.A.P. and van Leeuwen, J.P.T.M. (1998) Immunolocalization and quantification of noncollagenous bone matrix proteins in methylmethacrylate-embedded adult human bone in combination with histomorphometry, *Bone* **22**, 367-373.
4. Rodan, G.A. (1992) Introduction to bone biology, *Bone* **13**, S3-S6.
5. Yaszemski, M.J., Payne, R.G., Hayes, W.C., Langer, R. and Mikos, A.G. (1996) Evolution of bone transplantation: molecular, cellular and tissue strategies to engineer human bone, *Biomaterials* **17**, 175-185.
6. Aubin, J.E. and Liu, F. (1996) The steoblast lineage, in J.P.Bilezikian, L.G. Raisz and G.A. Rodan (eds.), *Principles of bone biology*, Academic Press, San Diego, USA, pp. 51-67.
7. Brown, K.L. and Cruess, R.I. (1982) Bone and cartilage transplantation in orthopaedic surgery, *J. Bone Joint Surg. Am.* **64**, 270-279.
8. de Boer, H.H. (1988) The history of bone grafts, *Clin. Orthop.*, 226-292.
9. Damien, C.J. and Parsons, J.R. (1991) Bone graft and bone graft substitutes: A review of current technology and applications, *J. Appl. Biomaterials* **2**, 187-208.
10. Coombes, A.G.A. and Meikle, M.C. (1994) Resorbable synthetic polymers as replacements for bone graft, *Clin. Mater.* **17**, 35-67.
11. Laurencin, C.T., Attawia, M. and Borden, M.D. (1999) Advancements in tissue engineered bone substitutes, *Curr. Opinion Orthop.* **10**, 445-451.
12. Lane, J.M., Tomin, E. and Bostrom, M.P.G. (1999) Biosynthetic bone grafting, *Clin. Orthop. Rel. Res.* **367S**, S107-S117.
13. Urist, M.R. (1965) Bone: formation by autoinduction, *Science* **150**, 893-899.
14. Cornell, C.N. and Lane, J.M. (1998) Current understanding of osteoconduction in bone regeneration, *Clin. Orthop.* **355**, S267-73.
15. Prolo, D.J. and Rodrigo, J.J. (1985) Contempory bone graft physiology and surgery, *Clin. Orthop.* **200**, 322-342.
16. Cowley, S.P. and Anderson, L.D. (1983) Hernias through donor sites for iliac-bone grafts, *J. Bone Joint. Surg. Am.* **65**, 1023-1025.
17. Arrington, E.D., Smith, W.J., Chambers, H.G., Bucknell, A.L. and Davino, N.A. (1996) Compilcations of iliac crest bone graft harvesting, *Clin. Orthop.* **329**, 300-309.
18. Coventry, M.B. and Tapper, E.M. (1972) Pelvic instability: a consequence of removing iliac bone for grafting. *J. Bone Joint Surg. Am.* **54**, 83-101.
19. Oklund, S.A., Prolo, D.J., Gutierrez, R.V. and King, S.E. (1986) Quantitative comparisons of healing in cranial fresh autografts, frozen autografts and processed autografts, and allografts in canine skull defects, *Clin. Orthop.* **205**, 269-291.
20. Oikarinen, J. and Korhonen, L.K. (1979) The bone inductive capasity of various bone transplanting materials used for treatment of expermental bone defects, *Clin. Orthop.* **140**, 208-215.
21. Anderson, M.L.C., Dhert, W.J.A., de Bruijn, J.D., Dalmeijer, A.J., Leenders, H., van Blitterswijk, C.A. and Verbout, A.J. (1999) Critical size defect in the goat's os ilium, *Clin. Orthop. Rel. Res.* **364**, 231-239.
22. Strong, D.M., Friedlaender, G.E., Tomford, W.W., Springfield, D.S., Shives, T.C., Burchardt, H., Enneking, W.F. and Mankin, H.J. (1996) Immunologic responses in human recipients of osseous and osteochondral allografts, *Clin. Orthop.* **326**, 107-114.
23. Oreffo, R.O.C. and Triffitt, J.T. (1999) Future potentials for using osteogenic stem cells and biomaterials in orthopedics, *Bone* **25**, 5S-9S.
24. Cook, S.D., Kay, J.F., Thomas, K.A. and Jarcjo, M. (1987) Interface mechanics and histology of titanium and hydroxyla coated titanium for dental implant applications, *Int. J. Oral Maxillofac. Implants* **2**, 15-22.
25. Cook, S.D., Thomas, K.A., Dalton, J.E., Volkman, T.H., Whitecloud, T.S. and Kay, J.F. (1992) Hydroxylapatite coating of porous implants improves bone ingrowth and interface attachment strength, *J. Biomed. Mater. Res.* **26**, 989-1001.
26. Spector, M. (1992) Biomaterial failure, *Orthop. Clin. North Am.* **23**, 211-217.

292

27. Klein, C.P., Patka, P. and den Hollander, W. (1989) Macroporous calcium phosphate bioceramics in dog femora: a histological study of interface and biodegradation, *Biomaterials* 10, 59-62.
28. Lemons, J.E. (1996) Ceramics: past, present, and future, *Bone* 19, 121S-128S.
29. Marcacci, M., Kon, E., Zaffagnini, S., Giardino, R., Rocca, M., Corsi, A., Benvenuti, A., Bianco, P., Quarto, R., Martin, I., Muraglia, A. and Cancedda, R. (1999) Reconstruction of extensive long-bone defects in sheep using porous hydroxyapatite sponges, *Calcif. Tissue Int.* 64, 83-90.
30. Wykrota, L.L., Wykrott, F.H.L. and Garrido, C.A. (2000) Long-term bone regeneration in large human defects using calcium-phosphate particulate, in J.E. Davies (ed.) *Bone Engineering*, em square incorporated, Toronto, Canada, pp. 516-565.
31. Daculsi, G. (1998) Biphasic calcium phosphate concept applied to artificial bone, implant coating and injectable bone substitute, *Biomaterials* 19, 1473-1478.
32. van Blitterswijk, C.A., Hesseling, S.C., Grote, J.J., Koerten, H.K. and de Groot, K. (1990) The biocompatibility of hydrixyapatite ceramic: a study of retrieved human middle ear implants, *J. Biomed. Mater. Res.* 24, 433-453.
33. Furlong, R.J. and Osborn, J.F. (1991) Fixation of hip prostheses by hydroxyapatite ceramic coatings, *J. Bone Joint Surg. Br.* 73, 741-745.
34. Salyer, K.E. and Hall, C.D. (1989) Porous hydroxyapatite as an onlay bone-graft substitute for maillofacial surgery, *Plast. Reconstr. Surg.* 84, 236-244.
35. Hing, K.A., Best, S.M., Tanner, K.E., Revell, P.A. and Bonfield, W. (1998) Histomorphological and biomechanical characterization of calcium phosphates in the osseous environment, *J. Eng. Medicine* 212, 437-451.
36. Marra, K.G., Szem, J.W., Kumta, P.N., DiMilla, P.A. and Weiss, L.E. (1999) In vitro analysis of biodegradable polymer blend/hydroxyapatite composites for bone tissue engineering, *J. Biomed. Mater. Res.* 47, 324-335.
37. Mendes, S.C., Reis, R.L., Bovell, Y.P., Cunha, A.M., van Blitterswijk, C.A. and de Bruijn, J.D. (2001) Biocompatibility testing of novel starch-based materials with potential application in orthopaedic surgery: a preliminary study, *Biomaterials* 22, 2057-2064.
38. Thomson, R.C., Yaszemski, M.J., Powers, J.M. and Mikos, A.G. (1995) Fabrication of biodegradable scaffolds to engineer trabecular bone, *J. Biomater. Sci. Polym. Ed.* 7, 23-38.
39. Goldstein, A.R., Zhu, G., Morris, G.E., Meszlenyi, R.K. and Mikos, A.G. (1999) Effect of osteoblastic culture conditions on the structure of poly(DL-lactic-co-glycolic acid) foam scaffolds, *Tissue Eng.* 5, 421-433.
40. Whang, K., Healy, K.E., Elenz, D.R., Nam, E.K., Tsai, D.C., Thomas, C.H., Nuber, G.W., Glorieux, F.H., Travers, R. and Sprague, S.M. (1999) Engineering bone regeneration with bioabsorbable scaffolds with novel microarchitecture, *Tissue Eng.* 5, 35-51.
41. Calvert, J.W., Marra, K.G., Cook, L., Kumta, P.N., DiMilla, P.A. and Weiss, L.E. (2000) Characterization of osteoblast-like behavior of cultured bone marrow stromal cells on various polymer surfaces, *J. Biomed. Mater. Res.* 52, 279-284.
42. Nichter, L.S., Yadzi, M., Kosari, K., Sridjaja, R., Ebramzadeh, E. and Nimni, M.E. (1992) Demineralized bone matrix polydioxanone composite as a substitute for bone graft: a comparative study in rats, *J.Craniofac. Surg.* 3, 63-69.
43. Peter, S.J., Lu, L., Kim, D.J. and Mikos, A.G. (2000) Marrow stromal osteoblast function on a poly(propylene fumarate)/β-tricalcium phosphate biodegradable orthopaedic composite, *Biomaterials* 21, 1207-1213.
44. Martin, I., Shastri, V.P., Padera, R.F., Yang, J., Mackay, A.J., Langer, R., Vunjak-Novakovic, G. and Freed, L.E. (2001) Selective differentiation of mammalian bone marrow stromal cells cultured on three-dimensional polymer foams, *J. Biomed. Mater. Res.* 55, 229-235.
45. Saad, B., Ciardelli, G., Matter, S., Welti, M., Uhlshmid, G.K., Neuenschwander, P. and Suter, U.W. (1998) Degradable and highly porous polyesterurethane foam as biomaterial: effects and phagocytosis of degradation products in osteoblasts, *J. Biomed. Mater. Res.* 39, 594-602.
46. Radder, A.M., Leenders, H. and van Blitterswijk, C.A. (1994) Interface reactions to PEO/PBT copolymers (polyactiveR) after implantation in cortical bone, *J. Biomed. Mater. Res.* 28, 141-151.
47. Deschamps, A.A., Claase, M.B., Sleijster, W.J., de Bruijn, J.D., Grijpma, D.W. and Feijen, J. (submmited) Design of segmental poly(ether ester) materials and structures for the tissue enginerring of bone, *J. Controlled release* .
48. Winter, G.D. and Simpson, B.J. (1969) Heterotopic bone formed in a synthetic sponge in the skin of young pigs, *Nature* 223, 88-90.
49. Ripamonti, U. (1991) The morphogenesis of bone in replicas of porous hydroxyapatite obtained from conversion of calcium carbonate exoskeletons of coral, *J. Bone Joint Surg. Am.* 73, 692-703.

293

50. Yang, Z.J., Yuan, H., Zou, P., Tong, W., Qu, S. and zhang, X.D. (1997) Osteogenic response to extraskeletally implanted synthetic porous calcium phosphate ceramics: an early stage histomorphological study in dogs, *J. Mater. Sci. Mater. Med.* **8**, 697-701.
51. Yuan, H., Yang, Z., Li, Y., Zhang, X.D., de Bruijn, J.D. and de Groot, K. (1998) Osteoinduction by calcium phosphate biomaterials, *J. Mater. Sci. Mater. Med.* **9**, 723-726.
52. Li, Y., Yuan, H. and Zhang, X. (1998) Calcium phosphate biomaterials:from osteoconduction to osteoinduction, Transactions of the Society for Biomaterials 1998, San Diego, CA; p. 428.
53. Yuan, H., de Bruijn, J.D., Zhang, X., van Blitterswijk, C.A. and de Groot, K. (2001) Bone induction by porous glaas ceramic made from bioglass ® (45S5), *J. Biom. Mater. Res.* **58**, 270-276.
54. Yuan, H., de Bruijn, J.D., Li, Y., Feng, J., Yang, Z., de Groot, K. and Zhang, X. (2001) Bone formation induced by calcium phosphate ceramics in soft tissue of dogs: a comparative study between porous α- TCP and β-TCP, *J. Mater. Sci. Mater. Med.* **12**, 7-13.
55. Yuan, H., Kurasshina, K., de Bruijn, J.D., Li, Y., de Groot, K. and Zhang, X. (1999) A preliminary study on osteoinduction of two kinds of calcium phosphate ceramics, *Biomaterials* **20**, 1799-1806.
56. Yang, Z., Yuan, H., Tong, W., Zou, P., Chen, W. and Zhang, X. (1996) Osteogenesis in extraskeletally implanted porous calcium phosphate ceramics: variability among different kinds of animal, *Biomaterials* **17**, 2131-2137.
57. Ripamonti, U. (1996) Osteoinduction in porous hydroxyapatite implanted in heterotopic sites of different animal models, *Biomaterials* **17**, 31-35.
58. Langer, R. and Vacanti, J.P. (1993) Tissue engineering, *Science* **14**, 920-926.
59. Lind, M. (1996) Growth factors: possible new clinical tools, *Acta Orthop. Scand.* **67**, 407-417.
60. Urist, M.R., DeLange, R.J. and Finerman, G.A. (1983) Bone cell differentiation and growth factors, *Science* **220**, 680-686.
61. Kim, K.J., Itoh, T. and Kotake, S. (1997) Effects of recombinant human bone morphgenetic protein-2 on human bone marrow cells cultured with various biomaterials, *J. Biomed. Mater. Res.* **35**, 279-285.
62. Puleo, D.A. (1997) Dependence of mesenchymal cell responses on duration of exposure to bone morphogenetic protein-2 in vitro, *J. Cell. Physiol.* **173**, 93-101.
63. Balk, M.L., Bray, J., Day, C., Epperly, M., Greenberger, J., Evans, C.H. and Biyibizi, C. (1997) Effect of rhBMP-2 on the osteogenic potential of bone marrow stromal cells from an osteogenesis imperfecta mouse (oim), *Bone* **21**, 7-15.
64. Takiguchi, T., Kobayashi, M., Suzuki, R., Yamaguchi, A., Isatsu, K., Nishihara, T., Nagumo, M. and Hasegawa, K. (1998) Recombinant human bone morphohenetic protein-2 stimulates osteoblast differentiation and suppresses matrix metalloproteinase-1 production in human bone cells isolated from mandibulea, *J. Periodont. Res.* **33**, 476-485.
65. Fromigué, O., Marie, P.J. and Lomri, A. (1998) Bone morphogenetic protein-2 and transforming growth factor-β_2 interact to modulate human bone marrow stromal cell proliferation and differentiation, *J. Cell Biochem.* **68**, 411-426.
66. Whang, K., Tsai, D.C., Aitken, M., Sprague, S.M., Patel, P. and Healy, K.E. (1998) Ectopic bone formation via rhBMP-2 delivery from porous bioabsorbable polymer scaffolds, *J. Biomed. Mater. Res.* **42**, 491-499.
67. Ripamonti, U., Ramoshebi, L.N., Matsaba, T., Tasker, J., Crooks, J. and Teare, J. (2001) Bone induction by BMPs/Ops and related family members in primates, *J. Bone Joint. Sur.* **83**, S116-S127.
68. Aspenberg, P., Jeppsson, C., Wang, J.S. and Boström, M. (1996) Transforming growth factor beta and bone morphogenetic protein 2 for bone ingrowth: a comparison using bone chambers in rats, *Bone* **19**, 499-503.
69. Boden, S.D., Martin, G.J., Morone, M., Ugbo, J.L., Titus, L. and Hutton, W.C. (1999) The use of coralline hydroxyapatite with bone marrow, autogenous bone graft, or osteoinductive bone protein extract for posterolateral lumbar spine fusion, *Spine* **24**, 320-327.
70. Wikesjö, U.M.E., Sorenson, R.G. and Wozney, J.M. (2001) Augmentation of alveolar bone and dental implant osseointegration: clinical implications of studies with rhBMP-2, *J. Bone Joint. Sur.* **83**, S136-S145.
71. Brekke, J.H. and Toth, J.M. (1998) Principles of tissue engineering applied to programmable osteogenesis, *J. Biomed. Mater. Res.* **43**, 380-398.

72. Boyne, P.J., Marx, R.E., Nevins, M., Triplett, G., Lazaro, E., Lilly, L.C., Alder, M. and Nummikoski, P. (1997) A feasibility study evaluating rhBMP-2/absorbable collagen sponge for maxillary sinus floor augmentation, *Int. J. Periodontics Restorative Dent.* **17**, 11-25.

73. Geesink, R.G., Hoefnagels, N.H. and Bulstra, S.K. (1999) Osteogenic activity of OP-1 bone morphogenetic protein (BMP-7) in a human fibular defect, *J. Bone Joint Surg. Br.* **81**, 710-718.

74. King, G.N., King, N., Cruchley, A.T., Wozney, J.M. and Hughes, F.J. (1997) Recombinant human bone morphogenetic protein-2 promotes wound healing in rat periodontal fenestration defects, *J. Dent. Res.* **76**, 1460-1470.

75. Mehler, M.F., Mabie, P.C., Zhang, D. and Kessler, J.A. (1997) Bone morphogenetic proteins in the nervous system, *Trends Neurosci.* **20**, 309-317.

76. Liu, P., Oyajobi, B.O., Russell, R.G.G. and Scutt, A. (1999) Regulation of osteogenic differentiation of human bone marrow stromal cells: interaction between transforming growth factor-β and 1,25(OH)$_2$ vitamin D$_3$ in vitro, *Calcif. Tissue Int.* **65**, 173-180.

77. Sumner, D.R., Turner, T.M., Purchio, A.F., Gombotz, W.R., Urban, R.M. and Galante, J.O. (1995) Enhancement of bone ingrowth by transforming growth factor-beta, *J. Bone Joint Surg. Am.* **77**, 1135-1147.

78. Martin, I., Muraglia, A., Campanile, G., Cancedda, R. and Quarto, R. (1997) Fibroblast growth factor-2 supports ex vivo expansion and maintenance of osteogenic precursors from human bone marrow, *Endocrinology* **138**, 4456-4462.

79. Zellin, G. and Linde, A. (2000) Effects of recombinant human fibroblast growth factor-2 on osteogenic cell populations during orthopic osteogenesis in vivo, *Bone* **26**, 161-168.

80. Nefussi, J.R., Brami, G., Modrowski, D., Oboeuf, M. and Forest, N. (1997) Sequential expression of bone matrix proteins during rat calvaria osteoblast differentiation and bone nodule formation in vitro, *J. Histochem. Cytochem.* **45**, 493-503.

81. Wada, Y., Kataoaka, H., Yokose, S., Ishizuya, T., Miyazono, K., Gao, Y.H., Shibasaki, Y. and Yamguchi, A. (1998) Changes in osteoblast phenotype during differentiation of enzymatically isolated rat calvaria cells, *Bone* **22**, 479-485.

82. Takushima, A., Kitano, Y. and Harii, K. (1998) Osteogenic potential of cultured periosteal cells in a distrated bone gap in rabbits, *J. Surg. Res.* **78**, 68-77.

83. Miura, Y. and O'Driscoll, S.W. (1998) Culturing periosyeum in vitro: the influence of different sizes of explants, *Cell Transplantation* **7**, 453-457.

84. Robey, P.G. and Termine, J.D. (1985) Human bone cells in vitro, *Calcif. Tissue Int.* **37**, 453-460.

85. Gundle, R., Joyner, C.J. and Triffitt, J.T. (1997) Interactions of human osteoprogenitors with porous ceramic following diffusion chamber implantation in a xenogenic host, *J. Mater. Sci. Mater. Med.* **8**, 519-523.

86. Druder, S.P. and Fox, B.S. (1999)Tissue engineering of bone. Cell based strategies, *Clin. Orthop.* **167**, S68-S83.

87. Doherty, M.J., Ashton, B.A., Walsh, S., Beresford, J.N., Grant, M.E. and Canfield, A.E. (1998) Vasular pericytes express osteogenic potential in vitro and in vivo, *J. Bone Miner. Res.* **13**, 828-838.

88. Lecoeur, L. and Ouhayoun, J.P. (1997) In vitro induction of osteogenic differentiation from non-osteogenic mesenchymal cells, *Biomaterials* **18**, 989-993.

89. Friedenstein, A.J., Chailakhyan, R.K. and Gerasimov, U.V. (1987) Bone marrow osteogenic stem cells: invitro cultivation and transplantation in diffusion chambers, *Cell Tissue Kinet.* **20**, 263-272.

90. Maniatopoulos, C., Sodek, J. and Melcher, A.H. (1988) Bone formation in vitro by stromal cells obtained from bone marrow of young adult rats, *Cell Tissue Res.* **254**, 317-329.

91. Owen, M. (1988) Marrow stromal stem cells, *J. Cell Sci.* **10**, 63-76.

92. Ohgushi, H. and Okumura, M. (1990) Osteogenic capacity of rat and human marrow cells in porous ceramics, *Acta Orthop. Scand.* **61**, 431-434.

93. Beresford, J.N. (1989) Osteogenic stem cells and the stromal system of bone and marrow, *Clin. Orthop.* **240**, 270-80.

94. Muschler, G.F., Boehm, C. and Easley, K. (1997) Aspiration to obtain osteoblast progenitor cells from human bone marrow: the influence of aspiration volume, *J. Bone Joint Sur.* **79**, 1699-1709.

95. Krebsbach, P.H., Kuznetsov, S.A. and Robey, P.G. (1999) Bone marrow stromal cells: characterization and clinical application, *Crit. Rev. Oral Biol. Med.* **10**, 165-181.

96. Bianco, P., Riminucci, M., Gronthos, S. and Robey, P.G. (2001) Bone marrow stromal cells: nature, biology, and potential applications, *Stem Cells* **19**, 180-192.

97. Gronthos, S., Graves, S.E. and Simmons, P.J. (1998) Isolation, purification and in vitro manipulation of human bone marrow stromal precursor cells, in J.N. Beresford and M.E. Owen (eds.), *Marrow stromal cell culture*, Cambridge University Press, Cambridge, UK, pp. 26-42.
98. Owen, M.E. (1998) The marrow stromal cell system, in J.N. Beresford and M.E. Owen (eds.), *Marrow stromal cell culture*, Cambridge University Press, Cambridge, UK, pp. 1-9.
99. Muraglia, A., Cancedda, R. and Quarto, R. (2000) Clonal mesenchymal progenitors from human bone marrow differentiate in vitro according to a hierarchical model, *J. Cell Science* **113**, 1161-1166.
100. Aubin, J.E. (2000) Osteogenic cell differentiation, in J.E. Davies (ed.) *Bone Engineering*, em square incorporated, Toronto, Canada, pp. 19-29.
101. Dennis, J.E., Merriam, A., Awadallah, A., Yoo, J.U., Johnstone, B. and Caplan, A.I. (1999) A quadripotential mesenchymal progenitor cell isolated from the marrow of an adult mouse, *J. Bone Miner. Res.* **14**, 700-709.
102. Pittenger, M.F., Makay, A.M., Beck, S.C., Jaiswal, R.K., Douglas, R., Mosca, J.D., Moorman, M.A., Simonetti, D.W., Craig, S. and Marshak, D.R. (1999) Multilineage potential of adult human mesenchymal stem cells, *Science* **284**, 143-146.
103. Bruder, S.P., Jaiswal, N. and Haynesworth, S.E. (1997) Growth kinetics, self-renewal, and the osteogenic potential of purified human mesenchymal stem cells during extensive subcultivation and following cryopreservation, *J. Cell Biochem.* **64**, 278-294.
104. Jaiswal, N., Haynesworth, S.E., Caplan, A.I. and Bruder, S.P. (1997) Ostegenic differentiation of purified, culture-expanded human mesenchymal stem cells in vitro, *J. Cell Biochem.* **64**, 295-312.
105. Pri-Chen, S., Pitaru, S., Lokiec, F. and Savion, N. (1998) Basic fibroblast growth factor enhances the growth and expression of the osteogenic phenotype of dexamethasone-treated human bone marrow-derived bone-like cells in culture, *Bone* **23**, 111-117.
106. Malaval, L., Modrowski, D., Gupta, A.K. and Aubin, J.E. (1994) Cellular expression of bone-related proteins during in vitro osteogenesis in rat bone marrow stromal cell cultures, *J. Cell Physiol.* **158**, 555-572.
107. Rickard, D.J., Kassem, M., Hefferan, T.E., Sarkar, G., Spelsberg, T.C. and Riggs, B.L. (1996) Isolatin and characterization of osteoblast precursor cells from human bone marrow, *J. Bone Miner. Res.* **11**, 312-324.
108. Fromigué, O., Marie, P.J. and Lomri, A. (1997) Differential effects of transforming growth factor β2, dexamethasone and 1,25-dihydroxyvitamin D on human bone marrow stromal cells, *Cytokine* **9**, 613-623.
109. Cooper, L.F., Harris, C.T., Bruder, S.P., Kowalski, R. and Kadiyala, S. (2001) Incipient analysis of mesenchymal stem-cell-dreived osteogenesis, *J. Dent. Res.* **80**, 314-320.
110. Komori, T., Yagi, H., Nomura, S., Yamaguchi, A., Sasaki, K., Deguchi, K., Shimizu, Y., Bronson, R.T., Gao, Y.H., Inada, M., Sato, M., Okamoto, R., Kitamura, Y., Yoshiki, S. and Kishimoto, T. (1997) Targeted disruption of cbfa1 results in a complete lack of bone formation owing to maturational arrest of osteoblasts, *Cell* **89**, 755-764.
111. Haynesworth, S.E., Baber, M.A. and Caplan, A.I. (1992) Cell surface antigens on human marrow-derived mesenchymal cells are detected by monoclonal antibodies, *Bone* **13**, 69-80.
112. Joyner, C.J., Bennett, A. and Triffitt, J.T. (1997) Identification and enrichment of human osteoprogenitor cells by using differentiation stage-specific monoclonal antibodies, *Bone* **21**, 1-6.
113. Simmons, P.J. and Torok-Storb, B. (1991) Identification of stromal cell precursors in human bone marrow by a novel monoclonal antibody, STRO-1, *Blood* **78**, 55-62.
114. Gronthos, S., Graves, S.E., Ohta, S. and Simmons, P.J. (1994) The STRO-1⁺ fraction of adult human bone marrow contains the osteogenic precursors, *Blood* **84**, 4164-4173.
115. Gronthos, S. and Simmons, P.J. (1995) The growth factor requirements of STRO-1-positive human bone marrow stromal precursors under serum-deprived condotions in vitro, *Blood* **85**, 929-940.
116. Oyajobi, B.O., Lomri, A., Hott, M. and Marie, P.J. (1999) Isolation and characterization of human clonogenic osteoblast progenitors immunoselected from fetal bone marrow stroma using STRO-1 monoclonal antibody, *J. Bone Miner. Res.* **14**, 351-361.
117. Stewart, K., Walsh, S., Screen, J., Jefferiss, C.M., Chainey, J., Jordan, G.R. and Beresford, J.N. (1999) Further characterization of cells expressing STRO-1 in cultures of adult human bone marrow stromal cells, *J. Bone Miner. Res.* **14**, 1345-1356.
118. Walsh, S., Jefferiss, C., Stewart, K., Jordan, G.R., Screen, J. and Beresford, J.N. (2000) Expression of the developmental markers STRO-1 and alkaline phosphatase in cultures of human

296

marrow stromal cells: regulation by fibroblast growth factor (FGF)-2 and relationship to the expression of FGF receptors 1-4, *Bone* **27**, 185-195.

119. Bennett, J.H., Joyner, C.J., Triffitt, J.T. and Owen, M.E. (1991) Adipocytic cells cultured from marrow have osteogenic potential, *J. Cell Science* **99**, 131-139.

120. Galotto, M., Campanile, G., Robino, G., Cancedda, F.D., Bianco, P. and Cancedda, R. (1994) Hypertrophic chondrocytes undergo further differentiation to osteoblast-like cells and participate in the initial bone formation in developing chick embryo, *J. Bone Miner. Res.* **9**, 1239-49.

121. Yoshikawa, T., Ohgishi, H., Dohi, Y. and Davies, J.E. (1997) Viable bone formation in porous hydroxyapatite: marrow cell-derived in vitro bone on the surface of ceramics, *Bio-Med. Mater. Eng.* **7**, 49-58.

122. Herbertson, A. and Aubin, J.E. (1995) Dexamethasone alters the subpopulation make-up of rat bone marrow stromal cell cultures, *J. Bone Miner. Res.* **10**, 285-294.

123. Hanada, K., Dennis, J.E. and Caplan, A.I. (1997) Stimulatory effects of basic fibroblast growth factor and bone morphogenetic protein-2 on osteogenic differentiation of rat bone marrow-derived mesenchymal stem cells, *J. Bone Miner. Res.* **12**, 1606-1614.

124. Vilamitjana-Amedee, J., Bareille, R., Rouais, F., Caplan, A.I. and Harmand, M.F. (1993) Human bone marrow stromal cells express an osteoblastic phenotype in culture, *In Vitro Cell Dev. Biol.* **29A**, 699-707.

125. Lennon, D.P., Haynesworth, S.E., Bruder, S.P., Jaiswal, N. and Caplan, A.I. (1996) Human and animal mesenchymal progenitor cells from bone marrow: identification of serum for optimal selection and proliferation, *In Vitro Cell Dev. Biol.* **32**, 602-611.

126. Reid, I.R. (1997) Glucocorticoid osteoporosis – mechanisms and management, *Eur. J. Endocrin.* **137**, 209-217.

127. Cheng, S-L., Yang, J.W., Rifas, L. and Zhang, U-F. (1994) Avioli LV, Differntiation of human bone marrow osteogenic stromal cells in vitro: induction of the osteoblast phenotype by dexamethasone, *Endocrinology* **134**, 277-286.

128. Kim, C-H., Cheng, S-L and Kim, G.S. (1999) Effects of dexamethasone on proliferation, activity, and cytokine secretion of normal human bone marrow stromal cells: possible mechanisms of glucocorticoid-induced bone loss, *J .Endocr.* **162**, 371-379.

129. Dieudonné, S.C., Kerr, J.M., Xu, T., Sommer, B., DeRubeis, A.R., Kuznetsov, S.A., Kim, I-S., Robey, P.G. and Young, M.F. (1999) Differential display of human marrow stromal cells reveals unique mRNA expression patterns in response to dexamethasone, *J. Cell Biochem.* **76**, 231-243.

130. Oreffo, R.O.C., Kusec, V., Romberg, S. and Triffitt, J.T. (1999) Human bone marrow osteoprogenitors express estrogen receptor-alpha and bone morphogenetic proteins 2 and 4 mRNA during osteoblastic differntiation, *J. Cell Biochem.* **75**, 382-392.

131. Leboy, P.S., Beresford, J.N., Devlin, C. and Owen, M.E. (1991) Dexamethasone induction of osteoblast mRNAs in rat marrow stromal cell cultures, *J. Cell Physiol.* **146**, 370-378.

132. Haynesworth, S.E., Goshima, J., Goldberg, V.M. and Caplan, A.I. (1992) Characterization of cells with osteogenic potential from human marrow, *Bone* **13**, 81-88.

133. Ohgushi, H., Okumura, M., Tamai, S., Shors, E.C. and Caplan, A.I. (1990) Marrow cell induced osteogenesis in porous hydroxyapatite and tricalcium phosphate: a comparitive histomorphometric study of ectopic bone formation, *J. Biomed. Mater. Res.* **24**, 1563-1570.

134. Yoshikawa, T., Ohgushi, H. and Tamai, S. (1996) Immediate bone forming capability of prefabricated osteogenic hydroxyapatite, *J. Biomed. Mater. Res.* **32**, 481-492.

135. Dennis, J.E., Konstantakos, E.K., Arm. D. and Caplan, A.I. (1998) In vivo osteogenesis assay: a rapid method for quantitative analysis, *Biomaterials* **19**, 1323-1328.

136. de Bruijn, J.D., van den Brink, I., Mendes, S.C., Dekker, R., Bovell, Y.P. and van Blitterswijk, C.A. (1999) Bone induction by implants coated with cultured osteogenic bone marrow cells, *Adv. Dent. Res.* **13**, 74-81.

137. Yoshikawa, T., Ohgishi, H., Nakajima, H., Yamada, E., Ichijima, K., Tamai, S. and Ohta, T. (2000) In vivo osteogenic durability of cultured bone in porous ceramics, *Transplantation* **69**, 128-134.

138. Ishaug-Riley, S.L., Crane, G.M., Gurlek, A., Miller, M.J., Yasko, A.W., Yaszemski, M.J. and Mikos, A.G. (1997) Ectopic bone formation by marrow stromal osteoblast transplantation using poly(DL-lactic-co-glycolic acid) foams implanted into the rat mesentery, *J. Biomed. Mater. Res.* **36**, 1-8.

139. Krebsbach, P.H., Kuznetsov, S.A., Satomura, K., Emmons, R.V.B., Rowe, D.W. and Robey, P.G. (1997) Bone formation in vivo: comparison of osteogenesis by transplanted mouse and human marrow stromal fibroblasts, *Transplantation* **63**, 1059-1069.

140. Anselme, K., Noël, B., Flautre, B., Blary, M-C., Delecourt, C., Descamps, M. and Hardouin, P. (1999) Association of porous hydroxyapatite and bone marrow cells for bone regeneration, *Bone* **25**, 51S-54S.

141. Solchaga, L.A., Dennis, J.E., Goldberg, V.M. and Caplan, A.I. (1999) Hyaluronic acid-based polymers as cell carriers for tissue-engineered repair of bone and cartilage, *Orthop. Res.* **17**, 205-213.

142. Kuznetsov, S.A., Krebsbach, P.H., Satomura, K., Kerr, J., Riminucci, M., Benayahu, D. and Robey, P.G. (1997) Single-colony derived strains of human marrow stromal fibroblasts form bone after transplantation in vivo, *J. Bone Miner. Res.* **12**, 1335-1347.

143. Yoshikawa, T., Ohgushi, H., Uemura, T., Nakajima, H., Ichijima, K., Tamai, S. and Tateisi, T. (1998) Human marrow cells-derived cultured bone in porous ceramics, *Bio-Med. Mater. Eng.* **8**, 311-320.

144. Lennon, D.P., Haynesworth, S.E., Arm, D.M., Baber, M.A. and Caplan, A.I. (2000) Dilution of human mesenchymal stem cells with dermal fibroblasts and the effects on in vitro and in vivo osteochondrogenesis, *Developmental Dynamics* **219**, 50-62.

145. Kadiyala, S., Jaiswal, N. and Bruder, S.P. (1997) Culture-expanded, bone marrow-derived mesenchymal stem cells can regenerate a critical-sized segmental bone defect, *Tissue Eng.* **3**, 173-185.

146. Bruder, S.P., Kurth, A.A., Shea, M., Hayes, W.C., Jaiswal, N. and Kadiyala, S. (1998) Bone regeneration by implantation of purified, culture-expanded human mesenchymal stem cells, *J. Orthop. Res.* **16**, 155-162.

147. Bruder, S.P., Kraus, K.H., Goldberg, V.M. and Kadiyala, S. (1998) The effect of implants loaded with autologous mesenchymal stem cells on the healing of canine segmental bone defects, *J. Bone Joint Surg.* **80-a**, 985-996.

148. Kon, E., Muraglia, A., Corsi, A., Bianco, P., Marcacci, M., Martin, I., Boyde, A., Ruspantini, I., Chistolin, P., Rocca, M., Giardino, R., Cancedda, R. and Quarto, R. (2000) Autologous bone marrow stromal cells loaded onto porous hydroxyapatite ceramic accelerate bone repair in critical-size defects of sheep long bones, *J. Biomed. Mater. Res.* **49**, 328-337.

149. Petite, H., Viateau, V., Bensaïd, W., Meunier, A., de Pollak, C., Bourguignon, M., Oudina, K., Sedel, L. and Guillemin, G. (2000) Tissue-engineered bone regeneration, *Nature Biotechnology* **18**, 959-963.

150. Louisia, S., Stromboni, M., Meunier, A., Sedel, L. and Petite, H. (1999) Coral grafting supplemented with bone marrow, *J. Bone Joint Surg.* **81-B**, 719-724.

151. de Bruijn, J.D., Yuan, H., Dekker, R., Layrolle, P., de Groot, K. and van Blitterswijk, C.A. (2000) Osteoinductive biomimetic calcium-phosphate coatings and their potential use as tissue-engineering scaffolds, in J.E. Davies (ed.) *Bone Engineering*, em square incorporated, Toronto, Canada, pp. 421-431.

152. Filshie, R.J.A., Zannettino, A.C.W., Makrynikola, V., Gronthos, S., Henniker, A.J., Bendall, L.J., Gottlieb, D.J., Simmons, P.J. and Bradstock, K.F. (1998) MUC18, a member of the immunoglobulin superfamily, is expressed on bone marrow fibroblasts and a subset of hematological malignancies, *Leukemia* **12**, 414-421.

153. Cortes, F., Deschaseaux, F., Uchida, N., Labastie, M.C., Friera, A.M., He, D., Charbord, P. and Peault, B. (1999) HCA, an immunoglobulin-like adhesion molecule present on the earliest human hematopoietic precursor cells, is also expressed by stromal cells in blood- forming tissues, *Blood* **93**, 826-837.

154. Bruder, S.P., Horowitz, M.C., Mosca, J.D. and Haynesworth, S.E. (1997) Monoclonal antibodies reactive with human osteogenic cell surface antigens, *Bone* **21**, 225-235.

155. Inui, K., Oreffo, R.O.C. and Triffitt, J.T. (1997) Effects of beta mercapto ethanol on the proliferation and differentiation of human osteoprogenitor cells, *Cell Biol. Int.* **21**, 419-421.

156. Phinney, D.G., Kopen, G., Righter, W., Webster, S., Tremain, N. and Prockop, D.J. (1999) Donor variation in the growth properties and osteogenic potential of human marrow stromal cells, *J. Cell Biochem.* **75**, 424-436.

157. Muschler, G.F., Nitto, H., Boehm, C.A. and Easley, K.A. (2001) Age and gender related changes in the cellularity of human bone marrow and the prevalence of osteoblastic progenitors, *J. Orthop. Res.* **19**, 117-125.

158. Bab, I., Passi-Even, L., and Gazit, D. (1988) Osteogenesis in in-vivo diffusion chamber cultures of human bone marrow cells, *Bone Mineral* **4**, 373-386.

159. Knight, S.M. and Gowen, M. (1992) The effect of age and sex on bone cell function, *Calcif. Tissue Int.* **50**, Suppl 1 A12.

298

160. D'ippolito, G., Schiller, P.C., Ricordi, C., Roos, B.A. and Howard, G.A. (1999) Age-related osteogenic potential of mesenchymal stromal cells from human vertebral bone marrow, *J. Bone Miner. Res.* **14**, 1115-1122.

161. Evans, C.E., Galasko, C.S. and Ward, C. (1990) Effect of donor age on the growth in vitro of cells obtained from human trabecular bone, J. Orthop. Res. **8**, 234-237.

162. Davies, J.E., Chernecky, B., Lowenberg, B. and Shiga, A. (1991) Deposition and resorption of calcified matrix in vitro by rat marrow cells, *Cells Mater.* **1**, 3-15.

163. de Bruijn, J.D., Davies, J.E., Flach, J.S., de Groot, K. and van Blitterswijk, C.A. (1992), in Tissue-Inducing Biomaterials L. Cima et al. (eds.) , Boston, USA, Res. Soc. Sp. Proc., vol 252, p. 63.

164. Yao, K.L., Todescan, R. and Sodek, J. (1994) Temporal changes in the matrix protein synthesis and m-RNA expression during mineralised tissue formation by adult rat bone marrow cells in culture, *J. Bone Miner. Res.* **9**, 231-240.

165. Bulstra, S.K., Geesink, R.G., Bakker, D., Bulstra, T.H., Bouwmeester, S.J. and van der Linden, A.J. (1996) Femoral canal occlusion in total hip replacement using resorbable and flexible cement restrictor, *J. Bone Joint Surg. Br.* **78**, 892-898.

166. Breitbart, A.S., Grande, D.A., Kessler, R., Ryaby, J.T., Fitzsimmons, R.J. and Grant, R.T. (1998) Tissue engineered bone repair of calvarial defects using cultured periosteal cells, *Plast. Reconst. Surg.* **101**, 567-574.

167. Perka, C., Schultz, O., Spitzer, R.S., Lindenhayn, K., Burmester, G.R. and Sittinger, M. (2000) Segmental bone repair by tissue-engineered periosteal cell transplants with bioresorbable fleece and fibrin scaffolds in rabbits, *Biomaterials* **21**, 1145-1153.

168. Weng, Y., Cao, Y., Silva, C.A., Vacanti, M.P. and Vacanti, C.A. (2001) Tissue engineered composites of bone and cartilage for mandible condylar reconstruction, *J. Oral Maxillof. Surg.* **59**, 185-190.

INJECTABLE BIODEGRADABLE MATERIALS FOR ORTHOPAEDIC TISSUE ENGINEERING*

J.S. TEMENOFF AND A.G. MIKOS
Department of Bioengineering, Rice University
6100 Main, Houston, TX, 77005-1892, U.S.A.

Abstract

The large number of orthopaedic procedures performed each year, including many performed arthroscopically, has led to great interest in injectable biodegradable materials for regeneration of bone and cartilage. A variety of materials have been developed for these applications, including ceramics, naturally-derived substances and synthetic polymers. These materials demonstrate overall biocompatibility and appropriate mechanical properties, as well as promote tissue formation, thus providing an important step towards minimally invasive orthopaedic procedures. This review provides a comparison of these materials based on mechanical properties, biocompatibility and regeneration efficacy. Advantages and disadvantages of each material are explained and design criteria for injectable biodegradable systems are provided.

1. Introduction

Every year in the United States, there are approximately 900,000 hospitalizations due to fractures (1). Of these, over 500,000 are fractures of the lower extremities (1), therefore requiring intervention to restore ambulation. In addition, 36 million Americans suffer from some form of arthritis (1). Although these maladies affect different tissues (bone and articular cartilage), such statistics demonstrate the vast need for new techniques to treat a wide variety of orthopaedic injuries.

When a lower extremity fracture occurs, immediate mechanical support of the bone is needed for continued limb loading. However, to reduce stress shielding and allow the fracture to heal completely, the support must be removed. Thus, a biodegradable support material would be ideal, as it would eliminate the need for a second surgery to remove the fixation device. Additionally, with the increasing popularity of arthroscopic procedures in orthopaedics, there has been great interest in fixation materials that are

*Reprinted from *Biomaterials* 21, J.S. Temenoff and A.G. Mikos, "Injectable biodegradable materials for orthopaedic tissue engineering.", pp. 2405-2412, Copyright (2000), with permission from Elsevier Science.

299

R.L. Reis and D. Cohn (eds.),
Polymer Based Systems on Tissue Engineering, Replacement and Regeneration, 299–312.
© 2002 *Kluwer Academic Publishers. Printed in the Netherlands.*

injectable as well as biodegradable. Currently, the most commonly-used injectable bone cement is poly(methyl methacrylate) (PMMA), but it suffers from the fact that it is not degraded and that its high curing temperatures can cause necrosis of surrounding tissue (2). Therefore, further development of alternative injectable materials is necessary, not only for fracture fixation, but also as cell carriers for tissue-engineered regeneration of bone and cartilage. This review will provide an overview of injectable biodegradable biomaterials for use in bone and cartilage regeneration, including design requirements for such materials, as well as comparisons of mechanical properties, biocompatibility and efficacy for each system. When developing a biomaterial for use in orthopaedics, there are several important requirements that must be satisfied:

Biocompatibility. Biodegradable materials for orthopaedic applications, as for any other application, must first be biocompatible. This means that the material must not elicit an unresolved inflammatory response nor demonstrate extreme immunogenicity or cytotoxicity. In addition, because it degrades *in vivo*, this must be true not only for the intact material and any of its unreacted components, but also for the degradation products (3).

Mechanical Properties. Especially important in orthopaedics are the initial mechanical properties of the biomaterial to be implanted. These properties must be as similar as possible to those of the tissue that is to be regenerated. As well as providing proper support in the early stages of healing, graded load transfer is needed later in the process for creation of replacement tissue that is identical to the original (4). While many mechanical properties should be considered for materials to be used in orthopaedics, including those in compression, tension and torsion, compressive properties are the most relevant for replacement of cancellous bone, while tensile properties are important for cortical bone (5). Compressive strength will be used to compare mechanical properties of the various materials covered in this review as it is the only parameter reported consistently throughout the literature.

Promotion of Tissue Formation. Properties such as amount of void space and degradation time should be chosen to encourage tissue growth and vascularization (if appropriate) within the material. It is important that the degradation rate be coupled to the rate of tissue formation so that the load-bearing capabilities of the tissue are not compromised (3).

Sterilizability. As with all implanted materials, orthopaedic materials must be easily sterilizable to prevent infection. The method of sterilization, however, must not interfere with the bioactivity of the materials or alter its chemical composition which could, in turn, affect its biocompatibility or degradation properties (4).

Furthermore, biodegradable materials that are injectable must posses additional properties:

Setting Time/Temperature Change. The material should set in several minutes to minimize the length of the procedure while allowing surgeons ample time for placement before hardening (4). If the setting reaction involves a temperature change, the increase or decrease should be as small as possible to reduce damage to the surrounding tissue. *Viscosity/Ease of Handling.* Ease of handling is of utmost importance for clinical use of any biomaterial. Therefore, viscous properties must be balanced between the need for the material to remain at the site of injection and the need for the surgeon to easily manipulate its placement.

Recent development of a variety of injectable materials, both ceramic-based and polymeric-based, have fulfilled many of these design criteria for very diverse orthopaedic applications. The results are promising and pave the way for a new era of orthopaedic materials that provide more and better tissue regeneration with less patient discomfort than other techniques currently employed.

2. Bone Tissue Engineering

Both polymeric materials and ceramics have been studied as a means to repair defects in bone. To facilitate comparison between these classes of materials, the focus of this review will be on those systems that have been well-characterized, especially in the area of ceramics where many combinations of calcium-phosphate materials have been tested.

2.1 CERAMIC MATERIALS

Since 1892, when Plaster of Paris was first used as a bone cement (6), ceramics have often been chosen to aid in fracture fixation and filling of bony defects. They promote bony ingrowth, are biocompatible and harden *in situ* (7). Further studies have shown that the most common types of calcium phosphate ceramics, hydroxyapatite (HA, $Ca_{10}(PO_4)_6(OH)_2$) and tricalciumphosphate (TCP, $Ca_3(PO_4)_2$) have different characteristics *in vivo*, although both forms have Ca/P ratios within the range known to promote bone ingrowth (1.50-1.67). In general, HA was found to be more osteogenic while TCP was degraded much faster (8,9).

Using this information, Daculsi et al. developed a biphasic calcium phosphate (BCP), a combination of HA and β-TCP (9). Particles of BCP (60/40 HA/β-TCP weight ratio) were incorporated at various concentrations in a 2% methyl cellulose carrier gel, so that the material was easily injectable. When implanted in rabbit distal femurs, none of the composites showed unresolved inflammation after ten weeks and bone ingrowth proceeded from the perimeter inward at a greater rate than in BCP blocks alone (10,11). This may be because the cellulose spaces the BCP particles apart, providing a ceramic scaffold with voids that favor bone and blood vessel formation (12,13). A study using particles of different sizes demonstrated that smaller grains result in faster bone formation, with 50% of the original BCP resorbed after 2 weeks in the small (40-80 μm) grain composite (10). However, the major drawback to this system is that it has no significant initial mechanical properties. If bone ingrowth were very rapid, this problem may eventually be overcome, but the lack of sufficient initial mechanical characteristics can also lead to difficulty in maintenance of the composite within the defect during surgery (10).

In response to these concerns, other researchers have chosen to investigate the use of ceramic pastes (7,14-19). This type of calcium phosphate ceramics sets *in-situ*, so that it is initially injectable but after several minutes undergoes non-exothermic setting to form materials with high compressive strength. These calcium phosphates can be divided into four types, depending on what is precipitated during setting: dicalcium phosphate dihydrate (DCDP, $CaHPO_4 \cdot 2H_2O$), calcium magnesium phosphate (CMP, $Ca_4Mg_5(PO_4)_6$), octacalcium phosphate (OCP, $Ca_8(HPO_4)_2(PO_4)_4 \cdot 5H_2O$), or calcium deficient hydroxyapatite (CDHA, $Ca_9(HPO_4)(PO_4)_5OH$) (20). Of these, the two types that have been studied most extensively are DCPD and CDHA.

A DCPD cement currently under investigation has an initial compressive strength of 25-35 MPa, higher than that of cancellous bone (5-10 Mpa (21)) (7,14). The cement is formed by mixing β-TCP, monocalcium phosphate monohydrate (MCPM, $Ca(H_2PO_4)_2 \cdot H_2O$), and calcium sulfate hemihydrate (CSH, $CaSO_4 \cdot 1/2H_2O$)) in an aqueous solution. The amounts of each component vary depending on the exact properties desired, but is generally about 40 wt.% β-TCP, 13 wt.% MCPM and 10 wt.% CSH (7,14). These are available in powder form and can be sterilized with γ-radiation (14).

Cylinders (4.7 mm diameter X 10 mm long) of this cement exhibited overall biocompatibility and were almost completely replaced by bone at 8 weeks after implantation in defects (5 mm diameter) in the distal femoral chondyle of rabbits (14). When injected, the material was easily shaped and formed a good interface with the existing bone (15). However, although this ceramic has acceptable mechanical properties and is quickly replaced by bone, DCPD cements have been found to turn acidic during setting, potentially inducing inflammation around the implantation site (20).

In contrast to the DCPD cements, CDHA cements remain neutral after setting (20). A material of this type is under development by several investigators (16-19, 22-25). The cement, a mixture of tetracalcium phosphate (TTCP, $Ca_4(PO_4)_2O$) and dicalcium phosphate anhydrous (DCPA, $CaHPO_4$), sets within 5-7 minutes in a sodium hydrogen phosphate solution. The components are sterilizable via γ-radiation and have demonstrated bone formation with no unresolved inflammatory response when implanted in rat tibiae for 8 weeks (17). Initial compressive strength was 10 MPa for specimens 6 mm in diameter by 3 mm high (16), but premature decay can ensue if the material contacts blood before completely setting (19). Also, this formulation shows no signs of degradation at 8 weeks (17). Addition of other materials such as alginate, chitosan and collagen is currently being investigated to resolve these problems (22-25).

A different type of CDHA cement has recently been developed with similar setting characteristics (26-30). A mixture of MCPM, TCP and calcium carbonate (CC, $CaCO_3$) in a sodium phosphate solution, this material has initial compressive strength of 10 MPa. After 12 hours hardening *in vivo*, the maximum compressive strength of 55 MPa and tensile strength of 2.1 MPa is attained (27). This value falls between that of cancellous (5-10 MPa) and cortical bone (130-220 MPa) in compression, but, in tension, the cement's properties are lower than those of native bone (5-10 MPa cancellous, 80-150 MPa cortical (21)).

This material can be sterilized by γ-radiation, shows no extended inflammatory response, and has been observed to cause bone replacement in canine proximal tibial metaphyseal and distal femoral metaphyseal defects (26). Because of these properties, the cement has recently been FDA approved for use in fixation of distal radial fractures (31). Clinical trials using the cement in conjunction with screw fixation for hip fractures improved load transfer and reduced screw cutout (29).

While this material is not currently approved for use in orthopaedics, encouraging clinical trials suggest that it may soon be employed throughout the body. However, it must be noted that in canines, full cement resorption did not occur during the 78 week experimentation period (26). Concerns about this lengthy resorption time have led to the development of another type of cement that sets endothermically when placed in the body (32). Calcium phosphate powders, sterilized through γ-radiation, were mixed in saline and implanted in femoral-slot defects in canines. Although initial mechanical properties of the material were not reported, nearly complete resorption was seen 1-2 months after implantation and histology showed lamellar or Haversian bone present in the defects after 12 weeks (32).

Despite such work on cements that are quickly resorbed, many existing calcium phosphate materials degrade very slowly (17-26), which can lead to decreased bone regeneration at the site of the implant. And, while these cements exhibit good biocompatibility (17,26,32) and perform well in compression (7,14,16,27), tensile strengths are still below those found in natural bone (27). In an effort to address these concerns, researchers have chosen to investigate polymeric materials for use in orthopaedic applications.

2.2 POLYMERIC MATERIALS

Polymers can offer some distinct advantages over ceramic materials. Like ceramics, they are injectable and harden *in situ*, but the mechanical properties and degradation times can be more easily tailored with polymers than with calcium phosphate materials (21). In addition, the widely varied polymer chemistry allows the possibility of functionalization to interact specifically with certain cell types. However, depending on the polymer formulation, they may be less biocompatible and more difficult to sterilize without damage than their ceramic counterparts.

To provide the necessary mechanical strength for use in orthopaedics, injectable polymers must be polymerized or cross-linked *in situ*. This curing is usually initiated either chemically or via the use of light. Recently, a group of photopolymerizable poly(anhydrides) with suitable strength has been developed to fill bony defects (21, 33-

304

35). These materials are polymers of sebacic acid (SA) alone, or copolymers of SA and 1,3-bis(p-carboxyphenoxy) propane (CPP), or 1,6-bis(p-carboxyphenoxy) hexane (CPH) (see Figure 1). The most effective means of photopolymerization for these polymers was found to be 1.0 wt.% camphorquinone (CQ) and 1.0 wt.% ethyl-4-N,N-dimethyl aminobenzoate (4EDMAB) with 150 mW/cm^2 of blue light. This system has been widely used in dentistry and allows penetration of the light to larger depths than UV systems because of the tendency of CQ to quickly photobleach (21,34).

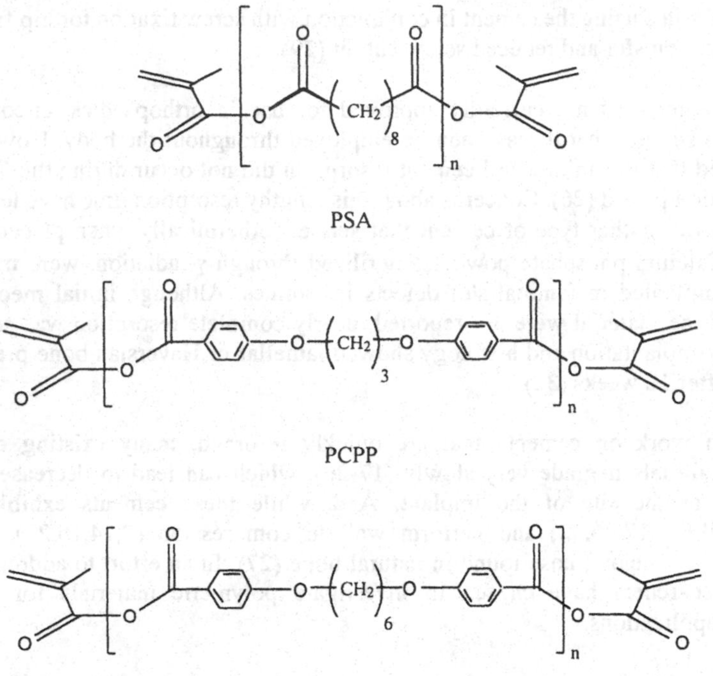

PSA

PCPP

PCPH

FIGURE 1. Diagram of the photopolymerizable dimethacrylated polyanhydrides currently under investigation for bone repair. PSA: poly(sebacic acid), PCPP: poly(1,3-bis(p-carboxyphenoxy)propane), PCPH: poly(1,6-bis(p-carboxyphenoxy)hexane) (21, 33).

Depending on the monomer(s) used, the mechanical properties as well as degradation time can be varied. In general, compressive strengths of 30-40 MPa and tensile strengths of 15-27 MPa were obtained, similar to those of cancellous bone (21-33). The tensile properties of these polymers are much higher than those of a CDHA ceramic (27), which may be important to prevent fractures in the time before the implants are replaced by host bone. Poly(sebacic acid) (PSA) degrades quickly (about 54 hours in saline), while poly(1,6-bis(p-carboxyphenoxy)hexane (PCPH) degrades much more slowly (estimated 1 year). Therefore, combinations of different amounts of SA with CPH would result in a polymer with degradation properties custom-designed for a specific application (21,33). Because these polymers are surface-eroding, they maintain bulk mechanical properties while undergoing degradation. With either PSA or PCPH, over 70% of the initial tensile modulus remained at 50% mass degradation (33). To further modulate either degradation time or mechanical properties, these polymers have been photopolymerized with particles or other linear polymers within them, making interpenetrating networks (IPNs). In this case, the additive within the IPN could alter the physical properties of the material to better suit the chosen application (33).

Minimal inflammatory response to the SA/CPP IPN was observed when implanted subcutaneously in rats up to 28 weeks. Loose vascularized tissue had grown into the implant at 28 weeks, with no evidence of fibrous capsule formation (35). No data have been reported on polymer sterilizability, heat generation during polymerization or polymerization time. However, a 12-week study using a 2.3 mm diameter full-thickness defect in the distal femur of rabbits showed good tolerance of the SA/CPP IPN and osseous tissue formation in the outer zone of some implants (35).

While this family of photopolymerizable polymers has many important characteristics for use in orthopaedics, dependence on light to polymerize the material may be impractical for use in deep crevices occurring in some bones. In these cases, the investigators suggest the use of a combination of photopolymerized and chemically polymerized materials (34).

In order to eliminate the need for the defect to be exposed to light and work toward a more minimally invasive surgery, other researchers are exploring polymers that, rather than polymerized, are chemically cross-linked in vivo. In this area, a promising candidate is poly(propylene fumarate) (PPF). PPF is an unsaturated linear polyester that can be crosslinked through the fumarate double bond (see Figure 2). The degradation products are propylene glycol, poly(acrylic acid-co-fumaric acid), and fumaric acid, a substance which occurs naturally as a part of the Kreb's cycle (36). Many methods to synthesize PPF have been explored, and each results in different polymer properties (see (37) for review of these methods) (37). Cross-linking usually occurs with either methylmethacrylate (MMA) (38,39) or N-vinyl pyrrolidone monomers (40-42) and benzoyl peroxide as a radical initiator. Depending on the ratio of initiator, monomer, and PPF, the curing time can range from 1-121 minutes (43). Although the cross-linking reaction is exothermic, it has been shown that temperatures never reach above 48°C during setting as compared to 94°C for a PMMA bone cement (43).

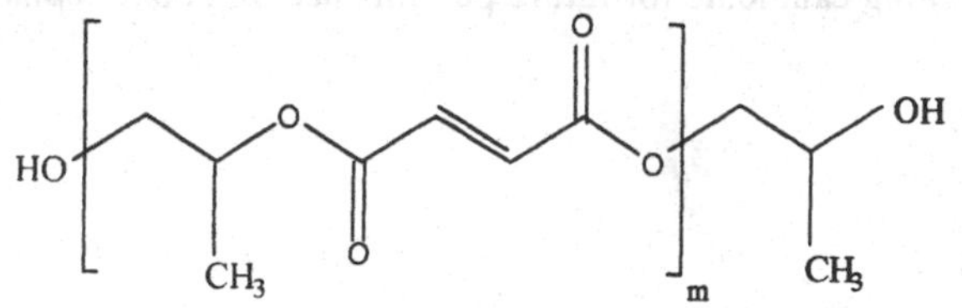

PPF

FIGURE 2. Poly(propylene fumarate) (PPF), an injectable polymer that can cross-link in situ with appropriate mechanical properties to fill defects in cancellous bone (37).

The mechanical properties of PPF can vary greatly according to the synthetic method and the cross-linking agents used. In order to improve these properties for use in orthopaedics, PPF is often combined with particles of ceramic materials such as β-TCP, calcium carbonate, or calcium phosphate (2,38,39,42). These composite materials exhibit compressive strengths from 2-30 MPa, which is appropriate for replacement of cancellous bone.

PPF degradation time is dependent on polymer structure as well as other components if it is part of a composite material. According to recent *in-vitro* studies, the time needed to reach 20% original weight ranged from near 84 (PPF/β-TCP composite) to over 200 days (PPF/CaSO$_4$ composite) (2,42). *In vivo*, β-TCP appears to act as a buffering agent, maintaining local pH and preventing accelerated PPF degradation (44). Although PPF undergoes bulk degradation, resulting in a decrease in overall mechanical properties with time, its compressive strength actually increases in the short term due to continued cross-linking. Therefore, larger strengths were seen after 3 weeks in saline (9.4 MPa) than when first placed in solution (2.6 MPa) (2).

PPF does not exhibit a deleterious long-term inflammatory response when implanted subcutaneously in rats. Initially, a mild inflammatory response was observed, and a fibrous capsule formed around the implant at 12 weeks (44). The PPF/β-TCP composite, sterilized by ultra-violet radiation, was implanted in rat tibiae for up to 5 weeks. At this time, the material was observed to be gradually replaced by bone from the perimeter inward (41). Recent *in vitro* work provides further information about the osteoconductive properties of PPF/β-TCP by demonstrating that the composite encourages attachment, proliferation and differentiated osteoblastic function of rat marrow stromal cells (45).

To improve the rate and extent of new bone formation, current research includes the addition of osteoblasts to the PPF construct (46). Another approach involves modification of the PPF with a GRGD (Gly-Arg-Gly-Asp) peptide sequence to encourage host cell attachment and migration once the construct has been injected (47). New studies with PPF composites include encapsulated growth factors (TGF-β1), which, rather than improving initial mechanical properties, act to direct cell migration and differentiation within the material (48-49). It is PPF's versatility, stemming from its ability to be easily modified, as well as its excellent mechanical properties, that makes this polymer an exciting candidate for future patients needing bone replacement.

3. Cartilage Tissue Engineering

Approaches to cartilage tissue engineering differ significantly from bone tissue engineering, although many of the same considerations for implantable replacements still apply. Like bone, cartilage must withstand compressive loads, but another important function of hyaline cartilage is frictionless movement in the joint (50,51). Therefore, the ability to withstand the shear forces at the joint surface is very important. However, mechanical properties of cartilage differ between joints and between different areas of the same joint (52-54) and the minimal necessary mechanical properties for new cartilage have not been determined. For these reasons, biochemical rather than mechanical analysis is often used to assess the extent of cartilage regeneration. In addition, cartilage does not undergo constant remodeling and often demonstrates poor regenerative capacity (55,56). Thus, unlike bone, most of the constructs implanted to repair cartilage have included a cellular component.

Because of the differences in needed mechanical properties and the lack of remodeling, ceramics are not appropriate for cartilage applications. In this section, a variety of both naturally-derived and synthetic polymers will be compared for use in cartilage repair, with a focus on biocompatibility and promotion of tissue formation as measures of implant efficacy.

Naturally-derived polymers, such as collagen, fibrin and hyaluronic acid, have often been used as carriers for cells to regenerate various tissues (3,57-60). Of these, the two that have been most widely studied for injectable cartilage applications are fibrin glues and alginate gels. Several researchers are exploring the option of injecting fibrinogen and thrombin to form a degradable fibrin mesh that can be used as a scaffold for chondrocytes (59,60). Because the patient's own fibrinogen and thrombin can be used, sterilizability, biocompatibility, and temperature change upon setting are not large concerns. When the cell-fibrinogen-thrombin mixture was injected into defects in horses, hyaline-like cartilage was formed, with more glycosaminoglycan (GAG) and type II collagen present at eight months than in defects that were left untreated (60).

Alginate, a derivative of seaweed, is another option for use in cartilage repair. A liquid, it is injected and crosslinked with calcium to prevent migration from the defect. Alginate can be steam sterilized and the cross-linking reaction does not adversely affect the surrounding tissue (61). Only mild inflammation was observed *in vivo*, but these studies were completed in nude mice and the elimination of the animals' immune system may have reduced the inflammatory response (62). Although histologic evaluation revealed the architecture of the newly formed tissue to be similar to that of native cartilage, there was little sign of alginate degradation after 12 weeks (61,62). An additional concern with the use of alginate is that some forms have been found to be immunogenic (63).

Because of possible complications with antigenicity and obtaining adequate amounts of natural polymers to fill large defects, other investigators have turned to the development of synthetic polymers, such as poly(ethylene oxide) (PEO) (see Figure 3). While sterilization methods were not described, Sims et al. injected a PEO gel with bovine articular chondrocytes into a nude mouse and did not observe any adverse inflammatory response (64). After 12 weeks, cartilage with histology similar to that of the epiphyseal plate and GAG content approaching that of natural bovine cartilage had been produced (64).

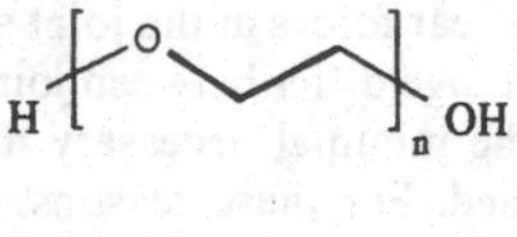

PEO

FIGURE 3. Chemical structure of poly(ethylene oxide) (PEO) or poly(ethylene glycol) (PEG), used as an injectable carrier for chondrocytes to repair cartilage (64).

Another synthetic injectable material currently under investigation is a PEO dimethacrylate (PEODM) that can be photopolymerized transdermally (see Figure 4). Initial results indicate the constructs can be polymerized in 3 minutes with no harm to imbedded chondrocytes. Specimens explanted from athymic mice at 2, 4, and 7 weeks show cartilage formation with increasing GAG and collagen content (65). While transdermal photopolymerization may not be possible for many orthopedic applications, a light source may be provided arthroscopically for some procedures. As in the case of bone, however, the problem of consistent photopolymerization in deep crevices remains.

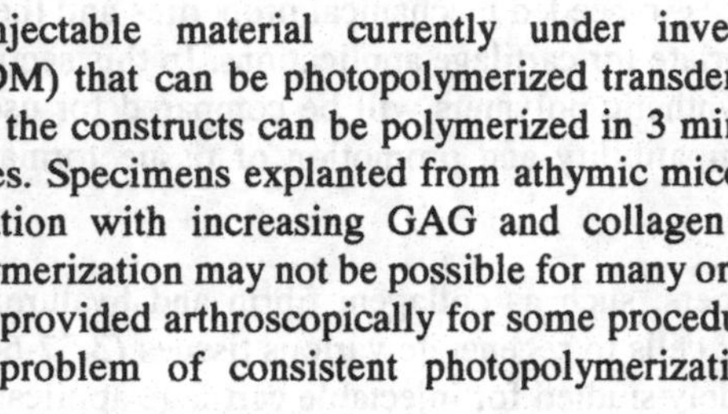

PEODM

FIGURE 4. Dimethacrylated poly(ethylene oxide) (PEODM), a polymer that can be transdermally photopolymerized with chondrocytes to form new cartilage (65).

An alternative may be the use of a material that chemically cross-links *in situ*. A poly(propylene fumarate-*co*-ethylene glycol) (P(PF-co-EG)) hydrogel has been developed for use in cardiovascular applications (66,67) and could be modified for use in a chondrocyte-polymer construct (see Figure 5). This material can be sterilized via ultra-violet radiation and has been found to be biocompatible, eliciting an initial inflammatory response in rats that recedes by 21 days, followed by fibrous capsule formation at 12 weeks (68,69). Cross-linking resulted in a slight temperature increase (from 37°C to 38.3°C), and short-term studies show no adverse effects of the reaction on endothelial cells injected concurrently with the polymer (70). These properties indicate that this material is a promising option for future work in cartilage repair.

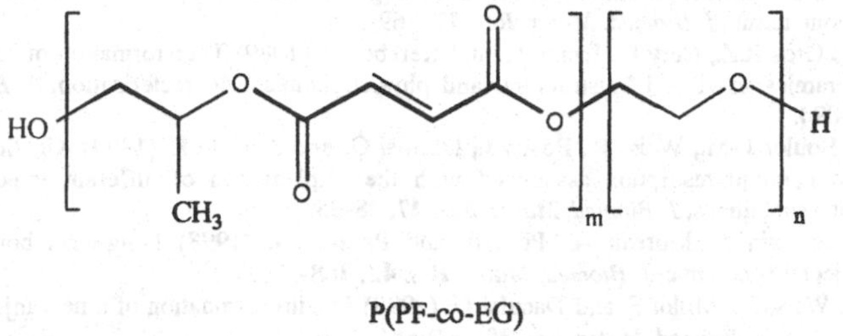

P(PF-co-EG)

FIGURE 5. Diagram of the structure of poly(propylene fumarate-co-ethylene glycol) (P(PF-co-EG)), a hydrogel currently studied for cardiovascular applications, but that could be modified for use in cartilage repair (66).

4. Conclusion

Although bone and articular cartilage are very diverse tissues providing different functions within the body, recent work has resulted in new injectable biomaterials with promise to repair both tissues. Like the tissues, the materials are also varied; they can be ceramic, naturally-derived, or based on synthetic polymers. They have different degradation characteristics and can contain imbedded cells or can be used alone. However, all are largely biocompatible and many demonstrate encouraging mechanical properties, an improvement over current injectables. In time, each material may find its own unique application. Taken together, these advances provide a means to lower cost as well as discomfort to persons undergoing a variety of orthopaedic procedures, thus signaling a significant step towards the widely-held ideal of inexpensive, minimally-invasive surgery.

References

1. American Academy of Orthopaedic Surgeons website. www.aaos.org.
2. Peter S.J., Nolley J.A., Widmer M.S., Merwin J.E., Yaszemski M.J., Yasko A.W., Engel P.S. and Mikos A.G. (1997) In vitro degradation of a poly(propylene fumarate)/b-tricalcium phosphate composite orthopaedic scaffold, *Tissue Eng.* **3**, 207-215.
3. Thomson R.C., Wake M.C., Yaszemski M.J. and Mikos A.G. (1995) Biodegradable polymer scaffolds to regenerate organs, *Adv. Polym. Sci.* **122**, 245-274.
4. Yaszemski M.J., Payne R.G., Hayes W.C., Langer R. and Mikos A.G. (1996) In vitro degradation of a poly(propylene fumarate)-based composite material, *Biomaterials* **17**, 2127-2130.
5. Yaszemski M.J., Payne R.G., Hayes W.C., Langer R. and Mikos A.G. (1996) Evolution of bone transplantation: Molecular, cellular and tissue strategies to engineer human bone, *Biomaterials* **17**, 175-185.
6. Frayssinet P., Gineste L., Conte P., Fages J. and Rouquet N. (1998) Short-term implantation of a DCPD-based calcium phosphate cement, *Biomaterials* **19**, 971-977.

310

7. Ikenaga M., Hardouin P., Lemaitre J., Andrianjatovo H. and Flautre B. (1998) Biomechanical characterization of a biodegradable calcium phosphate hydraulic cement: A comparison with porous biphasic calcium phosphate ceramics, *J. Biomed. Mater. Res.* **40**, 139-144.
8. Klein C.P.A.T., Dreissen A.A. and de Groot K. Biodegradation behavior of various calcium phosphate materials in bone tissue, *J. Biomed. Mater. Res.* **17**, 769-784.
9. Daculsi G., LeGros R.Z., Nery E., Lynch K. and Kerebel B. (1989) Transformation of biphasic calcium phosphate ceramics in vivo: Ultrastructural and physicochemical characterization, *J. Biomed. Mater. Res.* **23**, 883-894.
10. Gauthier O., Bouler J.-M., Weiss P., Bosco J., Daculsi G. and Aguado E. (1999) Kinetic study of bone ingrowth and ceramic resorption associated with the implantation of different injectable calcium-phosphate bone substitutes, *J. Biomed. Mater. Res.* **47**, 28-35.
11. Dupraz A., Delecrin J., Moreau A., Pilet P. and Passuti N. (1998) Long-term bone response to particulate injectable ceramic, *J. Biomed. Mater. Res.* **42**, 368-375.
12. Grimandi G., Weiss P., Millot F. and Daculsi G. (1998) In vitro evaluation of a new injectable calcium phosphate material, *J. Biomed. Mater. Res.* **39**, 660-666.
13. Gauthier O., Boix D., Grimandi G., Aguado E., Bouler J.-M., Weiss P. and Daculsi G. (1999) A new injectable phosphate biomaterial for immediate bone filling of extraction sockets: A preliminary study in dogs, *J. Periodontol.* **70**, 375-383.
14. Ohura K., Bohner M., Hardouin P., Lemaitre J., Pasquier G. and Flautre B. (1996) Resorption of, and bone formation from, new b-tricalcium phosphate-monocalcium phosphate cements: an in vivo study, *J. Biomed. Mater. Res.* **30**, 193-200.
15. Munting E., Mirtchi A.A. and Lemaitre J. (1993) Bone repair of defects filled with a phosphocalcic hydraulic cement: An in vivo study, *J. Mater. Sci. Mater. Med.* **4(3)**, 337-344.
16. Miyamoto Y., Ishikawa K., Fukao H., Sawada M., Nagayama M., Kon M. and Asaoka K. (1995) In vivo setting behaviour of fast-setting calcium phosphate cement, *Biomaterials* **16**, 855-860.
17. Miyamoto Y., Ishikawa K. Takechi M., Toh T., Yoshida Y., Nagayama M., Kon M. and Asaoka K. (1997) Tissue response to fast-setting calcium phosphate cement in bone, *J . Biomed. Mater. Res.* **37**, 457-464.
18. Miyamoto Y., Ishikawa K., Takechi M., Toh T., Yuasa T., Nagayama M. and Suzuki K. (1999) Histological and compositional evaluations of three types of calcium phosphate cements when implanted in subcutaneous tissue immediately after mixing, *J. Biomed. Mater. Res. Appl. Biomater.* **48**, 36-42.
19. Ishikawa K. and Asaoka K. (1995) Estimation of ideal mechanical strength and critical porosity of calcium phosphate cement, *J. Biomed. Mater. Res.* **29**, 1537-1543.
20. Driessens F.C.M., Boltong M.G., Bermudez O., Planell J.A., Ginebra M.P. and Fernandez E. (1994) Effective formulations for the preparation of calcium phosphate bone cements, *J. Mater. Sci. Mater. Med.* **5**, 164-170.
21. Anseth K.S., Shastri V.R. and Langer R. (1999) Photopolymerizable degradable polyanhydrides with osteocompatibility, *Nature Biotech.* **17**, 156-159.
22. Miyamoto Y., Ishikawa K., Takechi M., Toh T., Yuasa T., Nagayama M. and Suzuki K. (1998) Basic properties of calcium phosphate cement containing atelocollagen in its liquid or powder phases, *Biomaterials* **19**, 707-715.
23. Ishikawa K., Miyamoto Y., Kon M., Nagayama M. and Asaoka K. (1995) Non-decay type fast-setting calcium phosphate cement: Composite with sodium alginate, *Biomaterials* **16**, 527-532.
24. Ishikawa K., Miyamoto Y., Takechi M., Toh T., Kon M., Nagayama M. and Asaoka K. (1997) Non-decay type fast-setting calcium phosphate cement: Hydroxyapatite putty containing an increased amount of sodium alginate, *J. Biomed. Mater. Res.* **36**, 393-399.
25. Takechi M., Miyamoto Y., Ishikawa K., Toh T., Yuasa T., Nagayama M. and Suzuki K. (1998) Initial histological evaluation of anti-washout type fast-setting calcium phosphate cement following subcutaneous implantation, *Biomaterials* **19**, 2057-2063.
26. Frankenburg E.P., Goldstein S.A., Bauer T.W., Harris S.A. and Poser R.D. (1998) Biomechanical and histological evaluation of a calcium phosphate cement, *J. Bone Joint Surg.* **80-A**, 1112-1124.
27. Constanz B.R., Ison I.C., Fulmer M.T., Poser R.D., Smith S.T., VanWagoner M., Ross J., Goldstein S.A., Jupiter J.B. and Rosenthal D.I. Skeletal repair by in situ formation of the mineral phase of bone, *Science* **267**, 1796-1799.
28. Constanz B.R., Barr B.M., Ison I.C., Fulmer M.T., Baker J., McKinney L, Goodman S.B., Gunasekaren S., Delaney D.C., Ross J. and Poser R.D. (1998) Histological, chemical, and crystallographic analysis of four calcium phosphate cements in different rabbit osseous sites, *J. Biomed. Mater. Res. Appl. Biomater.* **43**, 451-461.

29. Goodman S.B., Bauer T.W., Carter D., Casteleyn P.P., Goldstein S.A., Kyle R.F., Larsson S., Stakewich C.J., Swiontkowski M.F., Tencer A.F., Yetkinler D.N. and Poser R.D. (1998) Norian SRS cement augmentation in hip fracture treatment, *Clin. Orthop. Rel. Res.* **348**, 42-50.

30. Kopylov P., Jonsson K., Thorngren K.G., and Aspenberg P. (1996) Injectable calcium phosphate in the treatment of distal radial fractures, *J. Hand Surg.* **21**, 768-771.

31. Food and Drug Administration (USA) website. www.fda.gov.

32. Knaack D., Goad M.E.P., Aiolova M., Rey C., Tofighi A., Chakravarthy P. and Lee D.D. (1998) Resorbable calcium phosphate bone substitute, *J. Biomed. Mater. Res. Appl. Biomater.* **43**, 399-409.

33. Muggli D.S., Burkoth A.K. and Anseth K.S. (1999) Crosslinked polyanhydrides for use in orthopedic applications: Degradation behavior and mechanics, *J. Biomed. Mater. Res.* **46**, 271-278.

34. Muggli D.S., Burkoth A.K., Keyser S.A., Lee H.R. and Anseth K.S. (1998) Reaction behavior of biodegradable, photo-cross-linkable polyanhydrides, *Macromolecules* **31**, 4120-4125.

35. Shastri V.R., Marini R.P., Padera R.F., Kirchain S., Tarcha P. and Langer R. (1998) Osteocompatibility of photopolymerizable anhydride networks, *Mat. Res. Soc. Symp. Proc.* **530**, 93-98.

36. He S., Timmer M.D., Yaszemski M.J., Yasko A.W., Engel P.S. and Mikos A.G. (2001) Synthesis of biodegradable poly(propylene fumarate) networks with poly(propylene fumarate)-diacrylate monomers as cross-linking agents and characterization of their degradation products, *Polymer* **42**, 1251-1260.

37. Peter S.J., Miller M.J., Yaszemski M.J. and Mikos A.G. (1997) Poly(propylene fumarate), in A.J. Domb, J. Kost and D.M. Wiseman (eds.), *Handbook of Biodegradable Polymers*, Harwood Academic, Amsterdam, pp. 87-97.

38. Frazier D.D., Lathi V.K., Gerhart T.N., Altobelli D.E. and Hayes W.C. (1995) In vivo degradation of a poly(propylene fumarate) biodegradable, particulate composite bone cement, *Mat. Res. Soc. Symp. Proc.* **394**, 15-19.

39. Frazier D.D., Lathi V.K., Gerhart T.N. and Hayes W.C. (1997) Ex vivo degradation of a poly(propylene glycol-fumarate) biodegradable particulate bone cement, *J. Biomed. Mater. Res.* **35**, 383-389.

40. Gresser J.D., Hsu S.-H., Nagaoka H., Lyons C.M., Nieratko D.P., Wise D.L., Barabino G.A. and Trantolo D.J. (1995) Analysis of a vinyl pyrrolidone/poly(propylene fumarate) resorbable bone cement, *J. Biomed. Mater. Res.* **29**, 1241-1247.

41. Yaszemski M.J., Payne R.G., Hayes W.C., Langer R.S., Aufdemorte T.B. and Mikos A.G. (1995) The ingrowth of new bone tissue and initial mechanical properties of a degrading polymeric composite scaffold, *Tissue Eng.* **1**, 41-52.

42. Kharas G.B., Kamenetsky M., Simantirakis J., Beinlich K.C., Rizzo A.-M.T., Caywood G.A. and Watson K. (1997) Synthesis and characterization of fumarate-based polyesters for use in bioresorbable bone cement composites. *J. Appl. Polym. Sci.* **66**, 1123-1137.

43. Peter S.J., Kim P., Yasko A.W., Yaszemski M.J. and Mikos A.G. (1999) Crosslinking characteristics of an injectable poly(propylene fumarate)/b-tricalcium phosphate paste and mechanical properties of the crosslinked composite for use as a biodegradable bone cement, *J. Biomed. Mater. Res.* **44**, 314-321.

44. Peter S.J., Miller S.T., Zhu G., Yasko A.W. and Mikos A.G. (1998) In vivo degradation of a poly(propylene fumarate)/b-tricalcium phosphate injectable composite scaffold, *J. Biomed. Mater. Res.* **41**, 1-7.

45. Peter S.J., Lu L., Kim D.J. and Mikos A.G. (2000) Marrow stromal osteoblast function on a poly(propylene fumarate)/b-tricalcium phosphate biodegradable orthopaedic composite, *Biomaterials* **21**, 1207-1213.

46. Payne R.G., Sivaram S.A., Babensee J.E., Yasko A.W., Yaszemski M.J. and Mikos A.G. (1998) Temporary encapsulation of rat marrow osteoblasts in gelatin microspheres, *Tissue Eng.* **4**, 497.

47. Jo S., Engel P.S. and Mikos A.G. (2000) Synthesis of poly(ethylene glycol)-tethered poly(propylene-co-fumarate) and its modification with GRGD peptide, *Polymer* **41**, 7595-7604.

48. Peter S.J., Lu L., Kim D.J., Stamatas G.N., Miller M.J., Yaszemski M.J. and Mikos A.G. (2000) Effects of transforming growth factor-b1 released from biodegradable polymer microparticles on marrow stromal osteoblasts cultured on poly(propylene fumarate) substrates, *J. Biomed. Mater. Res.* **50**, 452-462.

49. Lu L., Stamatas G.N. and Mikos A.G. (2000) Controlled release of transforming growth factor-b1 from biodegradable polymers, *J. Biomed. Mater. Res.* **50**, 440-451.

50. Cohen N.P., Foster R.J. and Mow V.C. (1998) Composition and dynamics of articular cartilage: Structure, function, and maintaining healthy state, *J. Orthop. Sports Phys. Ther.* **28**, 203-215.

51. Buckwalter J.A. and Mankin H.J. (1998) Articular cartilage: Tissue design and chondrocyte-matrix interactions, *AAOS Inst. Course Lect.* **47**, 477-486.

312

52. Athanasiou K.A., Rosenwasser M.P., Buckwalter J.A., Malinin T.I. and Mow V.C. (1991) Interspecies comparisons of in situ intrinsic mechanical properties of distal femoral cartilage, *J. Orthop. Res.* **9**, 330-340.

53. Athanasiou K.A., Agarwal A. and Dzida F.J. (1994) Comparative study of the intrinsic mechanical properties of the human acetabular and femoral head cartilage, *J. Orthop. Res.* **12**, 340-349.

54. Athanasiou K.A., Niederauer G.G. and Schenck R.C.J. (1995) Biomechanical topography of human ankle cartilage, *Ann. Biomed. Engr.* **123**, 697-704.

55. Coutts R.D., Sah R.L. and Amiel D. (1997) Effect of growth factors on cartilage repair, *AAOS Inst. Course Lect.* **46**, 487-494.

56. Buckwalter J.A. (1998) Articular cartilage: Injuries and potential for healing, *J. Orthop. Sports Phys. Ther.* **28**, 192-202.

57. Solchaga L.A., Dennis J.E., Goldberg V.M. and Caplan A.I. (1999) Hyaluronic acid-based polymers as cell carriers for tissue-engineered repair of bone and cartilage, *J. Orthop. Res.* **17**, 205-213.

58. Wakitani S., Goto T., Pineda S.J., Young R.G., Mansour J.M., Caplan A.I. and Goldberg V.M. (1994) Mesenchymal cell-based repair of large, full-thickness defects of articular cartilage, *J. Bone Joint Surg.* **76-A**, 579-592.

59. Sims C.D., Butler P.E.M., Cao Y.L., Casanova R., Randolph M.A., Black A., Vacanti C.A. and Yaremchuk M.J. (1998) Tissue engineered neocartilage using plasma derived polymer substrates and chondrocytes, *Plast. Reconstr. Surg.* **101**, 1580-1585.

60. Hendrickson D.A., Nixon A.J., Grande D.A., Todhunter R.J., Minor R.M., Erb H. and Lust G. (1994) Chondrocyte-fibrin matrix transplants for resurfacing extensive articular cartilage defects, *J. Orthop. Res.* **12**, 485-497.

61. Paige K.T., Cima L.G., Yaremchuk M.J., Vacanti J.P. and Vacanti C.A. (1995) Injectable cartilage, *Plast. Reconstr. Surg.* **96**, 1390-1400.

62. Paige K.T., Cima L.G., Yaremchuk M.J., Schloo B.L., Vacanti J.P. and Vacanti C.A. (1996) De novo cartilage generation using calcium alginate-chondrocyte constructs, *Plast. Reconstr. Surg.* **97**, 168-180.

63. Kulseng B., Skjak-Braek G., Ryan L., Andersson A., King A., Faxvaag A. and Espevik T. (1999) Transplantation of alginate microcapsules, *Transplantation* **67**, 978-984.

64. Sims C.D., Butler P.E.M., Casanova R., Lee B.T., Randolph M.A., Lee W.P.A., Vacanti C.A. and Yaremchuk M.J. (1996) Injectable cartilage using polyethylene oxide polymer substrates, *Plast. Reconstr. Surg.* **98**, 843-850.

65. Elisseeff J., Anseth K., Sims D., McIntosh W., Randolph M. and Langer R. (1999) Transdermal photopolymerization for minimally invasive implantation, *Proc. Natl. Acad. Sci.* **96**, 3104-3107.

66. Suggs L.J., Payne R.G., Yaszemski M.J., Alemany L.B. and Mikos A.G. (1997) Synthesis and characterization of a block copolymer consisting of poly(propylene fumarate) and poly(ethylene glycol), *Macromolecules* **30**, 4318-4323.

67. Suggs L.J., Kao E.Y., Palombo L.L., Krishnan R.S., Widmer M.S. and Mikos A.G. (1998) Preparation and characterization of poly(propylene fumarate-co-ethylene glycol) hydrogels, *J. Biomater. Sci. Polym. Edn.* **9**, 653-666.

68. Suggs L.J., Shive M.S., Garcia C.A., Anderson J.M. and Mikos A.G. (1999) In vitro cytotoxicity and in vivo biocompatiblity of poly(propylene fumarate-co-ethylene glycol) hydrogels, *J. Biomed. Mater. Res.* **46**, 22-32.

69. Suggs L.J., Krishnan R.S., Garcia C.A., Peter S.J., Anderson J.M. and Mikos A.G. (1998) In vitro and in vivo degradation of poly(propylene fumarate-co-ethylene glycol) hydrogels, *J. Biomed. Mater. Res.* **42**, 312-320.

70. Suggs L.J. and Mikos A.G. (1999) Development of poly(propylene fumarate-co-ethylene glycol) as an injectable carrier for endothelial cells, *Cell Transplantation* **8**, 345-350.

TISSUE ENGINEERING OF ELASTIC CARTILAGE BY USING SCAFFOLD/CELL CONSTRUCTS WITH DIFFERENT PHYSICAL AND CHEMICAL PROPERTIES

D.W. Hutmacher *,**, X. Fu[#], B.K. Tan[#], J-T.Schantz[&],
Department of Bioengineering*, Department of Orthopedic Surgery **,
National University Singapore; Department of Plastic Surgery, Singapore
General Hospital[#]; Laboratory for Biomedical Engineering [&], National
University Singapore

Abstract
In this study, we compared three scaffold/cell constructs with different physical and chemical properties, in their potential for tissue engineering elastic cartilage. Group I consisted of PCL scaffolds Group II custom-made non-woven made of PGA fibers in combination with agarose beads; collagen sponges were used in-group III. Chondrocytes were isolated via enzyme digestion from an ear cartilage biopsy of 2 year old male piglets. 250,000 cells in 30 µl were seeded into three different scaffold types. The specimens were then cultured for 1 week. The scaffold/cell constructs and controls were placed subcutaneously on the paravertebral fascia for 4 1/2 month. Explanted scaffolds were cut in cubical blocks and mechanically tested in phosphate buffered saline solution at 37^0C in accordance with ASTM F451-99a Standard Specification for Acrylic Bone Cement. After explantation the non-seeded PCL specimens (control) had compressive stiffness and 1 % offset yield strength in plane of 5.6 ± 0.4 MPa and out of plane 4.9 ± 0.5 MPa, respectively. In comparison, the scaffolds with a seeded specimens had compressive stiffness of 4.9 ± 0.5 MPa and 3.9± 0.3 MPa, respectively. The polymer molecular weight distribution was determined by gel permeation chromatography to study the in vivo degradation of the polymer matrix. Sections of explants were prepared using standard cryo-histochemical techniques and stained with anti-type II III, IX collagen antibodies, F-actin and anti-integrin beta, and I. There was considerably more organized extracellular matrix formation in group I, whereas group II and III exhibiting a loosely scattered patterns. Results of the toluidine blue, trichrome Goldner's staining and collagen II expression showed cartilage-like tissue formation in the PCL scaffolds whereas all the other matrices showed fibrous tissue in combination with calcification.

1. Introduction

Cartilage repair is a challenging clinical problem because once adult cartilage sustains damage, is it traumatic or pathologic, an irreversible, degenerative process

313

R.L. Reis and D. Cohn (eds.),
Polymer Based Systems on Tissue Engineering, Replacement and Regeneration, 313–332.
© 2002 *Kluwer Academic Publishers. Printed in the Netherlands.*

can occur. As autologous cartilage grafting has been used widely to reconstruct skeletal defects of the nose, ear and trachea because flexible cartilage lacks the ability to replace damaged tissue. However, this method has certain disadvantages, such as donor site morbidity, availability of only a limited amount of material and transplant degeneration leading to formation of fibrous tissue. [1,2]

Alternative methods for tissue engineering of cartilage have been investigated. Several groups have tissue engineered cartilage in animal models by using chondrocytes in combination with textile constructs which were made of the fast degrading polymers polyglycolic acid (PGA) [3-6] and copolymers of polylactic acid (PLA)/PGA[7-9]. However, the critical properties for a tissue engineered scaffold/cartilage graft to be used in plastic and reconstructive surgery include a degree of strength retention over time and a structural and mechanical equivalence to the different types of cartilage found in the body. Conventional scaffold technologies have limitations to design and fabricate of matrix architectures, which fulfill these biomechanical criteria. [10]

Recently, rapid prototyping technologies have emerged which can be used to manufacture scaffolds with more suitable structural and mechanical properties for bone and cartilage regeneration. [10] Hutmacher et al [11-13] designed and fabricated novel polycaprolactone (PCL) scaffolds by fused deposition modeling (Figure 1). The slow degrading matrices produced had a fully interconnected honeycomb-like pore architecture that revealed mechanical properties suitable for bone engineering. The PCL templates supported in vitro the adhesion, proliferation and differentiation of osteoblast-like cells and allowed the formation of mineralized extracellular matrix throughout the entire scaffold architecture. Implant studies in nude mice models showed that the PCL scaffold/osteoblast constructs lead to bone-like tissue formation. In this paper further experiments are reported which evaluate the potential of PCL scaffold/chondrocyte constructs to form elastic cartilage.

2. Material and Methods

2.1 SPECIMEN

Three different types with a total number of seven scaffolds per group were studied. Group I (Figures 2a) consisted of PCL scaffolds with a lay down pattern of $0/60/120^0$ and a porosity of 65% as described above. Group II (Figure 2b) were custom-made non-woven scaffolds made of PGA fibers with a porosity of 97% (Dexon II, Davis&Geck, NJ). Group III (Figure 2b) consisted of commercially available bovine collagen Type I sponge (Spongostan, Johnson&Johnson, NJ) with a porosity of 97%. Before cell seeding all scaffolds were sterilized by gamma irradiation to a dose of 2.5 Mrads.

2.2 CELL CULTURE

Unless otherwise noted, culture media supplements, and biochemical reagents were purchased from Gibco (Grand Island, NY) and Sigma (St. Louis, Missouri).

Cartilage pieces of 10x10 mm of the ears of two adult Yorkshire pigs were used for cell harvesting. Porcine ear chondrocytes were isolated by enzymatic digestion with 2-mg/mL collagenase type II (Gibco, Grand Island, NY) in DMEM cell culture medium (Gibco, Grand Island, NY). Cells of each ear were cultured on an individual base.

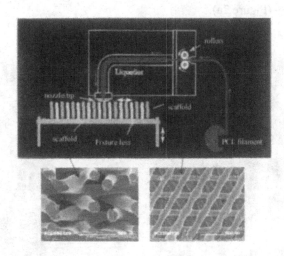

Figure 1. Graphical illustration of FDM process and freeze-fractured surfaces of a PCL scaffold with a 0/60/120° lay-down pattern and a porosity of 65% (side view and top view). View of PCL scaffold in the −z direction of the FDM build process showing a completely interconnected honeycomb pattern with triangular pores and large channel-like interconnections.

Chondrocytes were first cultured for multiplication under conventional monolayer culture conditions as previously described. [14] After confluent monolayers were achieved by the fourth passage; the cells were enzymatically lifted from the flask using 0.25% Trypsin/EDTA and counted using a hemocytometer. The cell mortality was less than 5% as shown by trypan blue staining and maintained a stable cell metabolism. Cell pellets were resuspended and aliquots of 30µl culture media containing 1.5×10^6 chondrocytes were mixed with 30µl of bovine collagen type I (Matrigel, BD, NY). The cell/hydrogel mixture was then injected into each of five scaffolds of each group. Two scaffolds of each group were not seeded with chondrocytes and served as controls.

Chondrocytes have been shown to elicit lymphocyte proliferation and migration in vitro. This may play a role in triggering a cellular response with a potential release of cytokines and other mediators of inflammation. It has been shown that the encapsulation of chondrocytes, using hydrogels, performed an immuno-protective role for the chondrocytes from surface antigens present in the host tissue. [15,16] Therefore, all three scaffold types were placed in a fibrin glue solution (Tissueseel Kit, Baxter AG, Austria) after seeding to form a capsule of 0.6

316

mm thickness. Subsequently, the seeded scaffolds were placed into an incubator to allow the cells to build adhesion plaques. After 2 hours, 5ml of complete medium was added to each well. The medium was renewed every third day. Cell-scaffold constructs were cultured for a period of 1 week prior to autogenic transplantation. The scaffold/tissue cell constructs were examined daily by light and scanning electron microscopy (Figure 2a).

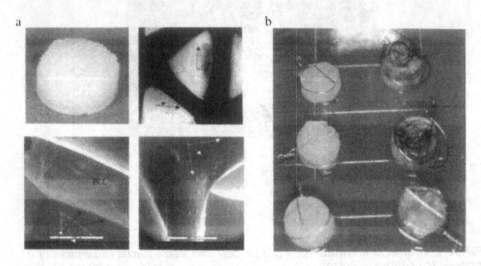

Figure 2a/b.   PCL scaffold/cell construct (Ø 12 × 10 mm) with a 65 % porosity and a lay-down pattern of 0/60/120° before implantation and 1 week of culturing (Figure 2a, left). The scaffolds were first coated with collagen and then seeded with chondrocytes. Finally, the seeded cells were embedded in fibrin glue solution. Inverted light microscopy (Figure 2a, right x 40) showed that the honeycomb like pores of the entire scaffold architecture was filled with a hydrogel cell composition.. A great number of cells maintained the round shape (arrows) and formed cell clusters, which is specific for chondrocyte phenotype. SEM images of a PCL scaffold/cell construct revealed that a great number of the chondrocytes showed their round three-dimensional phenotype (right, white arrows). A great number of cells established their filipodia and intercellular connections within the hydrogel matrix (left black arrows). Custom-made non-woven made of PGA fibres (right side) and collagen sponges (left side). Collagen matrices did loose already 20-30% of their original volume after 1 weeks in cell culture. Each scaffold/cell construct was encircled with a steel wire for easy identification in the soft tissue at the time of explantation.

2.3   SURGICAL PROCEDURE

Seeded and control scaffolds were implanted based on autogenic cell transplantation in two adult Yorkshire pigs as described below. The animals were housed in the

Animal Holding Facility/Department of Experimental Surgery, Singapore General Hospital, for the entire duration of the experiment. Housing and feeding was done according to standard animal care protocols. The studies were approved by the Animal Welfare Committee of the National University of Singapore/Singapore General Hospital and licensed by the National Institute of Health's Guide for Care and Use of Laboratory Animals.

The surgical procedure was carried out under general anesthesia. The animals were premedicated with ketamine, induced and orally intubated with pentobarbitane and maintained during the procedure with 1% halothane. The surgical regions were prepared with 1% cetrimide solution and 0.05% chlorhexidine acetate aqueous antiseptic solution. Lidocain (3%) was administered by local infiltration. Cultured scaffold/cell constructs and non-seeded scaffold (control) of each group were placed subcutaneously on the paravertebral fascia in two pigs. For easy identification in the soft tissue during retrieval, each scaffold/cell construct was encircled with a steel wire marker prior to transplantation (Figure 3). The wounds were carefully rinsed with 0,9% saline solution and closed with absorbable suture material. All pigs received amoxicillin (Ampicillin®) for five consecutive postoperative days. The pigs were sacrificed by an intravenous injection of euthanasia solution after 18 weeks and scaffold/issue constructs were harvested for analysis.

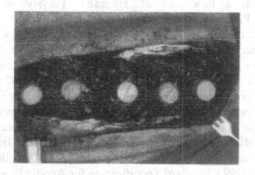

Figure 3. Intraoperative picture of the left paravertebral side of the pig 1 after placement of the scaffold/cell construct of group I to III. All scaffold/tissue constructs have a steel wire enclosure so that specimens could be easily identified after 4.5 month of implantation.

2.4   HISTOLOGY AND IMMUNOHISTOCHEMISTRY

The explanted PCL scaffold/tissue constructs were examined by scanning electron microscopy (SEM). For histology, specimens were fixed in 10% PBS buffered formalin (Merck, Germany) and embedded in paraffin and cut into 7 µm thick sections. Sections from each explant were cut in three different areas: the left and right margins and center. Qualitative assessment did not show major differences in histological appearance. Five micron cryostat sections were prepared from fresh tissue samples of each group, too.

Simultaneous detection of different collagen and cell nuclei was performed on central cryostat sections using the indirect immunofluoresence technique in combination with confocal laser microscopy (CLM). One set of cryostat sections was stained with anti-type II III, IX collagen antibodies, and I. Whereas on another set F-actin and anti-integrin β1 were analysed and counter stained with FITC or Rhodamine-conjugated secondary antibodies (Molecular Probes, Eugene, OR).

For the histological identification of cartilage-like tissue and calcified areas, von Kossa, Toludine blue and Trichrome Goldner, staining were carried out on 7 micron thick sections of paraffin embedded specimens. The expression of glycosaminoglycan (GAG) was viewed by Safranin-O stained sections via light microscopy.

2.5 PHYSICAL AND PHYSICO-CHEMICAL PROPERTIES

The polymer molecular weight distribution was determined by gel permeation chromatography (GPC) equipped with a differential refractor (Waters, Model 410, Milford, MA) and an absorbance detector refractor (Waters, Model 2690, Milford, MA). PCL samples in triplicates of each specimen were dissolved in tetrahydrofuran (THF) and eluted in a series of configurations through a Styragel columns refractor (Waters, Milford, MA) at a flow rate of 1mlmin⁻¹. Polystyrene standards (Polysciences, Warrington, PA) were used to obtain a calibration curve.

A thermal analysis differential scanning calorimeter (TA Instruments DSC 2910, New Castle, DE) was utilized to analyze the thermal transition of the PCL samples during melting to measure the crystallinity fraction. PCL samples in triplicates of each specimen were heated at a rate of 5°C/min from 25°C to 70° C in aluminum pans using nitrogen as purge gas. The crystallinity fractions were based upon an enthalpy of fusion value of 139.5J/g for 100% crystalline PCL as reported.[17] Mechanical properties of the PCL scaffolds and explanted scaffold/tissue constructs were tested in phosphate buffered saline solution (PBS) at 37 °C on an Instron 4502 Uniaxial Testing System and a 1-kN load-cell (Canton, MA) following the guidelines set in ASTM F451-99 a Standard Specification for Acrylic Bone Cement.

3. Results

3.1 MACROSCOPIC APPEARANCE

The animals recovered from anesthesia without complications and healed uneventfully. Immediately prior to explant, 18 weeks after implantation, areas of increased firmness could be palpated underneath the pig's skin. All scaffold/tissue constructs of group I to III could be easily identified due to the steel wire enclosure (Figure 3). In group I, there was no change in size of all implants to the original dimension of the PCL scaffold. At the time of explant, the specimens were surrounded by a thin fibrous tissue capsule with a light vascular network, but only minimal fibrotic reaction could be seen (Figure 4). After removal of the soft tissue capsule, the tissue inside the chondrocyte seeded PCL scaffolds had in some areas the gross appearance of elastic cartilage-like tissue whereas the non-seeded specimens had a firm soft tissue presentation. Mechanical stability of the PCL scaffold/chondrocyte construct was similar to elastic cartilage, with slight elasticity revealed by physical examination via probing with a pair of forceps. The tissue inside the non-seeded scaffolds appeared as firm and highly structured soft tissue.

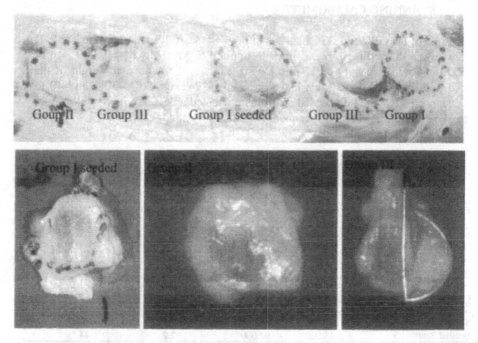

Figure 4. In situ picture of specimens before explantation. All scaffold/tissue constructs of group I to III could be easily identified after 4.5 month of implantation due to the steel wire enclosure. Group II specimens were found as a mass of tissue encased in a thick fibrous tissue whereas group I and III specimens revealed a thin fibrous tissue capsule with a vascular network, but only minimal fibrotic reaction. In group I, there was no change in size of chondrocyte seeded and non-seeded implants to the original dimension of the PCL scaffold. All specimens of group I revealed its original shape and volume whereas group II did loose 30 to 40% and group 40 to 60% of its original size, respectively.

In comparison, all tissue engineered scaffold/tissue construct of group II and III showed macroscopically a decrease in size and weight. Mechanical characteristics were clearly softer than in group I. The excised specimens of the custom-made non-woven made of PGA fibers were found as a mass of tissue encased in a thick fibrous tissue (Figure 4). The preparation of cross sections revealed that the textile/tissue constructs constituted of a thick outer layer of whitish tissue with a yellowish internal segment. At the time of explant the collagen foam based specimens were found as distinct masses of tissue, each surrounded by a thin fibrous tissue capsule with a vascular network with minimal fibrotic and inflammatory reaction (Figure 4). After removal of the tissue capsule, specimens had the gross appearance of a highly structured muscle like tissue.

3.2 GAS PERMEATION CHROMATOGRAPHY AND DIFFERENTIAL SCANNING CALORIMETRY

The results of the GPC and DSC analysis of the PCL scaffolds are shown in Table 1. GPC analysis showed the samples prior to implant to have Mw of 140.000, Mn of 78.5000 and polydispersity of 1.7. Following explant, the molecular weight of the seeded and non-seeded scaffolds had decreased on average 20 to 30% and 40 to 60%, respectively. The non-seeded scaffolds presented a decrease in crystallinity of up to 30% whereas the seeded specimens revealed only minor changes.

TABLE 1. Molecular weight distribution and DSC results of the PCL specimens before and after implantation

PCL sample	M_n	M_w	Polydispersity M_w/M_n	Crystallinity fraction (%)
Scaffold after sterilisation	78.500	140.000	1.8	55.8
Non-seeded scaffold 1 (control)	27.500	89.000	3.2	35.2
Non-seeded scaffold 2 (control)	29.000	90.000	3.1	37.1
Seeded Scaffold 1	54.500	96.000	1.8	49.1
Seeded Scaffold 2	54.000	113.000	2.1	54
Seeded Scaffold 3	45.000	97.000	2.2	44.5
Seeded Scaffold 5	54.500	108.000	2.0	54.4
Seeded Scaffold 6	44.500	99.000	2.2	45.8

3.3 MECHANICAL TESTING

After 4.5 month in vivo, the collagen sponge and custom-made textile construct retained only approximately 25% and 55% of their original volume, respectively. Therefore, it was not possible to perform mechanical testing on the specimens of group II and III. Figure 5 shows the typical stress-strain curve for a PCL scaffold/tissue construct. It comprises of three distinct regions: a linear-elastic region followed by a plateau of roughly constant stress, leading into a final region of steeply rising stress.

Statistical analysis of the results was used to compare the stiffness and the compressive 1 % offset yield strength. Before seeding and transplantation PCL scaffolds with a 0/60/120° lay-down pattern had a compressive stiffness and 1 %

offset yield strength in plane of 29.4 ± 4.0 MPa and out of plane 23.2 ± 3.4. MPa. After explantation the non-seeded specimens (control) had compressive stiffness and 1 % offset yield strength in plane of 5.6 ± 0.4 MPa and out of plane 4.9 ± 0.5 MPa. In comparison, the seeded scaffolds had in plane and out of plane compressive stiffness of 4.9 ± 0.5 MPa and 3.9± 0.3 MPa, respectively.

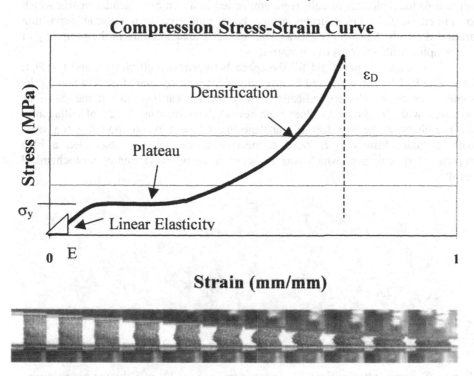

Figure 5. Typical stress-strain curve and video images of PCL scaffold/tissue construct under compression loading in a simulated physiological environment. The modulus of such specimens was measured as the gradient of the stress-strain curve at a strain level of 10%. The offset yield strength is defined at 1% strain. As strain increases, first the scaffold/tissue construct undergoes a densification process. When the rods and struts of the scaffold in combination with tissue inside the fully interconnected pore network are crushed the stress level rises quickly.

3.4   HISTOLOGY

Cell seeded PCL scaffold sections showed typical histological features of native pig cartilage with collagen rich ECM and lacunae embedding roundly configured singular chondrocytes. Occasionally bicellular chondrocytes were seen (Figure 6). All cell seeded PCL specimens had similar histology with confluent nodules of cartilage rimmed by fibrous tissue suggestive of a perichondrium-like layer. The fibrous capsule around the PCL scaffold/cartilage construct was free of foreign body giant cells, suggesting absence of chronic inflammation.

The PCL scaffold seems to provide the new developing tissue a porous and fully interconnected architecture, which gives the seeded chondrocytes sufficient space to proliferate and produce extracellular matrix. In the Toluidine blue staining (Figure 6) focal clusters of cells were embedded in a rich extracellular matrix which is stained purple. This indicates proteoglycan and glycosaminoglycan deposition around the cells. However, explants also included areas, mainly in the periphery of the samples, with a fibrous tissue matrix.

Collagen bundles did fill the space between the cell clusters and the PCL bars. The PCL material showed a close alignment with cellular structures having the same color of the clusters indicating newly formed cartilage-like tissue. Sections prepared with Trichrome Goldner stain revealed intertwining bands of collagen at the periphery of the new formed cartilage-like tissue. Several spotted green areas with cellular clusters in a loose connective tissue network resembled a high deposition of collagen, which was in accordance with the immunohistochemical results.

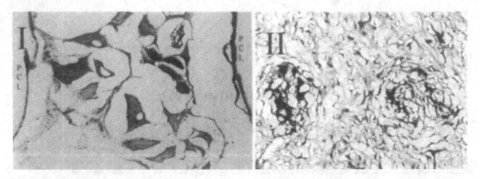

Figure 6. Representative light microscopic images (x100) of Toludine blue staining of specimens of group I (left) and II (right). Section of tissue obtained with a PCL scaffold/cell construct showing lacunae with singular and bicellular chondrocytes. In contrast, the section obtained from the PGA textile/cell construct showed only fibrous and calcified tissue formation.

The tissue developed in the custom-made non-woven PGA mesh presented an irregular woven network with several clusters of spindle shaped cell conglomerates. These appear in a dark blue color in the Toluidine blue staining (Figure 7), which indicated calcification of fibrous cartilage tissue. The polymer matrix was not detectable in the entire specimen and there seemed to be no organized structure in the newly build tissue. Strong fibrous tissue ingrowth from the surrounding soft tissue of the subcutaneous area was visible in all PGA mesh specimens.

The Trichrome Goldner staining showed a large reticular fibrous network of calcified ECM. An extensive network of fibrous encapsulating lacuna of loosely packed cell conglomerates was revealed in the histomorphologic structure of the spongostan specimens (Figure 8). Size and shape of the cells might lead to the conclusion to define them as adipocytes. This is in accordance with the macroscopical findings where a fatty character of these tissue constructs was detected.

The fibrous network is dark purple to blue in the Toluidine blue staining as mineralized and hyperthrophic tissue. The Trichome Goldner staining of the group III specimens appeared similar to the group II specimen with the difference that the light green character of the some areas is very prominent indicating high collagen deposition. A meandering network of collagen fibres could also be seen. Histological sections of the explants of group II had similar features that showed broad areas of fibrous tissue containing blood vessels and adipose tissue formation. Lymphohistiocytic cells infiltrated the fibrous area and occasionally foreign body giant cell formation was detected.

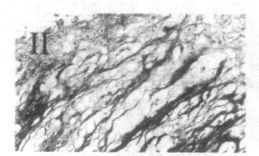 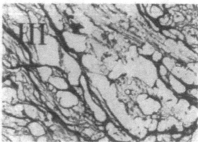

Figure 7. Representative light microscopically images (x100) of Trichrome Goldner staining of specimens group II, and III. A large reticular fibrous network of calcified ECM is detected in the PGA mesh specimens. An extensive network of fibrous encapsulating lacuna of loosely packed cell conglomerates was revealed in the histomorphologic structure of the spongostan specimens. Size and shape of the cells lead to the conclusion to define them as adipocytes. This is in accordance with the macroscopical findings where a fatty character of these tissue constructs was detected.

The thin fibrous tissue layers encapsulating transplants of group I and III contained abundant small vessels whereas the group II specimens showed an interconnected vascular network. The specimens of group I contained few vessels in areas peripheral to the PCL scaffold/tissue construct. In the central portion of the PCL scaffold architecture no vessels could be detected whereas inside the transplants of group II a high number and group III a small number of vessels, were observed throughout the entire specimen.

3.5   SCANNING ELECTRON MICROSCOPY

For both the chondrocyte seeded and non-seeded PCL scaffolds the entire volume of the polymer matrix was filled with tissue (Figure 10). However, as demonstrated by

324

histology only the scaffolds seeded with chondrocytes exhibited cartilage-like tissue formation. The PCL scaffolds cultured for 1 week with chondrocytes did show a direct attachment of tissue at the polymer surface (Figure 10, upper row). At high magnification cell clusters as well as a dense ECM formation could be observed. The PCL surface showed signs of hydrolytic attack via surface erosion after 18 weeks in situ. The non-seeded specimens revealed a slight detachment of the ingrown tissue (Figure 10, lower row). The PCL appears much smoother indicating slower erosion of the polymer surface. The original scaffold architecture of the explanted specimens of group II and III could not be detected via SEM and only scattered tissue structures were observed. This was an indication that both the PGA textile and the bovine collagen type 1 matrices were completely resorbed and metabolized.

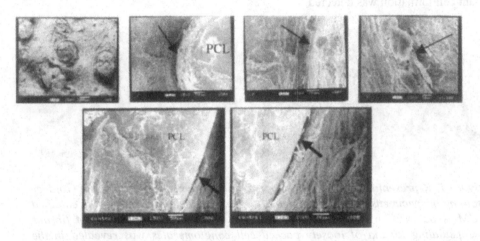

Figure 8. Representative SEM micrographs of PCL scaffold/tissue construct after 4 1/2 month of implantation. The entire volume of a scaffold measuring $\varnothing$ 12 × 10 mm was filled with a dense tissue network. Non-seeded matrices (lower row) revealed no direct contact between the tissue and the polymer surface (arrows) whereas the chondrocytes seeded scaffolds (upper row) showed a tight surface/tissue interface (arrows). The seeded PCL surface appears to be attacked by surface erosion whereas the non-seeded specimen shows a smoother pattern.

3.6   IMMUNOHISTOCHEMISTRY

Collagen types II, VI, IX, X, and XI are found in the extracellular matrix (ECM) of elastic cartilage, although 80-90% of the collagen's present are type II. In vivo, type I is representatively found in mesencymal tissues, but in cartilage, type I collagen is found mainly in the perichondrium. [1] Figure 11 and the Table 2 show Confocal Laser Microscopy images along with collagen analysis data. This semi-quantitative analysis revealed that all specimens presented types I, II, III and IX collagen. However, all collagen types were most strongly expressed inside the PCL scaffolds.

PCL specimens prepared by immunohistochemical cryo-sectioning showed the brightest expression of F-actin (Figure 12) for the PCL scaffold/chondrocyte constructs. The cryostat-sections of group II and III showed no expression of integrin b1 whereas group I specimens showed positive staining (Figure 13).

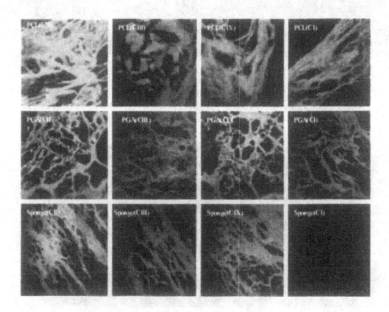

Figure 9. Representative confocal laser microscopy images (all x100) of the immunohistochemical collagen assay. The semi-quantitative analysis revealed the strongest expression of type II, III and IX collagen of the tissue inside the PCL scaffolds (see also Table 2)

TABLE 2. Semi-quantitative analysis of the ECM collagen's in the tissue engineered constructs

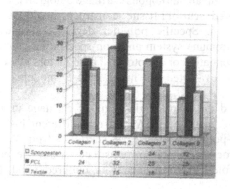

Figure 10. Representative fluorescence microscopy images (all x 100) of F-actin stained cryo-sections of the harvested tissue engineered constructs of group I-III.

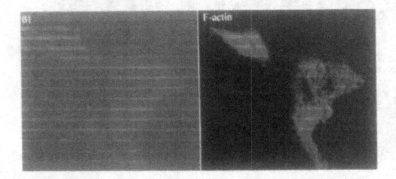

Figure 11. Representative fluorescence microscopy images of to subsequently microtomed cryo-sections. The cartilage-like tissue inside the PCL scaffold architecture stained positively for F-actin and anti-integrin β1.

4. Discussion

Studying tissue engineering concepts in an immunocompetent model is essential towards the development of human autologous cartilage autografts. Theoretically, an autologous, tissue engineered scaffold-tissue construct should be able to avoid problems related to rejection. Specific polymer degradation and resorption by products might trigger the immune system and non-specific inflammatory reactions could result in cell re-differentiation or apoptosis. These factors made it necessary to study concepts previously evaluated in nude mouse models, using a larger immunocompetent model.

Cao et al [11] studied the tissue engineering of cartilage in a pig model by using a non-woven scaffold made of PGA and two different hydrogels: calcium alginate and polyethylene-polypropylene. Five scaffolds from each group were seeded with $1x10^7$ chondrocytes, which were isolated from the external ear via enzymatic digestion. Scaffold-cell constructs were cultured for 1 week and then implanted subcutaneously on the ventral surface of each pig. All specimens were harvested and studied via histology 6 weeks post operatively. The PGA textile-pig chondrocytes construct did show 30 to 40% cartilage formation. However, a strong, localized non-specific inflammatory and foreign body reaction was detected, resulting in the formation of 60 to 70% of fibrous tissue. The authors concluded that the results differed from those in their previous nude mice experiments [3] due to the immunological reaction induced by the degradation by-products of the PGA.

Sutures made of PGA have been used in surgery for more than two decades and it is known that the fast degradation and resorption kinetics of PGA can elicit a local inflammation due to the production of acidic by-products during metabolization in the 'Krebs cycle'. [18] A strong and clinically detectable foreign body reaction to PGA pins and screws, used clinically for fracture fixation, has been reported by several groups. [19-21] Therefore, it is important from a scaffold matrix point of view to choose a polymeric material whose degradation and resorption rates minimizes stimulation of inflammation in immunocompetent animal models.

The PCL scaffold was designed to maintain the interconnected pore network for at least 4 to 6 months. In this way, the neocartilage has sufficient time to mature without biomechanical overloading. In the study presented constructs engineered from autologous chondrocytes in combination with PCL scaffolds made by FDM, developed inside the matrix cartilage-like tissue, evident both grossly and histologically. Fibrin flue coated PCL scaffolds/cell constructs showed fibrous encapsulation without any significant pathologies. In contrast both collagen type I foam matrix and a laboratory-made PGA textile construct showed significant inflammatory responses and foreign body cell reactions. The collagen scaffold led to formation of fibrous tissue with sparse ill defined nodules of cartilage. Similarly, as reported initially by Cao et al [15], it was found that the implantation in an immunocompetent animal model of a custom-made non-woven construct made of PGA fibres in combination with cultured chondrocytes resulted in a mass of tissue encased within a thick fibrous capsule.

Cartilage is a form of connective tissue composed exclusively of chondrocytes with a highly specialized ECM. Like the dense regular connective tissue of the tendon and ligament, it is an avascular and the matrix serves as route for the diffusion of substances between blood vessels in the surrounding connective tissue and chondrocytes. The matrix of cartilage is solid and firm though somewhat pliable, which accounts for the special resilient properties of cartilage. The critical properties for a tissue engineered scaffold-chondrocyte-ECM construct to be used in plastic and reconstructive surgery, include a degree of strength retention over time, and a structural and mechanical equivalence to the different types of cartilage found in the body.

Cao et al [3] used a nonwoven PGA mesh in combination with bovine chondrocytes to tissue engineer an ear. The textile construct was shaped into a human infant ear by using a plaster mold and heat. Dipping the fragile scaffold into a polylactide solution stabilized the 3D geometry. However, the textile construct had to be stented externally for 4 weeks because the scaffold was not able to resist the contraction forces of healing. In contrast, this study in piglets could show that the chemical and physical properties of PCL scaffolds are able to counteract the contracting forces of the soft tissue for up to 18 weeks.

An understanding of the structure of tissue engineered constructs needs to be closely associated with an assessment of the biomechanical properties. A number of groups [22-24] has reported on the mechanical properties of tissue engineered articular cartilage. However, groups that regenerated elastic cartilage [7,9,15,16] did not quantify the mechanical properties of their tissue engineered specimens. From the mechanical characterization of the explanted PCL scaffold-tissue constructs under a simulated physiological environment, it can be concluded that the tissue engineered autogenous cartilage graft retained a compressive stiffness similar to that of human elastic cartilage, 18 weeks after transplantation. In contrast, non-wovens made of PGA fibers and a collagen Type I sponges of bovine origin were used, the explanted specimens did not reveal any mechanical comparative properties.

Collagen type II is known to bind a high amount of carbohydrate groups, allowing more interactions with water than some other types. Types IX and XI, along with type II, form fibrils that interweave to form a mesh. This organization provides tensile strength as well as physically entrapping other macromolecules. Although the exact function of types IX and XI are unknown, type IX has been observed to bind superficially to the fibers and extending into the inter-fiber space to interact with other type IX molecules, possibly acting to stabilize the mesh structure of the ECM. [25] The specific analysis of collagen of the explants by using the indirect antibody-immunofluoresence technique revealed the highest collagen type II expression in the PCL scaffold architecture.

It is known for more then two decades that chondrocytes isolated from articular cartilage lose their polygonal morphology and temporarily the capacity for typical three-dimensional (3D) cartilaginous matrix synthesis when cultured as monolayers in culture flask or Petri dishes.[26,27] Hence, such cells cultured in a 2D environment are defined as dedifferentiated chondrocytes because they can retain some of their chondrocytic differentiation markers.

Immunohistochemical analysis of chondrocytes isolated from human septal cartilage revealed that the re-expression of cartilage-specific collagen type II occurred in cells that had been cultured first in monolayer flask and then seeded onto the 3D matrices. [16] Hence, cells maintained part of their chondrocyte phenotype and produced an extracellular matrix containing an enhanced ratio of type II collagen to type I collagen. The de-differentiation of the expanded chondrocytes and especially the complete loss of collagen type II expression is a crucial issue of concern when tissue engineering cartilage in the honeycomb PCL architecture. Hence, the dedifferentiation and chondrogenic potential of in 3D matrices seeded cells and their phenotypic changes was a critical feature to be investigated in this study. Although, a noticeable expression of collagen type I was found in the PCL matrix, a cartilage-like tissue with a collagen type II containing ECM was developing within the PCL scaffold architecture.

The collagen assay of the natural pig cartilage as positive control also revealed a small amount of collagen type I. In addition, it cannot be ruled out that bovine collagen type I gel, which was used for the cell seeding was measured, even though a pig specific collagen type I antibody was used. Other groups also reported similar problems encountered when using commercially available collagen specific antibodies to analyze the ECM production of chondrocytes. [9, 28] However, it remains to be established whether this collagen type I is produced by dedifferentiated chondrocytes in the PCL matrix where chondrocytes have already re-differentiated and synthesized collagen type II, or if the dedifferentiating cells can coexpresses both collagen types concurrently. Further *in vitro* and *in vivo* experiments are planned to study this phenomenon.

Changes in the cellular form that are caused by the mechanical and surface properties of the ECM serve to switch cells between growth and differentiation programs. Integrins transmit mechanical stresses across the cell surface and the F-actin fibers of the cytoskeleton. Because of this mechanical interaction and the resulting biochemical processes, cells spread to varying degrees depending on the ability of the ECM to resist mechanical tension that cells exert through integrins within their focal adhesion sites. Beta-1 integrin plays a major role in cell attachment and was believed to be involved in mediating the interactions of chondrocytes with their environment. [29] The regulatory biomechanical network provides chondrocytes with the capacity to respond to the environment they are placed in. For example, in a non-deformable synthetic polymer substrate the cells might switch to an adhesion mode that provides firm anchorage. [25,30] In contrast, pliable substrates such as collagen based scaffolds, allow a significant translocation of the ECM together with the integrins. It may be concluded from this study that a rigid PCL matrix, when compared to a pliable PGA textile construct and a collagen foam, lead to the steady immobilization of adhesion complexes and focal contacts on the scaffold surface as well as the high tension at the cytoskeleton-membrane interface.

Pitt et al. [31] was among the first who observed that in vivo biodegradation of non-porous PCL specimens proceeded in two stages. The first stage of the degradation process involves non-enzymatic, random, hydrolytic ester cleavage and its duration is determined by the initial molecular weight of the polymer as well as its chemical structure. The second phase is characterized by the onset of weight loss. GPC was employed to assess the molecular weight distribution as function of the degradation kinetics. The Mw and Mn of the non-seeded and seeded PCL scaffolds decreased significantly over 18 weeks. Perhaps the best representation of the change in molecular weight is visible in the polydispersity index (PI). High PI values as measured for the non-seeded scaffolds correspond to a conglomerate of chains spanning a wide range of molecular weights. The sterilized scaffolds initially had a low PI value and during implantation the longer molecular chains were cut by hydrolysis into small fragments of wide ranges, which resulted in an increased PI. In contrast, in the implanted seeded samples the chains broke into shorter chains, which resulted in a smaller increase in PI.

Crystallinity is known to play an important role in determining both permeability and biodegradability of PCL because during the earlier stages of degradation the bulk crystalline phase is inaccessible to water and small biological molecules, such as enzymes and lipids. [32] DSC analysis of non-seeded explants showed a change of crystallinity of up to 60%, whereas the seeded samples showed no significant changes. It may be concluded from this study that the degradation and resorption mechanism of highly porous PCL scaffolds exhibited different characteristics from solid PCL films and rods. However, the entire mass loss over time of the tissue engineered scaffold-cell constructs has to be documented to define the degradation and resorption characteristics of highly porous PCL specimens. This is presently under investigation in a number of studies, which aim to tissue engineer bone and cartilage.

5. Conclusion

In vitro engineered scaffold-cell constructs were studied for their ability to generate elastic cartilage in an immunocompetent animal model. From the analysis of the histological, immunohistochemical, mechanical, and physico-chemical results of the tissue inside the scaffolds, it can be concluded that a novel PCL matrix showed its potential to support elastic cartilage-like tissue formation whereas a textile construct made of PGA and a bovine collagen type I foam allowed only the formation of fibrous and calcified tissue. A number of studies are in progress to further evaluate the potential of by FDM fabricated PCL scaffolds in cartilage tissue engineering.

ACKNOWLEDGEMENT
The authors would like to acknowledge Mr. Kim Cheng Tan for the support of fabricating the PCL scaffolds and Dr. Arthur Brandwood for reviewing the manuscript.

6. References

1. Buckwalter, J.A., Mankin, H.J. (1997) Articular Cartilage Part II: Degeneration and osteoarthrosis, repair, regeneration, and transplantation, *The Journal of Bone and Joint Surgery* **79**, 612-632

2. Sittinger M., Bujía J., Rotter N., Reitzel D., Minuth W.W., Burmester G.R. (1996) Tissue engineering and autologous transplant formation: practical approaches with resorbable biomaterials and new cell culture techniques. *Biomaterials* **17**, 237 -242.

3. Cao, Y., Vacanti ,J.P., Paige, K.T., Upton, J., Vacanti, C.A. (1997) Transplantation of chondrocytes utilizing a polymer-cell construct to produce tissue-engineered cartilage in the shape of a human ear. *Plast. Reconstr. Surg* **100**, 297-302,

4. Vacanti, C.A., Vacanti, J.P. ,(1997) Bone and cartilage reconstruction. In: Lanza R, Langer R, Chick W, editors. *Principles of tissue engineering. New York: R.G. Landes Co.*, 619-31.

5. Freed, L.E., Hollander, A.P., Martin, I., Barry, J.R., Langer, R., Vunjak-Novakovic, G. (1998) Chondrogenesis in a cell-polymer-bioreactor system, *Experimental Cell Research* **240**, 58-65.

6. Puelacher, WC., Kim, SW., Vacanti, JP., Schloo, B., Mooney, D., Vacanti, CA. (1994) Tissue-engineered growth of cartilage: the effect of varying the

concentration of chondrocytesseeded onto synthetic polymer matrices. *Int J Oral Maxillofac Surg* **23**, 49-53.

7. Aigner, J., Bujía, J., Hutzler, P., Kastenbauer, E. (1997) Distribution and viability of cultured human chondrocytes in a three-dimensional matrix as assessed by confocal laser scan microscopy. *In Vitro Cell. Dev. Biol.* 18, 407-409.

8. Sittinger, M., Reitzel, D., Dauner, M., Hierlemann, H., Hammer, C., Kastenbauer, E., Planck H., Burmester, G.R., Bujia, J. (1996) Resorbable polyesters in cartilage engineering: affinity and biocompatibility of polymer fiber structures to chondrocytes. *J Biomed Mater Res* **33**,57-63.

9. Rotter, N., Aigner, J., Naumann, A., Planck, H., Hammer, C., Burmester, G., Sittinger M., (1998) Cartilage reconstruction in head and neck surgery: comparison of resorbable polymer scaffolds for tissue engineering of human septal cartilage, *Journal of Biomedical Materials Research* **42**,.347-356.

10. Hutmacher, D.W. (2000) Polymeric Scaffolds in Tissue Engineering Bone and Cartilage. *Biomaterials 21*, 2529-2543.

11. Hutmacher, D.W., Zein, I., Teoh, S.H., Ng, K.W., Schantz, J.T., Leahy, J.C. (2000) Design and Fabrication of a 3D Scaffold for Tissue Engineering Bone, in C. M. Agrawal, J. E. Parr and S. T. Lin (eds.), *Synthetic Bioabsorbable Polymers for Implants, STP 1396*, American Society for Testing and Materials, West Conshohocken, PA **b**,152-167.

12. Zein I, Hutmacher DW, Teoh SH, Tan KC (2002). Poly(ε-caprolactone) Scaffolds Designed and Fabricated by Fused Deposition Modeling. Biomaterials (in press).

13. Hutmacher, DW., Schantz, JT., Zein, I., Ng, KW., Tan, KC., Teoh, S H. (2001) A Mechanical Properties and Cell Cultural Response of Polycaprolactone Scaffolds Designed and Fabricated via Fused Deposition Modeling. *J. Biomed. Mater Res.* **55**, 1-15.

14. Fu, B., Tan, BK., Lee, ST., Foo, CL, Sun, DF. Aw, SE. (2000). Effect of Fibrin Glue Coating on the Formation of New Cartilage. *Transplanation Proceedings* **32**, 210-217.

15. Cao, Y., Rodriguez, A., Vacanti, M., Ibarra, C., Arevalo, C., Vacanti, C. (1998) Comparative study of the use of poly(glycolic acid), calcium alginate and pluronics in the engineering of autologous porcine cartilage. *J Biomater Sci Polym Edn* **9**,475-87.

16. Haisch, A., Groger, A., Gebert, C., Radke, C., Ebmeyer, J., Sittinger, (2001) M. Designing Artificial Perichondrium: Immune- protection and Stabilization by Macroencapsulation for Transplantation of Tissue- Engineered Cartilage. *Transplantation* in press.

17. Perrin, D.E., English J.P., (1998) Polycaprolactone, in A.J. Domb, J. Kost, D.M. Wiseman (eds.), *Handbook of Biodegradable Polymers*, Harwood Academic Publishers, 63-77.

18. Vert, M., Li, S.M., Spenlehauer, G., Guerin P., (1992) Bioresorbability and biocompatibility of aliphatic polyesters, *Journal of Materials Science: Materials in Medicine* 3, 432-446.

19. Rehm, K.E., Claes, L., Helling, H.J., Hutmacher, D., (1994) Application of a Poly-lactide Pin. An Open Clinical Prospective Study. in K.S. Leung, L.K. Hung, P.C.

Leung (eds), *Biodegradable Implants in Fracture Fixation, World Scientific,* Hong Kong, 54.

20. Bergsma, EJ., Rozema, FR., Bos, RM., Brujn, W. (1993) Foreign Body Reactions to Resorbable Poly (L-lactide) Bone Plates and Screws Used for the Fixation of Unstable Zygomatic Fractures. *J Maxillofac Surg* **51**,666-670.

21. Böstmann, O., Hirvensalo, E., Mäkinen, ., Rokkanen, P. (1990) Foreign body reactions to fracture fixation implants of biodegradable synthetic polymers. *J Bone Joint Surg* **72-B**, 592.

22. Freed, L.E., Vunjak-Novakovic, G., (1997) Tissue culture bioreactors: chondrogenesis as a model system, in R.P.Lanza , R. Langer, W.L. Chick (eds), *Principles of Tissue Engineering,* Landes Co., Austin, 151-165.

23. Ma, P.X., Langer, R., (1999) Morphology and mechanical function of long-term in vitro engineered cartilage, *Journal of Biomedical Materials Research* **44**, 217-221.

24. Lohmann, C.H., Schwartz, Z., Niederauer, G.G., Carnes, Jr. D.L., Dean, D.D., Boyan, B.D., (2000) Pretreatment with platelet derived growth factor-BB modulates the ability of osteochondral resting resting zone chondrocytes incorporated into PLA/PGA scaffolds to form new cartilage in vivo, *Biomaterials* **21**, 49-61.

25. Zamir, E., Katz, M., Posen, Y., Yamada, K., Katz, B.Z., Lin, S., Lin, D., Bershadsky, A., Geiger, B., (2000) Dynamics and segregation of cell-matrix adhesion in cultured fibroblasts. *Nature Cell Biology* **2**, 34.

26. von der Mark, K., Gauss, V., von, der Mark, H., Muller, (1977) P. Relationship between cell shape and type of collagen synthesized as chondrocytes lose their cartilage phenotype in culture. *Nature* **267**, 531-532.

27. Takigawa, M., Shirai E., Fukuo K., Tajima, K., Mori, Y., Suzuki, F. Chondrocytes dedifferentiated by serial monolayer culture from cartilage nodules in nude mice. *Bone Miner* **2**:449-462.

28. Haisch, A., Groger, A,, Radke, C,, Ebmeyer, J,, Sudhoff, H,, Grasnick, G,, Jahnke, V, Burmester, GR,, Sittinger, M. (2000) Macroencapsulation of human cartilage implants: pilot study with polyelectrolyte complex membrane encapsulation. *Biomaterials* **21**, 1561-66.

29. Bouchet, BY., Colon, M., Polotsky, A., Shikani, AL., Hungerford, DS., Frondoza, C., Beta-1 integrin expression by human nasal chondrocytes in microcarrier spinner culture. *J Biomed Mat Res* **52**, 716-724.

30. Lamba, N.M.K., Baumgartner, J.A., Cooper, S.L. (1998) Cell-Synthetic Surface Interactions. in C.W. Patrick Jr., A.G. Mikos, L.V. McIntire (eds), *Frontiers in Tissue Engineering.* Elsevier Science Inc., New York, 107-120.

31. Pitt, C.G., Chasalow, F.I., Hibionada, Y.M., Klimas, D.M., Schindler, A., (1981) Aliphatic polyesters I: the degradation of poly(ε-caprolactone) in vivo, *Journal of Applied Polymer Science,* **26**,.3779-3787.

32. Pitt, C.G., (1990) Poly(ε-caprolactone) and its Copolymers, in R. Chasin, R. Langer (eds.), *Biodegradable Polymers as Drug Delivery Systems,* Marcel Dekker, New York, 71-120.

CRANIOFACIAL BONE TISSUE ENGINEERING USING MEDICAL IMAGING, COMPUTATIONAL MODELING, RAPID PROTOTYPING, BIORESORBABLE SCAFFOLDS AND BONE MARROW ASPIRATES

D.W. Hutmacher [*,**], D. Rohner [#,@], V. Yeow[#], S.T. Lee[#], A. Brentwood [***], J-T.Schantz[&],
Department of Bioengineering[*], Department of Orthopedic Surgery [**],
National University Singapore; Department of Plastic Surgery, Singapore
General Hospital[#]; Laboratory for Biomedical Engineering [&], National
University Singapore[@], Department of Reconstructive Surgery,
University Hospital Basel, Switzerland, [***] Graduate School of
Biomedical Engineering, University of New South Wales, Australia

Abstract
The clinical goals for craniofacial skeletal reconstruction are multifaceted. Aesthetic
and functional considerations often dictate the use of moldable implant materials.
However, in most cases these three-dimensional shaped transplants must also provide
immediate structural integrity. In addition, to minimize periimplant morbidity, the host-
graft interface should not produce an immunological or inflammatory response. Bone
tissue engineering has emerged as a potential method to address the problems of
autogenic bone grafting as well as allo- and xenoplastic materials. We report the three
month results of a study in an immuno-competent model, the aim of which was to treat
complex craniofacial defects.

1. Introduction

Successful craniofacial surgical experience with autograft bone has made it the material
against which all others are measured. Unfortunately, autografted bone is limited in
amount and desired morphology. In addition, the use of a patient's own bone is
associated with donor-site morbidity and graft resorption. When autogenous tissue is
not available, or its use is limited because of defect size or shape, allogenic, xenogenic,
and a variety of man-made biomaterials are used for craniofacial reconstruction.
Metallic, ceramic, and synthetic polymer materials are readily available. [1-3]

Reconstructive surgery utilizing bone substitutes such as hydroxyapatite, bioactive glass
ceramics, stainless steel and titanium implants has benefited from many advances in
biomaterials research and biomedical engineering over the last 20 years. [4] These
advances, however, have often occurred without clinical input in their early
development. Recently, a number of research teams realized that complex skeletal
reconstruction will require interdisciplinary interaction such that clinically oriented
problems will be introduced at the outset into the development of biomaterials and
medical devices. Kermer et al [5] were among the first which presented clinical data
which showed that medical imaging, finite element analysis, computer aided analysis

333

R.L. Reis and D. Cohn (eds.),
Polymer Based Systems on Tissue Engineering, Replacement and Regeneration, 333–354.
© 2002 *Kluwer Academic Publishers. Printed in the Netherlands.*

and rapid prototyping can be presented as an integrated approach that can be used for the modeling and simulation of hard tissue structures and the design of complex-shaped implants. In the late 1990's several groups reported the great clinical value of stereolithographic models for preoperative diagnosis of craniofacial deformities and planning of surgical corrections. Holck et al. 1999 [6] concluded that in complex cases stereolitography offered highly accurate models of the bony orbit for preoperative evaluation, surgical planning and teaching and could act as a template for custom-made implant manufacturing.

Hoffmann et al [7] designed ceramic implants on the basis of stereolitography models and prefabricated them using a commercially available dental copy-milling machine. Five patients received customized implants, which had been inserted for reconstruction of the lamina papyracea, zygomatic complex and infraorbital floor and rim. The authors conclude that based on the good aesthetic and functional results, with significantly reduced operating times and morbidity in all cases,that this technique should be the first choice for the reconstruction of complex orbital fractures. Eufinger et al. [8] published a clinical report in which 50 successful applications with individually prefabricated titanium implants - produced with computer-aided design (CAD) and computer-aided manufacturing (CAM) -are presented.

Ono et al [7] used pre-formed hydroxyapatite implants in the reconstruction of very large and complex- cranial bone defects in nine patients. The size, shape, and curvature of the hydroxyapatite ceramic implants were determined based on high-precision, full-scale models fabricated through a laser lithographic molding method using computed tomographic data. The authors reported that this method allowed the fabrication of hydroxyapatite ceramic implants of shapes that accurately matched the area of bone defect, allowing for a minimum of adjustment during the operation even with a complex implantation site. However, in one patient a crack in the hydroxyapatite implant was reported after an external traumatic injury. The authors concluded that the mechanical properties of a ceramic-based implant have to be improved. In addition, hydroxyapatite implant types are not designed to degrade and resorb, and the potential for replacement by host tissue does not exist.

The aim of the study of Erickson et al [8] was to determine the possible benefits from stereolithographic (SL) models produced from computed tomography (CT) and magnetic resonance imaging (MRI) data. Surgeon's opinions about the use of SL models for diagnosis, treatment planning, preoperative trials and practice, use within the operating room during surgical procedures, and for the construction of custom titanium implants or surgical devices were recorded and analyzed. The author concluded that surgeons using RP models in their surgical treatment regimes found them beneficial for diagnosis, planning, as a reference during surgery, and in the fabrication of custom implants and surgical devices that afforded surgical solutions previously not available. Overall, patients were believed to have received better care, because the surgeons had more knowledge of their unique anatomy before surgery. Through the use of these anatomical models, the patients experienced shorter surgical procedures, with more predictable results.

Recently, the application of rapid prototyping technologies has been extended to design and fabricate scaffolds for bone tissue engineering. [10-14] Hutmacher et al [15-18] designed and fabricated novel scaffolds by fused deposition modeling. The Poly Caprolactone (PCL) and Poly Caprolactone/Hydroxyapatite (PCL/HA) matrices produced had a fully interconnected honeycomb-like pore architecture that revealed mechanical properties suitable for bone engineering. The PCL templates supported in vitro the adhesion, proliferation and differentiation of osteoblast-like cells and bone-marrow-derived mesencymal stem cells. In those studies the formation of mineralized extracellular matrix throughout the entire scaffold architecture was seen after induction with osteoblastic media. In vivo studies in nude mice and rabbit models showed that the PCL scaffold/cell constructs lead to bone-like tissue formation in combination with a vascular network. Based on these findings Hutmacher and his group have developed a clinically driven hard tissue engineering program which utilizes medical imaging, computational modeling, rapid prototyping, bioresorbable scaffolds, and innovative transplantation surgery to treat complex craniofacial bone defects. [19] The objective of the presented study was to study the feasibility of this concept for the reconstruction of orbital wall defects in a pig model.

2.   Material and Methods

2.1   COMPUTATIONAL MODELING AND SCAFFOLD FABRICATION

In a preliminary laboratory anatomical study in pig cadavers, three different methods were evaluated to design a custom-made and defect-specific scaffold. As a defect model the orbital reconstruction was chosen because it represents a complex case in craniofacial surgery and the anatomy of the orbit of the pig is comparable to that of the human. [20] A defect of 1.5 x 2.5 cm at the medial wall of both orbits was surgically created. The defect was cast with a silicone material (Zerosil Supersoft®, Drewe AG, Germany). The silicone cast was used to fabricate a mold by using fast hardening dental cement. After this procedure the head was dissected from the body and CT scanned in the Radiology Department. The scaffolds designed and fabricated by the three methods described below were then tested for their accuracy of fit to the orbital bone defect. For this purpose both a rapid prototyped models manufactured based on the CT data of the skull as well as the anatomically processed bone structure of the pig head were used.

Three scaffold manufacture methods were used:

stereolithography based upon CT data,

stereolithography based upon laser surface scanning and

shape moulding.

2.1.1 Image based Scaffold Engineering using CT scan data

This method involved Computer Tomography (CT) in combination with modeling software to acquire the STL file for the scaffold design and fabrication. The starting point was a helical high-resolution computed tomogram of the orbital regions using a spin-echo CT unit (General Electric, New Jersey). The axial and coronal sections had a resolution of 1.6 mm per section. The borders of the cortical bone were automatically outlined and digitized using the standard software package of the CT scanner. The database was then exported and a virtual three-dimensional model of the

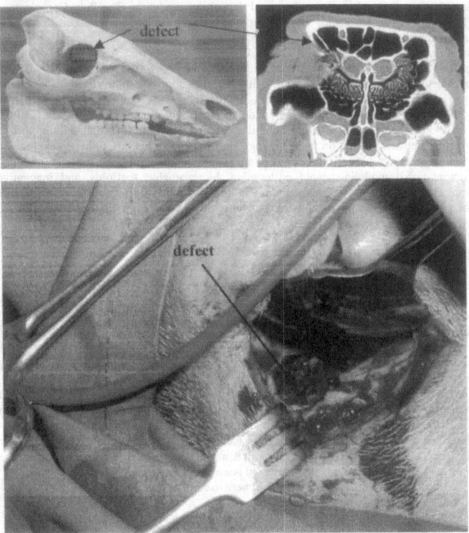

Figure 1. The anatomical model of the pig head shows the intact orbital architecture and the defect area is outlined. The CT scan in coronal projection (upper right) and the intraoperative picture show the structure and volume of the orbital defect (arrows).

bone was mathematically computed in a commercial modeling software (Mimics™, Materialise, Belgium). The CT scans were displayed as a series of superimposed images. The engineer and surgeon inspected the images in detail on the computer screen. The sections accommodating the defect were chosen to model the scaffold. The scan view chosen was the anterior-posterior and on average the defect spanned 12 layers (Figure 3-5). The area of the defect was masked out layer by layer in a different color. The 12 masks were then combined and . The modeled defect was exported as a solid 3D image in form of a stl. file which was used to design and fabricate polycaprolactone (PCL) scaffolds as described by Hutmacher et al.[14] Briefly, the FDM method was used to fabricate three-dimensional objects from computer generated solid or surface models and these models were used to derive software inputs to a customised stereolithographic apparatus designed for production oof porous scaffolds from layered polymer beams. Porous scaffold architecture was created directly in Stratasys' QuickSlice (QS) software. The head speed, fill gap, and raster angle for every layer was programmed through the QS software and saved as a .SLC (Slice) format file. A lay-down pattern of 0/60/120°, and gap spacing of 0.020" (0.508mm) were programmed to obtain a honeycomb-like pattern of triangular pores. The 2D slice data were converted into QS's .SML (Stratasys Machine Language) format file, which automatically generated the build paths based on the input parameters for each slice layer. The .SML data was sent to the FDM machine to fabricate the defect specific and custom-made scaffolds.

2.1.2 Laser Surface Scanner

For capturing surface details and morphologies, laser surface scanning systems are the method of choice for many commercial engineering-based applications. The potential of laser scanning of the surface and morphology of the bone defect represented in the dental cement mold was evaluated. The utilized scanner (Renishaw Cyclone Gloucestershire, UK) employs a low energy laser that points vertically downwards on the subject. It makes multiple passes in the x-direction in order to plot the contour of the exposed surface. The bone defect inside the mold was scanned three times and the acquired data were stored in form of a stl. file. This format was used to design and fabricate PCL scaffolds via FDM as described above.

2.1.3 Shape Molding

The third method utilized a low cost molding technique. By FDM, prefabricated PCL sheets (100 x 1000 x 4 mm, 65% porosity) were cut to scaffolds, which were 20% to 30 % larger, than the approximate size of defect represented in the dental cement mold. The mold was placed into saline solution bath at a temperature of 45°C. The sheets were molded to the shape of the defect by forming them into the cast. To solidify the defect morphology the mold in combination with the shaped PCL scaffold were cooled using iced saline solution. The custom-made PCL scaffold, which was formed to fit the contour of the bone defect, was then removed from the mold.

2.2   SURGERY I (DEFECT)

Ten pigs were housed in the animal holding facility at the Department of Experimental Surgery, Singapore General Hospital, for the entire duration of the experiment. Housing and feeding were according to standard animal care protocols. The study was approved by the Animal Welfare Committee of the Singapore General Hospital and was licensed by the National Institute of Health's Guide for Care and Use of Laboratory Animals. All

338

animals were premedicated with ketamine, induced and orally intubated with pentobarbitane and maintained with 1-% halothane. The periorbital region was prepared

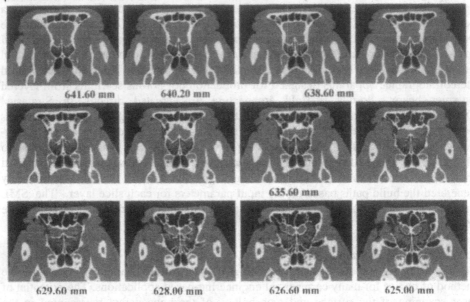

Figure 2. Coronal CT scans of the orbital defect computed in the Mimics software. The defect is outlined with a white mask layer by layer. It can be estimated that the defect spans a total of approximately 17 mm coronal direction.

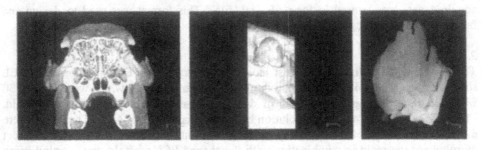

Figure 3. Three-dimensional computation of the modeled defect (white) and the pig skull (yellow). The computed mask of defect is exported as a stl. file. The screen picture on the right shows the general 3-D view of the external (solid) scaffold shape.

with 1% centrimide solution and 0,05% chlorhexidine acetate aqueous antiseptic solution. The transconjunctival approach with lateral cantothomy was provided as access to the orbit. A defect of 1.5x2.5 cm at the medial wall of both orbits was surgically created (Fig. 1). The defect was cast in situ with a sterile silicone material (Zerosil Supersoft®, Drewe AG, Germany) (Fig. 2). The silicone cast was used to produce a mold using dental cement (see above). The wound was rinsed with 0.9% saline solution and the incision closed in two layers with absorbable sutures (Vicryl®, Johnson&Johnson, NJ).

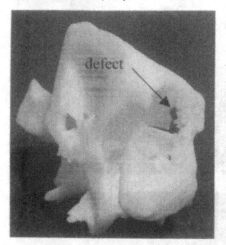

Figure 4. A full-scale model of the defect site was manufactured by Rapid Prototyping to evaluate the accuracy of the scaffold design and fabrication by using the three methods described above. It was demonstrated that PCL scaffolds (right), which would follow the three-dimensional contour of the defect site could be manufactured by utilizing medical imaging, computational modeling and FDM.

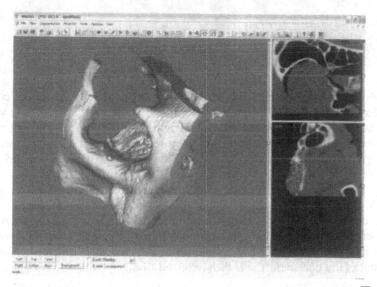

Figure 5. Screen picture of the in Mimics computed 3D CT scan of the pig skull after placement of a custom made PCL scaffold. The PCL scaffold follows and fills the contour of the complex anteriomedial orbital wall defect.

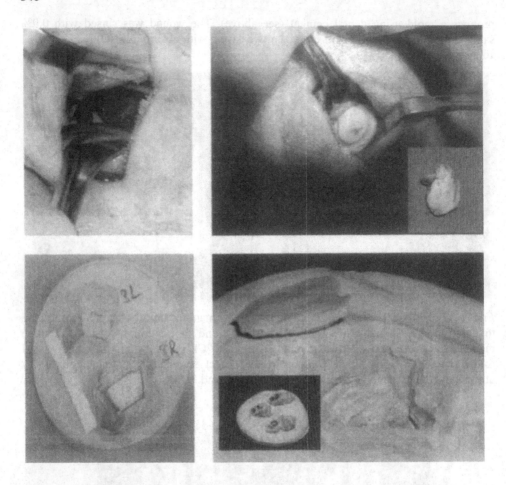

Figure 6. Defect in the posterior medial orbit upper (left). The defect cast with a standard silicone material. The outline and volume of the defect was clearly visible and cut out accordingly. The inset (left, lower right hand corner) give a picture of the osteomized fragment that were surgically removed. The defect geometry was then cast into a negative mold by using dental cement (lower right). By FDM prefabricated PCL sheets (100 x 1000 x 4 mm, 65% porosity) were cut to scaffolds, which were 20% to 30 % larger, then the approximate size of defect represented in the dental cement mold. The mold was placed into saline solution bath at a temperature of 45 °C. The sheets were molded to the shape of the defect by forming them into the cast. To solidify the defect morphology the mold in combination with the shaped PCL scaffold were cooled by using iced saline solution. The custom-made PCL scaffold, which was formed to fit the contour of the bone defect, was then removed from the mold (lower right).

2.3 SURGERY II (RECONSTRUCTION)

Simulating the clinical situation, three to seven days after the defect surgery (phase I) the reconstruction surgery (phase II) was carried out. Three different implant techniques were analyzed for the reconstruction of the orbital defect. The technique for the reconstruction was randomized for each orbit and none of the animals had the same technique in the right and the left orbit. The animals were again premedicated and anesthetized as previously described during the second phase surgery. Group 1 was used as control group and therefore the defect was not reconstructed. In group 2 non-porous PLLA/PDLA 70/30 sheets (60x40x 0.5 mm) were fabricated by compression molding as described by Hutmacher et al. [21] The rather stiff bioresorbable sheet was cut to size and placed by an onlay technique. Stable fixation was achieved by using a bioresorbable pins (Resorpin®, Geistlich, Germany). All defect-specific PCL scaffolds used in the study were manufactured by the shape molding process. Economical and logistic constraints did not allow CT scanning of ten live pigs. In group 3 non-coated PCL scaffolds were press-fitted into the defect site without additional fixation. Following the concept of Dean et al. [22] , the PCL scaffolds in group 4 were stored for 30 minutes in heparinised bone marrow aspirate before placement. The bone marrow (20ml) was harvested from the iliac crest using a trephine needle under aseptic conditions. The wounds were carefully rinsed with 0,9% saline solution and closed in two layers with absorbable sutures (Vicryl, Johnson&Johnson, NY). All pigs received amoxicillin (Ampicillin®) for five consecutive postoperative days. The pigs were sacrificed by an intravenous overdose of barbiturate solution after three months and the orbital regions processed for analysis as described below.

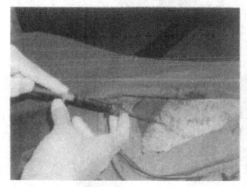

Figure 7. The bone marrow (20ml) was harvested from the iliac crest using a trephine needle under aseptic conditions (left). PCL scaffolds in group 4 were stored for 30 minutes in heparinised bone marrow aspirate before surgical placement (right).

2.4 HISTOLOGY

The entire orbital regions of all 10 pigs were dissected and were fixed in 3.5% neutral buffered formaldehyde (Merck, Germany). Specimens were decalcified in 30% formic

342

acid for 2 weeks and embedded in paraffin after dehydration in ascending concentrations of ethanol and xylene. Paraffin section (5-7 μm) were de-paraffinized and stained with H&E, Toluidine blue, von Kossa, Trichrome Goldner and Safranin O/Fast Green.

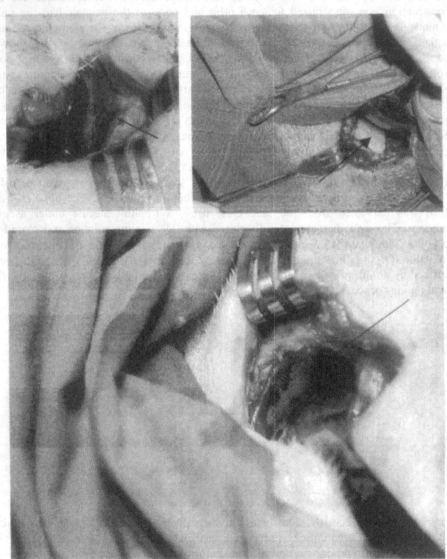

Figure 8. Reconstruction of the anteriomedial wall defects by using a bone-marrow coated PCL scaffold (upper left), a non-coated PCL scaffold (upper right), and non-porous PLLA/PDLA 70/30 sheets (lower). Shape moulded PCL scaffolds (upper left) did exactly follow the contour of the defect site (arrow). The fixation of the rather stiff foil with a bioresorbable pin did keep the bended PLLAPDLA 70/30 sheet in place (arrow).

2.5   RADIOGRAPHIC ANALYSIS

Macro and micro structural assessment was performed on the dissected orbital regions 3 month after reconstructive surgery. Plain antero-posterior microradiographs were taken at the defect sites using mammography equipment at 28kV and 32mAs (Mammomat 3000 equipped with an Rh beam, Siemens, Erlangen, Germany).

The scaffold/tissue constructs were scanned using a computer tomography scanner (Somatom Plus 4 volume zoom, Siemens, Germany) for non-destructive assessment of the bone mineral density. The tissue density of the reconstructed areas was semi-quantitatively determined in every scanned plane (0.5mm slice thickness) with a standard imaging software (Voxel Q, version 4.1, Macroni, Italy).

2.6   GEL PERMEATION CHROMATOGRAPHY

The polymer molecular weight distribution was determined by gel permeation chromatograph equipped with a differential refractor (Waters, Model 410, Milford, MA) and an absorbance detector refractor (Waters, Model 2690, Milford, MA). Triplicates of specimen from each group were dissolved in tetrahydrofuran (THF) and eluted in a series of configurations through a styragel columns refractor (Waters, Milford, MA) at a flow rate of $1mlmin^{-1}$. Polystyrene standards (Polyscience, Warrington, PA) were used to obtain a calibration curve.

2.7   DIFFERENTIAL SCANNING CALORIMETRY

A differential scanning calorimeter (TA Instruments DSC 2910, New Castle, DE) was utilized to analyze the thermal transition of the samples during melting to determine the crystallinity fraction of the PCL scaffolds. Triplicates from each group were heated at a rate of $5°$ C/min from $25°$ C to $70°$ C in aluminum pans using nitrogen as purge gas. The crystallinity fractions were based upon an enthalpy of fusion value of 139.5J/g for 100% crystallinity PCL as reported.

3.   Results

3.1   LABORATORY STUDY

Preoperative planning and surgical reconstruction of complex craniofacial defects, such as the bony region of the orbita often requires CT and/or MRI images. Several commercial software programs are available to read and manipulate CT and MRI data sets. To be able to design and fabricate a custom-made and defect specific scaffold it is necessary to represent the computer-modeled mask in a data format which can me imported and read by the software of the rapid prototyping machine. The most common format is known as stereolitography (.stl) which defines the surfaces of a computerized object as connected triangles. The Mimics software was able to read the CT data and allowed the user to generate a solid model of the defect site, which could be exported in the form of a .stl file. Based on the obtained data sets it was concluded that a resolution of 1.6 mm per section might be not sufficient in case larger and even more complex defects had to be precisely modeled. In such cases a resolution of 0.5 mm would be recommended.

Although, for logistical reasons, the presented animal study was not based on using the CT scan technique, the results of this laboratory based study showed the feasibility of using image based scaffold engineering to design and fabricates PCL matrices, which follow the contour of the defect site.

The laser scanning of the dental cement mold only allowed representation of the external geometry of the orbita defect in the form of dense point clouds, which had to be manually connected and compiled. Hence, the stl. data sets obtained were insufficient to design and fabricate accurate scaffolds. Even coloring of the defect side of the cast with a special paint which is normally used in industrial applications did not significantly improve the scanning process.

PCL has an unusually low glass transition temperature of –60°C, among the aliphatic polyesters used for tissue engineering applications. It exists in a rubbery state at room temperature, and also has a low melting temperature of 60°C. Therefore it is an ideal material to be used in shape molding when preprocessed into porous sheets. It could be shown that by this method fabricated PCL scaffolds filled out precisely the complex shaped defect site. Therefore, it was decided to use this inexpensive and straightforward technique in the animal model presented in this paper.

3.2   ANIMAL STUDY

3.2.1 Macroscopic Appearance

Postoperative healing was uneventful in all the pigs. After three months of healing there were no signs of infection. The defects in the group I (untreated controls) showed coverage with fibrous tissue and herniation of periorbital tissue. In group II, the entire defect was covered with the PLLA/PDLA 70/30 sheet, which was securely in place and embedded in a thick with fibrous tissue capsule. Macroscopically, there were no signs of infection or inflammatory reaction. The PCL specimens of group III and IV were all integrated in the host bone and covered by thin fibrous soft tissue. No signs of infection or inflammatory reaction could be detected.

3.2.2 Histology

The animals were sacrificed 3 month post impantation and specimens were processed and analyzed as described above. Typical cross-sections of the orbits were processed in a coronal plane and included on one side of the defect the orbital cavity and on the other side the nasal cavity. A gross light microscopical inspection of all histological sections was performed. For the group I specimens the orbital defect was bridged with fibrous soft tissue. The surface towards the nasal cavity was lined with a thin epithelium layer. The border zone of the defect showed minimal signs of new bone formation. There was no detection of giant cells. In group II the non-porous polymer foils did act as a template so that the nasal surface was lined with a thick fibrous tissue capsule. Within this fibrous tissue some giant cells could be detected. A significant inflammatory

template so that the nasal surface was lined with a thick fibrous tissue capsule. Within this fibrous tissue some giant cells could be detected. A significant inflammatory reaction was not present. There was little new bone formation starting from the border zone of the defect.

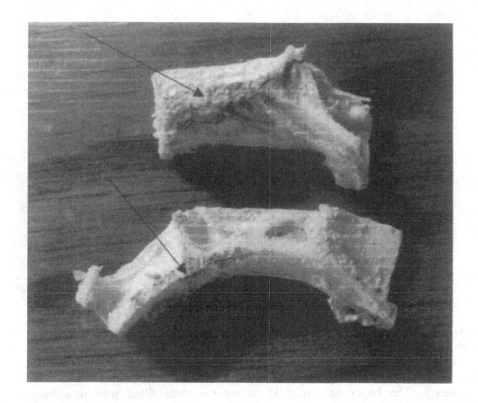

Figure 9. Gross morphological pictures of a bone-marrow coated scaffold construct (upper specimen) and a PLLA/PDLA 70/30 foil (lower specimen retrieved 3 month after implantation. It can been seen that the porous PCL scaffold (arrow) was able to fill out the defect site whereas the foil (arrow) did only bridge the lacking bone volume.

In group III the PCL scaffolds fitted precisely into the defect site. Most of the honeycomb matrix architecture was filled with fibrous tissue and only at the nasal side was new bone formation detected. Giant cells were present at the interface between fibrous tissue and the PCL scaffold. However, this giant cell reaction did not result in a clinically detectable foreign body reaction. In group IV the defects were filled up with the scaffolds in the same manner than in group III. The pore architecture that was oriented towards the nose showed formation of new bone, too. Qualitative analysis revealed that the amount seemed to be higher than the new bone in the non-coated scaffold. It was difficult to differentiate between new bone that was originating from the host and new bone that was starting to grow within the scaffold induced by the bone marrow coating. Giant cell formation without signs of septic or aseptic inflammation was seen in this group, too.

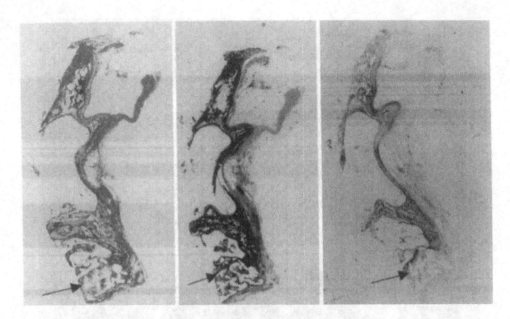

Figure 10. Histological photographs of a bone-marrow coated scaffold construct. Prior to histological analysis the orbits were cut in to half's. The typical cross-sections of orbits were done in a coronal plane and included on one side of the defect the orbital cavity and on the other side the nasal cavity. The PCL scaffold filled out the defect completely. The honeycomb scaffold architecture was filled with new bone in combination with islets of fibrous tissue. In general, on the nasal side larger amounts of new bone formation was detected. However, bone formation was also regular seen in the orbital cavity (see also radiographs) which most likely were induced by the transplanted bone marrow. Randomly, mononuclear cells were seen at the interface between fibrous tissue and the PCL matrix.

3.2.3 *Radiographic Analysis*

Postmortem dorsal and lateral soft radiographs were taken to evaluate the new bone formation and integration/ displacement of the PCL scaffolds and PLLA/PDLA 70/30 foil. The cortical or cancellous radiological structure on the defect edges was detectable in all samples. The control defects and the PLLA/PDLA 70/30 foil showed no regeneration but at the edges of some defects new mineralized tissue was observed. In contrast, PCL scaffold groups showed mineralized tissue areas inside the pores of the polymer matrix. Qualitative assessment revealed larger radiopaque areas in the bone marrow coated specimens, which was in accordance with the histological results.

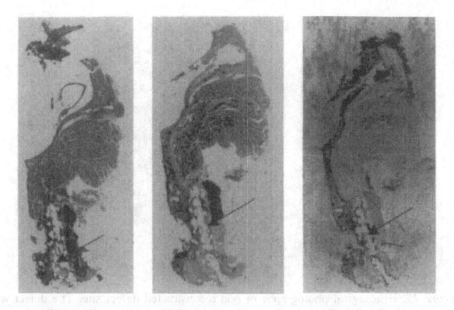

Figure 11. Histological photographs of a non-coated PCL scaffold. . The PCL matrix did follow the contour and occupied the defect volume in the same manner as the bone-marrow-coated group. In most cases, new bone formation was detected which originated from the nasal side. In two animals induction of new bone was seen also in the orbital cavity.

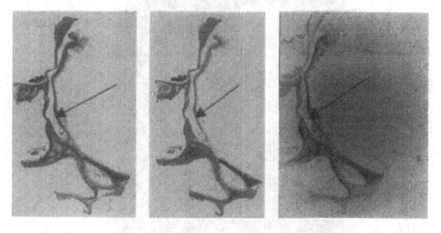

Figure 12. The defect was completely bridged with the PLLA/PDLA 70/30 foil (arrows). The nasal surface was lined with epithelium. The polymer sheet that has been dissolved during the decalcifying process showed a thick fibrous tissue capsule. Within this fibrous tissue some mononuclear giant cells could be detected A clinically relevant inflammatory reaction was not present. There was little new hard tissue formation starting from the boney border zone of the defect.

348

Figure 13. Histological photographs of non-reconstructed defect site. The defect was bridged with fibrous soft tissue. The surface towards the nasal cavity was lined with a thin epithelium. The border zone of the defect showed no signs of new bone formation and no inflammatory reaction was detectable.

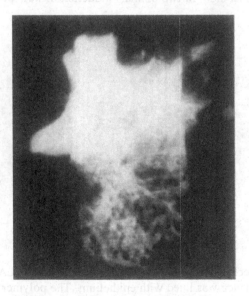

Figure 14. Radiograph of a bone-marrow coated PCL scaffold reveals that a large volume of mineralized tissue is growing on the interface to the host bone. Parts of the porous matrix towards the orbital cavity is also filled with calcified tissue.

The computer tomography data further confirmed the x-ray examination. Analysis of the new bone formation scan by scan revealed a higher percentage of mineralized tissue in the bone marrow coated group (data not shown). The 3D reconstruction of the CT scans exhibited that the PCL scaffolds were firmly incorporated into the host bone and that bone formation started also in the center of the PCL scaffolds as also seen in the histology.

3.2.4 Gas Permeation Chromatography and Differential Scanning Calorimetry

GPC analysis showed that prior to implantation the ethanol sterilized PCL scaffolds had a Mw of 145.000, Mn of 77.000, polydispersity of 1.7 and a crystallinity of 56%. The molecular weight of the coated and non-coated PCL scaffolds decreased on average 20 to 30% and the crystallinity 10 to 20% after 3 month of implantation. The ethanol sterilized PLLA/PDLA 70/30 foils had a Mw of 335.000, Mn of 161,000 polydispersity of 2.1 and a crystallinity of 32%. Three month post implantationem a reduction of Mw to 144.000 and Mn to 103,000 (polydispersity 1.4) and an increase in crystallinity to 39% was detected.

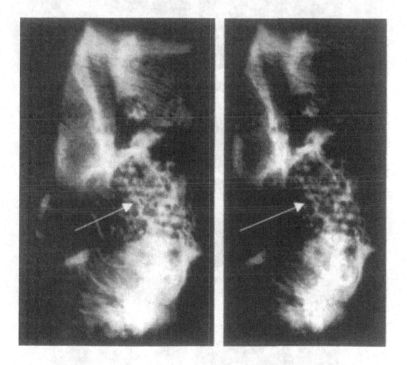

Figure 15. Direct radiographical comparison of group IV (left) and III (right) specimens of the same animal show larger amounts of calcified tissue in the bone-marrow coated PCL scaffolds (arrow).

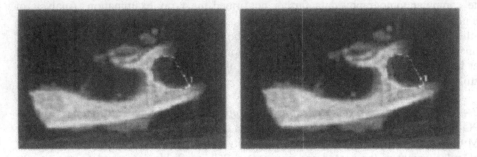

Figure 16. CT scans of non- coated PCL scaffold group shows only minimal calcified tissue inside the scaffold architecture (yellow marking)

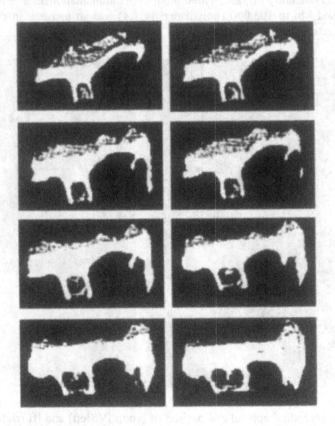

Figure 17. CT scans of bone-marrow coated PCL scaffold group shows layer by layer the formation of mineralized tissue within the scaffold architecture

4.   Discussion

In the last two decades remarkable progress has been made in the development of surgical techniques for major cranioplastic reconstruction. Although these sophisticated techniques have revolutionized craniofacial surgery from a clinical point of view, they may have already reached a number of their practical limits. The limitations in solving the remaining – and somewhat difficult - cranioplastic surgery problems may be approached from the perspective of the nature of the graft material with which the surgeon works. [23] Today, a great number of materials are available for craniofacial bone reconstruction. [1] These include autogenous and allogenic bone, non-absorbable implants such as silicone, polyamide, hydroxyapatite and polyethylene; titanium implants; bioresorbable implants and membranes made of synthetic polymers. It is obvious that with this many choices the optimal material for craniofacial reconstruction has yet to be found.

Successful surgical experience with autograft bone has made it the material against which all others are measured. [24,25] Unfortunately, autograft bone is limited in desired morphology. In addition, the use of patient's own bone is associated with donor-site morbidity and graft resorption. Metallic, ceramic, natural and synthetic polymer materials are readily available. All metals, most ceramics, and many polymers are not designed to degrade and resorb, and the potential for replacement by host tissue does not exist. The application of bioresorbable foil and sheets showed minimal potential for new bone growth and insufficient remodeling adjacent to the reconstructed area. [1]

Until recently the application of rapid prototyping technologies been restricted to surgical planning and simulation for reconstructive and plastic surgery to the fabrication of implants and medical devices. Lately, a number of research groups have fabricated scaffolds for bone tissue engineering by using a variety of RP methodologies because they have certain advantages when compared to conventional scaffold fabrication techniques. [12,14] FDM is currently being used by a Tissue Engineering Group at the National University of Singapore, to design and fabricate scaffolds of various porosity and micro-architecture. The honeycomb-like designs fulfill a number of requirements for bone tissue engineering such as allowing the cells to form a mineralized ECM in combination with a vascular network throughout the entire matrix architecture. FDM fabricated PCL matrix showed its potential to be used as scaffold material for the tissue engineering of bone in several in vitro studies and small animal models. [17,18]

The study in an immunocompetent large animal model presented in this paper revealed that after three months of healing both the bone marrow coated and non-coated PCL scaffold had an ingrowth of new bone that was significantly higher than compared to the amount of new bone along the surface of the PLLA/PDLA 70/30 sheets. It was shown first by Dean et al that porous scaffolds coated by bone marrow can induce new bone formation due to the attachment and entrapment of proteins and cells on the surface and inside the pore architecture. Our results confirm that bone marrow coating results in a higher amount of bone formation when compared to non-coated specimens.

The accumulation of small and non-randomly distributed bone islets in the non-coated scaffold group might be related to the filling of the porous PCL matrix with blood after implantation. It is known from the concept of guided bone regeneration (GBR) that a stable blood clot formation promotes next to the host bone hard tissue formation. Since the physiological wound healing is associated with wound contraction and shrinkage of the hematoma it is crucial for a successful GBR treatment that defect filler is used next to the membrane barrier. It is a necessary condition that the porous biomaterial has the ability to retain the blood clot for a sufficient period of time. Whang et al [26] postulated that the stabilization of the blood clot by matrices with pore sizes smaller then 50 microns would significantly enhance bone regeneration through the induction potential of growth factors within the blood and bone marrow. However, a rabbit study by Fisher et al [27] which used scaffolds with similar pore sizes did not show bone regeneration, although the filling of the scaffolds with blood after implantation was noted. However, the group the protein adsorption characteristics of the polymer used (not reported by Fisher et al.). It that this material characteristic might cane an important factor in bone biocompatibility and may have influenced the outcome of their study. [14]

Qualitatively, mono-nucleated cells of lymphocytic-monocytic type could be detected more often in the decalcified connective tissue in both PCL scaffold groups when compared to the nonporous PLLA/PDLA 70/30 sheets. The constant paucity, the small number of poly-nucleated cells, absence of phagocytosis and the few capillary vessel suggest that these mono-nucleated cells may not be due to inflammation and/or reaction to the presence of the bioresorbable polymer constructs. They possibly represent mono-nucleated elements such as those, which are normally found in the interstices and connective stroma in all tissues and organs.

Recently, Hutmacher [19] reported the initial results of a clinical pilot study, which is utilizing a bioresorbable scaffold in combination with cells. This report represents the first clinical effort to bring bone tissue engineering concepts based on the design and fabrication of patient specific bone grafts by applying medical imaging, computational modeling, bioresorbable scaffolds and osteoblast-like cells into the field of reconstruction of large, complex-formed cranial bone defects.

5.   Conclusion

Autogenous bone is currently the standard graft material for craniofacial reconstruction, but its shortcomings are numerous. Donor site morbidity, material scarcity, graft caliber and handling characteristics, and graft resorption are significant clinical problems. A logical solution would be the availability of autogenic bone graft, which require no or minimal donor site, would be precisely matched to the defect, highly biocompatible, and would be integrated and remodeled by the host soft and hard tissue. Such grafting materials do not yet exist.   This study indicated that the use of polycaprolactone scaffolds for the reconstruction of craniofacial defects was superior regarding to an already clinically used PLLA/PDLA 70/30 foil. The soft and hard tissue response of the PCL scaffolds in the presented pig model further confirmed that defect-specific scaffolds processed by FDM are a promising alternative to conventional scaffold materials and techniques. The minimal cellular encapsulation of the PCL scaffolds indicates that the highly regular honeycomb-like pore architecture allows fast graft vascularisation and subsequent formation of new host bone. The clinically uneventful course showed that the PCL might be in the future a potential scaffold material in the field of craniofacial surgery. The 12 months results of an ongoing study will reveal if this type of PCL scaffold can completely regenerate a bone defect of this size.

6. References

1. Hammer B., Kunz C., Schramm A., de Roche R., Prein J. (1999) Repair of complex orbital fractures: technical Problems, state-of-the-art solutions and future perspectives. Ann Acad Med Singapore 28:687-91

2. Grant M.P., Iliff N.T., Manson P.N.(1997) Strategies for the treatment of enopththalmos. Clin Plast Surg 24:539-50

3. Eufinger H., Wehmöller E., Machtens L., Heuser A., Harders D.; (1995) Reconstruction of craniofacial bone defects with individual alloplastic implants based on CAD/CAM-manipulated CT-data. J Craniomaxillofac Surg 23:175-81

4. Hutmacher D., Kirsch A., Ackermann K.L., Huerzeler M.B. (1998) Matrix and Carrier Materials for Bone Growth Factors - State of the Art and Future Perspectives. In: Stark GB, Horch R, Tancos E(eds). Biological Matrices and Tissue Reconstruction, Springer Verlag, Heidelberg, Germany; p197-206

5. Deckard C. and Beamann J (1987). Advances in Selective Laser Sintering. In: Proceedings of the 14th Conference on Production Research Technology: University of Michigan, 447-452

6. Holck D.E., Boyd E.M. Jr, Ng J., Mauffray R.O. (1999) Benefits of stereolithography in orbital reconstruction. Ophthalmology Jun;106:1214-8

7. Hoffmann J., Cornelius C.P., Groten M., Probster L., Pfannenberg C., Schwenzer N. (1998) Orbital reconstruction with individually copy -milled ceramic implants. Plast Reconstr Surg Mar; 101(3):604-12

8. Hoffmann J., Cornelius C.P., Groten M., Probster L., Pfannenberg C., Schwenzer N. (1998) Orbital reconstruction with individually copy -milled ceramic implants. Plast Reconstr Surg Mar; 101(3):604-12

9. Ono I., Tateshita T., Satou M., Sasaki T., Matsumoto M., Kodama N. (1997). Treatment of Large Complex Cranial Bone Defects by Using Hydroxyapatite Ceramic Implants. Plastic Reconstruction Surgery Aug;104(2):339-49

10. Erickson D.M., Chance D., Schmitt S., Mathis J. (1999) An opinion survey of reported benefits from the use of stereolithographic models. J Oral Maxillofac Surg. Sep;57(9):1040-3.

11. Feinberg S.E., Hollister S.J., Halloran J.W., Chu T.M., Krebsbach P.H.(2001) Image-based biomimetic approach to reconstruction of the temporomandibular joint. Cells Tissues Organs.; 169(3):309-21.

12. Hutmacher D.W. (2000) Polymeric Scaffolds in Tissue Engineering Bone and Cartilage. Biomaterials 21, pp 2529-2543

13. Marra K.G., Campbell P.G., Dimilla P.A., Kumta P.N., Mooney M.P., Szem J.W., Weiss L.E. (1999) Novel three dimensional biodegradable scaffolds for bone tissue engineering. Mater Res Soc Symp Proc 550:155-160.

14. Porter N.L., Pilliar R.M., Grynpas M.D. (2001) Fabrication of porous calcium polyphosphate implants by solid freeform fabrication: a study of processing parameters and in vitro degradation characteristics. J Biomed Mater Res. Sep 15;56(4):504-15.

354

15. Hutmacher, D.W., Zein I., Teoh S.H., Ng K.W., Schantz J.T., Leahy J.C. Design and Fabrication of a 3D Scaffold for Tissue Engineering Bone, In: Synthetic Bioabsorbable Polymers for Implants, STP 1396, C. M. Agrawal, J. E. Parr and S. T. Lin, Eds., American Society for Testing and Materials, West Conshohocken, PA, 2000. p152-167
16. Zein I, Hutmacher DW, Teoh SH, Tan KC (2002). Poly(ε-caprolactone) Scaffolds Designed and Fabricated by Fused Deposition Modeling. Biomaterials 23, 1169-1185
17. Hutmacher, DW., Schantz, JT., Zein, I., Ng, KW., Tan, KC., Teoh, S H. (2001) A Mechanical Properties and Cell Cultural Response of Polycaprolactone Scaffolds Designed and Fabricated via Fused Deposition Modeling. J. Biomed. Mater Res. 55, 1-18. Schantz, JT., Hutmacher, DW., Ng, KW., Lim T.C., Chim H. Teoh, S H. (2002) Induction of ectopic bone formation by using human periosteal cells in combination with a novel scaffold technology. Cell Transplant (in press)
19. Hutmacher D.W., Lauer G. (2002). Grundlagen und aktuelle Anwendungen des Tissue Engineering in der Mund-, Kiefer,- und Gesichtschirurgie. Implantologie (in press)
20. Farkas L.G., Munro I.R., Vanderburg B.M.(1976) Quantitative assessment of the morphology of the pig head as a model in surgical experimentation. Part 1: methods of measurement. Can J Comp Med 40:397
21. Hutmacher DW, Kirsch, A., Ackermann, KL, Huerzeler, MB. A Tissue Engineered Cell Occlusive Device for Hard Tissue Regeneration – A Preliminary Report. Int J Periodontics Restorative Dent 2001, 21:48-59
22. Dean D., Topham N.S., Rimnac C., Mikos A.G., Goldberg D.P., Jepsen K., Redtfeldt R., Liu Q., Pennington D., Ratcheson R. (1999). Osseointegration of preformed polymethylmethacrylate craniofacial prostheses coated with bone marrow-impregnated poly (DL-lactic-co-glycolic acid) foam. Plast Reconstr Surg. Sep;104(3):705-12.
23. Vacanti C.A., Vacanti J.P. (1997) Bone and cartilage reconstruction. (In: Lanza R., Langer R., Chick W. editors). Principles of tissue engineering. New York: R.G. Landes Co. p. 619-31.
24. de Roche R., Kuhn A., de Roche-Weber P., Gogolewski S., Printzen G., Geissmann A., De Jager M., Hammer B., Prein J., Rahn B.(1998) Experimental reconstruction of the sheep orbit with biodegradable implants. Mund Kiefer Gesichtschir May;2 Suppl 1:117-20
25. Rozema F.R., Bos R.R., Pennings A.J., Jansen H.W. (1990) Poly(L-lactide) implants in repair of defects of the orbital floor: an animal study. J Oral Maxillofac Surg 48:1305-1309, discussion 1310
26. Whang K., Healy K.E., Elenz D.R. et al. (1999). Engineering bone regeneration with bioabsorbable scaffolds with novel microarchitecture. Tissue Eng. 5(1):8-16
27. Fisher J. Vehof J., Dean D. et al. (2002). Soft and Hard Tissue Response to photocrosslinked poly (propylene fumarate) scaffolds in a rabbit model. J Biomed Mater. 59:547-556

Tissue Engineering and Regeneration of Other Tissues

BIODEGRADABLE POLYMERS AS SCAFFOLDS FOR TISSUE ENGINEERING AND AS TISSUE REGENERATION INDUCERS

Y. IKADA
Institute for Frontier Medical Sciences, Kyoto University, Shogoin, Sakyo-ku, Kyoto 606-8507, Japan

1. Introduction

The most powerful treatment to cure diseases may be medication, that is, drug administration to the patients. However, drugs are no more effective when large part of a tissue has been severely damaged or an organ has irreversibly lost its function. In these cases, either artificial organ or organ transplantation is at present the first choice for reconstruction of the defective or lost organ. Unfortunately, these therapeutic methods are not always effective, but have several problems that are difficult to solve. For instance, the number of organ donors is quite smaller than that of the patients waiting for the organ to be transplanted. Complications of immuno-suppresive agents are also trouble for the organ recipients. Also, current artificial organs are required to improve the poor biocompatibility and the insufficient ability to replace defective organs.

Recently, a new therapeutic means has emerged to complement this reconstructive surgery supported by the artificial organs and the organ transplantation. That is the regenerative medicine which is characterized by tissue regeneration mostly using autologous cells[1]. If a new tissue is regenerated from the cells of patient herself or himself to replace a defective tissue, this regenerated tissue is absolutely free of immune rejection and foreign-body reaction, because the tissue is autologous. Therefore, the regenerative medicine is often called ultimate ideal means for reconstructive medicine.

R.L. Reis and D. Cohn (eds.),
Polymer Based Systems on Tissue Engineering, Replacement and Regeneration, 357–370.
© 2002 *Kluwer Academic Publishers. Printed in the Netherlands.*

358

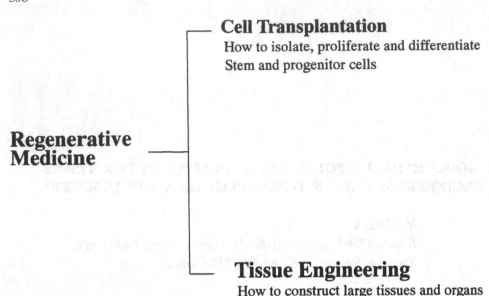

Cell Transplantation
How to isolate, proliferate and differentiate
Stem and progenitor cells

**Regenerative
Medicine**

Tissue Engineering
How to construct large tissues and organs
from differentiated cells

Fig. 1 Two Subgroups of Regenerative Medicine

1.1. WHAT IS TISSUE ENGINEERNG?

Regenerative medicine can be divided into two categories; cell transplantation
and tissue engineering, as shown in Fig.1. The difference of tissue engineering
from cell transplantation is ascribed to the requirement of artificial scaffold for
cell growth and differentiation. Bone marrow cells are transplanted to patients
suffering from leukemia or other severe hematopoetic diseases. This bone
marrow transplantation does not need any artificial scaffold, since the bone
marrow of patient has stroma which serves as the scaffold for the transplanted
hematopoetic progenitor cells. Other cells currently applied for transplantation
includes cardiac myocytes, pancreatic cells, and fetal neural cells.

Most of tissues comprise cells and extracellular matrix. When a tissue is
largely lost, the extracellular matrix is also lost. In such a case cell infusion is
ineffective to regenerate the tissue, as the infused cells will be diffused out
from the diseased area due to the absence of extracellular matrix. This is a
major reason to provide a scaffold for tissue regeneration. Such regenerative
medicine that utilizes scaffold is called tissue engineering.

1. **Scaffold (artificial extracellular matrix)**
   from porous, resorbable biomaterials for cell   attachment,
   proliferation and differentiation
2. **Cells**
   Autologous, allogenic or xenogenic; stem, progenitor, or
   differentiated
3. **Cell growth factors**
   for promotion of cell proliferation and differentiation
4. **Carriers of growth factors**
   for their sustained release
5. **Barriers**
   for making space by preventing fibrous tissue ingrowth

Fig 2. Materials Necessary for Tissue Engineering

1.2. ELEMENTS REQUIRED FOR TISSUE ENGINEERING

As mentioned above, tissue engineering needs cell and scaffold for regeneration of a new tissue[2,3]. In addition, signaling factors such as cell growth factors are required for tissue engineering. In some cases the signaling factors necessary for tissue regeneration are secreted from the cells, but sometimes the secretion is not enough to give prompt signaling to interacting cells. In such a case we should provide signaling factors exogenously. The fundamental elements necessary for tissue engineering are given in Fig.2.

When aqueous solution of growth factors is administered to the site of tissue regeneration, most of the growth factor molecules will be quickly diffused out from the administered site. This suggests that some effective methods should be exploited to maintain the growth factor at an effective concentration for a certain period of time in the vicinity of the site for regeneration. Scaffold also requires the most adequate design to regenerate the target tissue by selecting the best three-dimensional structure, pore size, and bioabsorption rate.

Therefore, it is apparent that tissue engineering is the typical interdisciplinary science and technology requiring cooperation among biomaterial scientists, cell biologists, and medical doctors.

Table 1   Functions of Biomaterials Used for Tissue Engineering

1. Scaffold	Substrate for cell proliferation and differentiation
2. Space Maker	Maintenance of the space for tissue regeneration
3. Barrier	Prevention of fibrous tissues infiltration
4. Pattern regulator	Patterning of regenerated tissues
5. Depot	Sustained release of growth factors
6. Nutrient supplier	Oxygen and nutrients supply
7. Immuno-isolator	Suppression of immune attack from the host

2. Scaffold

As pointed out above, tissue engineering is characterized by scaffold, which is also called artificial extracellular matrix. This matrix biomaterial is necessary for tissue engineering, because the cells regenerating a tissue should be provided with a proper biological environment. Therefore, the scaffold to be used for tissue engineering has to fulfill a number of functions such as summarized in Tab.1.

2.1. REQUIREMENTS FOR SCAFFOLD

The most important requirements for scaffold are to be biodegradable, porous, and cell-adhesive.

2.1.1. *To be biodegradable*

Any biomaterial supporting tissue engineering should disappear from the site of action, when the tissue regeneration is completed, because it is no more needed. Otherwise, the biomaterial still remaining there would disturb the tissue regeneration or induce foreign-body reactions later. Most ideally, scaffold should disappear from the regeneration site, in harmony with formation of new extracellular matrix, but it is too difficult to control the biodegradation of scaffold in such a fashion. Although the time course of tissue regeneration is not monotonous but very complicated, many man-made scaffolds have proved to work well for tissue regeneration.

Table 2   Biodegradable Materials Used for Tissue Engineering

Materials	Examples
1. Natural polymers	Fibrin, Collagene, Gelatine, Hyalurotane, Matrigel®
2. Synthetic polymers	Polyglycolide (PGA), Glycolide-lactide copolymers (PGLA), Polylactides (PLA; PLLA and PDLLA), Ethylene oxide block copolymers with PLA or propylene oxide chains
3. Inorganic materials	Tricalcium phosphate, Calcium carbonate, Non-sintered hydroxy-apatite

Most of scaffolds are currently made from biodegradable biomaterials listed in Tab.2. As can be seen, there are very few inorganic biodegradable materials. Although hydroxyapatite is not absorbable at the physiological pH, this biomaterial can be used as scaffold for bone regeneration, because the hydroxyapatite used as scaffold will be integrated into the regenerated bone tissue. Use of β–tricalcium phosphate(β-TCP) seems to be preferred to hydroxyapatite when used as scaffold, because of higher solubility.

Polyglycolide(PGA) is often used as scaffold, similar to other synthetic aliphatic polyesters, but it should be kept in mind that the glycolic acid produced as a result of quick biodegradation of PGA will lower the local pH at the regeneration site. This will lead to subacute inflammation, when a large amount of PGA scaffold is applied to the location, where the flow of body fluids is not high enough. In any case, accumulation of biodegradation by-products at a limited place should be avoided to prevent the pH decrease there. Poly-2-hydroxyalkanoates such as poly-2-hydroxybutyrate(PHB) are easy in fabrication to desired shape, but can undergo biodegradation only in the earth environments (in soil and sea) through enzymes of microorganisms which are missing in our body. Therefore, these polyesters cannot be used for patients.

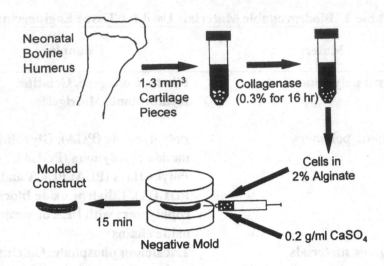

Fig.3 A Construct Made from Alginate, CaSO₄, and Chondrocites by Injection

Biopolymers such as collagen, gelatin, hyaluronate, chitin, and alginate are also used as scaffold. Fig.3 illustrates injection molding of an alginate-CaSO₄ composite together to chondrocytes[4]. The mixture, formed upon adding aqueous solution of alginate to CaSO₄ solution, is still flowable liquid, but gradually sets to a gel as a result of salt crosslinking of alginic acid with Ca^{2+} ion. This gel will be resorbed in the body as the Ca ions are slowly replaced with Na ions existing at a high concentration in the body.

Biological tissues harvested from animals or cadavers have been also used as scaffold. Generally, the harvested tissues are subjected to treatment with surfactant to remove all the cellular components, as they might induce immune reactions. The major component of the purified tissues is fibers of type I collagen. Such biological scaffolds can be obtained from gastrointestine, blood vessel, heart valve, and amniotic membrane. Fibrous encapsulation formed around a subcutaneous implant also produces a collagenous tissue which is applicable as scaffold[5].

2.1.2. *To Be Porous*

Following selection of a biodegradable material with the most appropriate molecular weight and chemical composition, this material is processed so as to have desired shape and dimension in accordance with the requirement for the tissue to be regenerated. The feature common to almost all the scaffolds is their porous structure, which is required since scaffolds should encourage cells as many as possible to enter into the inside of scaffold. Too large pore size of scaffold cannot effectively accept numerous cells, while cells cannot go into scaffold if the pore is too small in comparison with the cell size. The optimum pore size ranges from several ten μm to several hundred μm, depending on the nature (vascular or avascular) of the tissue to be regenerated. Scaffold of high porosity is not recommended, as the mechanical strength of such a scaffold becomes low due to its high porosity.

One problem facing at the scaffold fabrication is to maintain excellent mechanical properties including strength and modulus. For instance, scaffolds to be used for ligament regeneration require high tensile strength and sufficient rubbery elasticity, as they will be subjected to cyclic elongation and contraction to promote formation of ligament with high tensile strength. However, it is quite difficult to address this biodynamic demand if we select the scaffold from currently available synthetic biomaterials, because these materials are not so tough and pliable as natural ligament, cartilage, and blood vessel walls.

Porous structure can be produced with various methods. Most simply, porous scaffolds are made with the freeze-drying method from solutions of scaffold materials. Fig.4 shows a scaffold prepared with the freeze-drying method for regeneration of heart valve[6]. Microscopic phase-separation is also a means to produce porous scaffold. Widely used is the porogen method where a scaffold material is mixed with porogen microparticles, followed by extraction of the porogen from the solidified mixture. Inorganic salts like NaCl are the most common porogen used for fabrication of porous scaffolds. Woven and non-woven products made from biodegradable fibers also produce porous scaffolds with high reproducibility. An example is the scaffold for blood vessel regeneration prepared from a fabric of glycolide-lactide copolymer[7].

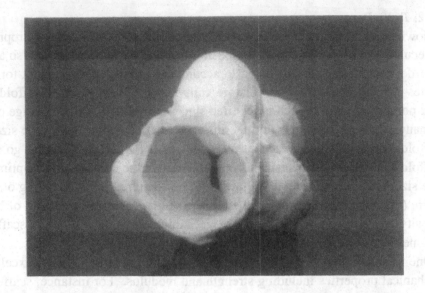

Fig.4   A Scaffold Made from Lactide-ε-Caprolactone
Copolyme for Heart Valve Regeneration

In addition to the methods mentioned above, modern microtechnologies such as laser and three-dimensional photo-polymerization processing have been applied to scaffold fabrication in recent years.

2.1.3. *To Be Cell-Adhesive*

Cells should adhere to a substrate when they divide to multiply the cell number. This indicates that the surface of scaffolds should be cell-adhesive. Generally, both too hydrophilic and too hydrophobic surface do not allow cell adhesion, while surface with well-balanced hydrophobicity-hydrophilicity promotes cell adhesion. In addition, many attempts have been made to modify the surface of scaffolds by immobilizing cell-adhesive proteins including collagen, fibronectin, and laminin, since immobilization of cell- adhesive proteins onto synthetic surface gives the best substrate for cell adhesion. However, it is not always necessary to immobilize these expensive proteins or oligopeptides such as RGDS onto scaffolds, because cell-adhesive proteins present in the culture medium or secreted from cells will deposit onto the man-made scaffolds unless their surface is neither too hydrophobic nor too hydrophilic.

In this respect, scaffolds made from collagen are applicable without any surface modification and, in many cases, simple coating of scaffolds with collagen is sufficient to make the surface cell-adhesive.

2.2. ROLES OF SCAFFOLD

Scaffold plays several important roles in tissue engineering, as already shown in Tab.1. To provide cells with substrate for their adhesion is a most important function of scaffold. Keeping space for tissue regeneration is another important role of scaffold. Tissue regeneration becomes impossible if the space for tissue regeneration is not available due to previous occupation by collagenous tissues such as scar. It should be stressed that the growth activity of fibroblasts which often invade into the space reserved for tissue regeneration is remarkably high in comparison with other cells. The primary role of barrier membranes used for regeneration of periodontal tissues and alveolar bone is to protect the space for regeneration from epidermal ingrowth to the site. The protected space is initially filled with blood clot, which functions as natural scaffold for tissue regeneration.

Another role of scaffold is to regulate the final shape of the tissue regenerated. As the cell growth and extracellular matrix formation cannot proceed beyond the scaffold presence, tissue regeneration is normally restricted to the original shape of the scaffold, so far as scaffold biodegradation takes place in harmony with the tissue regeneration.

When a scaffold is implanted exactly at the place, where the lost tissue existed, the scaffold should work, in principle, as the alternative of the lost tissue. If it is a load-bearing tissue, the implanted scaffold also should bear load[8]. One example is regeneration of long bone. In this case, a hydroxyapatite rod is a possible candidate for the scaffold, but porous hydroxyapatite is too brittle to support high load. Much tougher scaffold will be needed for this long-bone regeneration, unless an extracorporeal fixation device is used.

1. **Injection of plasmid DNA**
2. **Transplantation of cells modified with gene technology**
3. **Use of DDS technology**

Fig. 5 Methods for sustained Release of Growth Factors

3. Supply of Growth Factors

Generally, the number of cells which can be obtained from patients is not large enough, especially when patients are aged. A promising way to overcome this poor cell supply is to make use of cell growth factors which are capable of multiplying cell number. Some growth factors can induce tissue regeneration without cell seeding. The well-known examples include capillary and arteriole regeneration by basic fibroblast growth factor(bFGF) and vascular endothelial growth factor(VEGF) and bone regeneration by bone morphogenetic protein(BMP) and transforming growth factor-β (TGF-β).

When these signaling factors are administered to the site for regeneration by one shot in aqueous solution, they will quickly be diffused out and remain there at such a very low concentration that is no more effective for tissue regeneration.

To circumvent this situation, several methods have been proposed as tissue regeneration inducers, as shown in Fig.5. One of them is application of gene technology to introduce the gene responsible for biosynthesis of signaling factors to cells. It has been reported that simple administration of the naked plasmid DNA of VEGF to the defective cardiac tissue leads to angiogenesis around the administered site, resulting in recovery of ischemic heart diseases[9]. Cellular modification by gene recombination also facilitates synthesis and secretion of growth factors.

Another method to keep the growth factor concentration high enough for tissue regeneration at the site of regeneration is to utilize drug delivery systems(DDS). This DDS technology uses polymeric carriers for sustained release of physiologically-active substances such as growth factors[10]. A problem associated with this system is poor availability of carriers suitable for

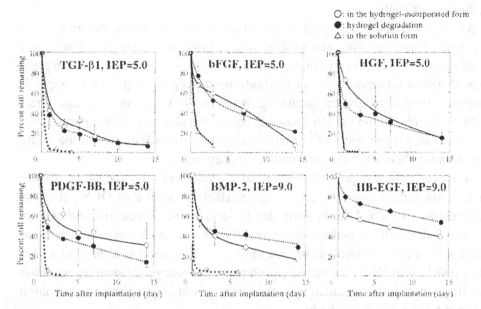

Fig.6 In Vivo Release Profiles of Various Growth Factors from Crosslinked Gelatin Carrier

growth factors.   This is because deactivation of protein drugs very readily occurs when bioactive proteins such as growth factors are incorporated into hydrophobic polymeric carriers including glycolide-lactide copolymers, unless proteins are mixed physically at low temperature with carriers without using any organic solvents.

An attempt to avoid this protein deactivation during formulation of carrier-protein composites is to impregnate bioactive protein molecules into a preformed, biodegradable carrier.   For this purpose, crosslinked collagen and hydroxyapatite were often selected as the carrier, but insignificant sustained release of growth factors was observed. Probably, collagen cannot trap growth factors as both collagen and growth factors are positively charged, while protein adsorption to hydroxyapatite is too high to release the adsorbed protein molecules. This suggests that negatively charged hydrogels will be able to trap basic growth factors and to release the trapped proteins upon biodegradation of the hydrogels. Indeed, this idea was realized when bFGF was impregnated into crosslinked gelatin hydrogel with negative charges. Fig.6 shows the percentage of still trapped bFGF as a function of time when gelatin hydrogel with incorporated bFGF was subcutaneously implanted in mice, along with the results for other growth factors[11].

4. Source of Cells

As demonstrated above, even the sustained release of growth factors alone at the site of regeneration can initiate some tissue regeneration without any scaffold. It is also possible that scaffold alone without both cell seeding and growth factor incorporation can facilitate tissue engineering. An example is dermal tissue regeneration occurring when only a collagen sponge sheet without fibroblast and growth factors is put on the skin site, where part of full-thickness skin tissue was lost[12]. In this case many fibroblasts are recruited into the collagen sheet from the surrounding healthy tissue.

However, cells are generally seeded to scaffold and, in some cases, growth factors are further impregnated into the scaffold in a form of DDS. If tissue regeneration is carried out in vitro, that is, in clean room, scaffold and cell are indispensable, often together with growth factors. Most of tissue regeneration studies have been carried out in vivo until now, because it is at present a very hard task to provide all the materials and tools necessary for tissue regeneration in vitro, whereas the so-called natural healing power assists tissue regeneration when it takes place in our body.

As the cell source for tissue engineering, heterogeneous cells are used instead of autologous cells when the latters are not available, but heterogeneous cells are always associated with immune rejection. It seems probable that implanted neural cells are less attacked by the self-defence system, similar to fetal cells, but the level of their immune attack should be examined in more detail before their clinical application.

The best cell source is autologous cells from patients, but some problems are involved in harvesting these cells. One of them is difficulty in obtaining a sufficient amount of cells from patients. Generally, the cells in the final differentiation stage or very close to it do not multiply to a large extent, although we have to seed a sufficiently large number of cells to the scaffold. Normally, the cell number is increased by culture prior to implantation. If the cell density is not high enough, it will take long time until tissue regeneration comes to completion or it will fail.

Therefore, extensive studies have been performed on mesenchymal stem cell as this cell has high potential with respect to cell multiplication and differentiation into a variety of blastic cells. Somatic stem cells, which can differentiate into matured cells to produce various tissues, are also known in recent years[13], but these stem cells are much more difficult to isolate than the mesenchymal stem cell, probably because of lower cell density. If the mesenchymal stem cell is readily isolated from the bone marrow or fatty tissue of patients, this will tremendously contribute to clinical medicine. As is well known, embryonic stem(ES) cell can produce all the tissues[14] and hence is called totipotent cell, but it should be kept in mind that ES cell is allogeneic unless it is obtained from a cloned embryo.

References

1. Petit-Zeman, S. (2001), Regenerative medicine, *Nature Biotechnology*, **19**, 201-206.
2. Lanza, R.P., Langer, R., and Vacanti, J. (eds.),(2000), Principles of Tissue Engineering, Second Edition, Academic Press, San Diego.
3. Caplan, A.I. (2000), Tissue engineering, designs for the future: New logics, old molecules, *Tissue Eng.*, **6**, 1-8.
4. Chang, S.C.N., Rowley, J.A., Tobias, G., Genes, N.G., Roy, A.K., Mooney, D.J., Vacanti, C.A., and Bonassar, L.J. (2001), Injection molding of chondrocyte/alginate constructs in the shape of facial implants, *J. Biomed. Mater. Res.*, **55**, 503-511.
5. Tsukagoshi, T., Yenidunya, M.O., Sasaki, E., Suse, T., and Hosoda, Y. (1999), Experimental vasucular graft using small-caliver fascia-wrapped fibrocollagenous tube: short-term evaluation, *J. Reconstr. Microsurg.*, **15**, 127-131.
6. Shinoka, T., Imai, Y., Hibino, N., Watanabe, M., Matsumura, T., Kosaka, Y., Konuma, T., Toyama, S., Yamakawa, M., Ikada, Y., and Morita, S. (2000), Development of blood vessels by tissue engineering using autologuous cells, *Artificial Organs* (in Japanese), **29**, S-25.
7. Niklason, L.E., Abbott, W., Gao, J., Klagges, B., Hirschi, K.K., Ulubayram, K., Conroy, N., Jones, R., Vasanawala, A., Sanzgiri, S., and Langer, R. (2001), Morphologic and mechanical characteristics of engineered bovine arteries, *J. Vasc. Surg.*, **33**, 628-638.
8. Kim, B-S. and Mooney, D.J. (2000), Scaffolds for engineering smooth muscle under cyclic mechanical strain conditions, *Trans. ASME*, **122**, 210-215.
9. Isner, J.M. and Asahara, T. (2000), Angiogenesis and vasculogenesis as therapeutic strategies for postnatal neovascularization, *J. Clin. Invest.*, **103**, 1231-1234.
10. Ikada, Y. (1998), Preface, *Adv. Drug. Delivery Reviews*, **31**, 183-184.
11. not yet published.

370

12. Suzuki, S., Matsuda, K., Maruguchi, T., Nishimura, Y., and Ikada, Y.(1995). Further applications of "bilayer artificial skin", *British J. Plastic Surgery*, **48**, 222-229.

13. Orlic, D. Kajstura, J. Chimenti, S., Jakoniuk, I., Anderson, S.M., Li, B., Pickel, J., Mckay, R., Nadal-Ginard, B., Bodine, D.M., Leri, A., and Anversa, P. (2001), Bone marow cells regenerate infarcted myocardium, *Nature*, **410**, 701-705.

14. Colman, A. (2001), Stem cells -the long march forward to the clinic-, Drug Discovery World Summer, 66-71.

SCAFFOLDING ARTERIAL TISSUE

D. COHN AND G. MAROM

*Casali Institute of Applied Chemistry, The Hebrew University of Jerusalem,
Jerusalem 91904, Israel*

1. Introduction

The treatment of diseased or injured arteries has been a key area of contemporary surgery and a focus of intensive research for more than one hundred years. The interposition of a jugular vein segment into the carotid artery, conducted by Carrel as early as 1908 [1], initiated the use of veins for the replacement of artery sections. Among other important contributions to modern cardiovascular surgery, it is worth mentioning the pioneering work performed by Tuffier [2] using paraffin-lined silver tubes to replace damaged arteries, and the study conducted by Blakemore and coworkers [3] on the utilization of non-sutured anastomoses using Vitalium tubes. It is worth stressing, though, that it was the work conducted during the early 1950s, that changed dramatically the world of vascular surgery, by introducing textile vascular grafts. Of special importance was the landmark work published by Blakemore and Voorhees in 1951, who demonstrated the clinical applicability of Vinyon N, as the first fabric arterial prosthesis [4].

The continuously increasing clinical demand for improved arterial prostheses has focused a great deal of interest on studies of both the key parameters which determine the biological performance of blood-contacting implants [5-8] and on the development of new superior prostheses [9-12]. The different strategies pursued to engineer clinically successful vascular grafts varied over the years, but can be generally categorized into three main classes: (a) those who developed fully artificial substitutes, entirely based on synthetic polymers, (b)

371

R.L. Reis and D. Cohn (eds.),
Polymer Based Systems on Tissue Engineering, Replacement and Regeneration, 371–390.
© 2002 *Kluwer Academic Publishers. Printed in the Netherlands.*

those who focused on developing totally biological constructs and, (c) hybrid systems comprising both and capitalizing on Tissue Engineering working concepts.

2.   Fully synthetic vascular grafts.

The performance of an artificial blood vessel is determined by diverse factors, the most crucial of which pertains to the thrombogenicity of its luminal surface. Protein adsorption and cell adhesion processes play a fundamental role in determining the hematological response elicited by the blood-contacting surface [13-15]. It is due to its effect on these phenomena, that the biomaterial surface largely dictates the thrombogenicity and long-term patency of prostheses implanted in the cardiovascular system. Despite numerous attempts to address the problem, the need for a suitably hemocompatible prosthesis remains unfulfilled.

The transmural ingrowth of vascularized perigraft tissue, a phenomenon which, evidently, necessitates a porous prosthesis, is considered to be indispensable for the development of a healthy neointimal lining and an overall successful healing process [16-18]. Unfortunately, highly porous grafts are not universally implantable, their utilization being seriously restricted, mainly due to the risk of severe intraoperative hemorrhage through the interstices of the prosthesis. It is clear, therefore, that new improved prostheses should combine minimal blood loss at implantation, with high healing porosity.

The dynamic mechanical regime under which arterial tissue is called upon to function, comprises three fundamental components: stresses normal to the surface, tangential shear stresses and cyclic elongation. Under this dynamic loading regime, natural arteries display a substantially more compliant dimensional response than the stiff Dacron and Gore-Tex prostheses in clinical use. Compliance of blood vessels can be defined, in a simplified way, as the fractional change in diameter *per* unit change in pressure. Because of the difference between the mechanical behavior of the native artery and the graft, different diametric fluctuations are created in the two anastomosed vessels, during the systolic/diastolic sequence [19, 20]. It has been claimed that whereas axial compliance mismatch affects flow patterns at the distal anastomosis of the graft [21], a mismatch in the hoop compliance causes flow-induced shear stresses on platelets and endothelial cells, and results in higher anastomotic stresses [22, 23]. The significant distensibility difference existing between the graft and the host artery, causes intimal hyperplasia, thrombotic phenomena and substantially reduced patencies [24-29]. In light of this compliance mismatch, attempts to develop a new generation of arterial prostheses displaying enhanced pulsatility, have been conducted [30-37].

It is apparent, therefore, that regardless of their widespread clinical use, contemporary arterial prostheses still pose serious problems. Due to the complexities of their biological performance, designing arterial prostheses remains a challenge which contemporary synthetic vascular grafts have met only partially.

3. Fully biological vascular grafts.

The working hypothesis on which this approach is based, states that tissue-engineered arterial prostheses consisting entirely of biological materials, will display superior properties [38]. This, mainly, due to the ability of these tissue constructs to heal and remodel at the site of implantation, in response to environmental stimuli. The work conducted by Weinberg and Bell [39] in the mid eighties, producing the first fully biological arterial prosthesis, represented a major milestone in the field. Their graft comprised only collagen gels, smooth muscle cells, endothelial cells and fibroblasts. Regrettably, the tissue structures generated exhibited utterly unacceptable mechanical properties. Even when wrapped with a reinforcing PET mesh, the negligible burst strength of the constructs prevented their implantation in animals. Other laboratories followed the same conceptual framework, but the mechanical performance of the prostheses produced were improved only marginally [40, 41].

Typically, these fully biological vascular grafts were built aiming at mimicking the structure of natural vessels. Arteries are complex, multi-layered tubular tissues, comprising collagen and elastin fibers, smooth muscle, ground substance, *vaso vasora* and endothelium, as shown schematically in Fig. 1. Due to the directionality and spatial conformation of their fibrous components, they display an anisotropic, non-Hookean dimensional response under physiological loading. While seeding endothelial cells on the luminal surface of the prosthesis was expected to render it nonthrombogenic [42, 43], collagen, smooth muscle cells and fibroblasts were viewed as the basic components required to generate a multi-layered functional arterial wall.

Figure 1 A schematic view of an artery.

Based on this concept, step-by-step processes were used to generate the different layers of the arterial wall [40, 44]. The fabrication procedure started by pouring a smooth muscle cells-suspended collagenous solution into a mold, followed by a maturation period and by an endothelial cells seeding stage. Due to its poor mechanical properties, the tissular construct produced had to be reinforced with an elastomeric knit to avoid rupture, even under pressures well below arterial levels.

L'Heureux and coworkers [38] described what they claimed was the first entirely biological vascular graft that succeeded to attain satisfactory mechanical properties. The biological tubular structure they produced, comprised three distinct layers, remarkably akin to the intima, media and adventitia present in native arterial tissue. The authors assigned the enhanced strength of this biological graft, to the exceptionally high level of organization of its collagenous adventitial layer.

4.   Tissue engineered vascular grafts.

Tissue engineering was defined by Langer and Vacanti [45] as "an interdisciplinary discipline that applies the principles of engineering and the life sciences toward the development of biological substitutes that restore, maintain or improve tissue function". Even though several definitions of the field of Tissue Engineering have been proposed in recent years, proper recognition should be given to Guthrie who, as early as 1919, had the foresight to make the following remarkable statement: "To restore and maintain function, an implanted segment needs only temporarily to restore mechanical continuity and serve as scaffolding or bridge for the laying down of an ingrowth of tissue derived from the host" [46].

The emerging field of Tissue Engineering and its nascent application in the vascular field, represents a major breakthrough in arterial prostheses design both conceptually as well as technologically. The objective of vascular Tissue Engineering is to induce regeneration of functional arterial tissue, by providing the appropriate three-dimensional scaffolding construct on which cells will be able to grow, differentiate and generate new tissue. Clearly, the composition and mechanical properties of the materials, strongly affect the ability of the system to actively promote the regeneration of autologous functional tissue. In addition, the macrostructural characteristics of the scaffold, play also a fundamental role in determining the type of cells and tissue components present in the new tissue.

The template's ultimate task is to provide a gradually disappearing, temporary construct for the generation of viable new tissue. Therefore, if autologous tissue is to regenerate and replace the scaffold, biodegradability is one of its indispensable attributes. It is also necessary for the template to perform as an adhesive substrate for cells, promoting their growth and differentiation, while retaining cell function. Also, for a scaffold to perform successfully, it is required to be biocompatible, to display the right porosity and to be mechanically suitable.

In general terms, Tissue Engineering can be classified into *in vitro* and *in vivo* types. While the former concentrates on the *ex vivo* generation of tissues from cells removed from a donor site, the latter aims at regenerating functional tissue at the site of implantation, by the combined action of biomolecules and cells, *in situ*. Since most of the work conducted to date has focused on structures formed under *in vitro* conditions, the authors will concentrate initially on tissue engineered vascular grafts generated *ex vivo*.

The study published in 1999 by Niklason *et al* [47] illustrates very eloquently the tremendous progress made in this field, during recent years. The authors described the *in vitro* generation of functional arteries under pulsating loading conditions, using a polyglycolic acid biodegradable scaffold. In accordance with theoretical considerations [48] as well as ample experimental findings [49-54], the grafts fabricated under a mechanical pulsatile regime, displayed much improved histological features and mechanical properties, when compared to their non-pulsated counterparts. Also, the enriched composition of the culturing media used, proved to strongly affect the behavior of the tissue formed. While controls cultured in standard medium were very weak, rupturing below 300 mm Hg, the pulsated vessels grown in an enriched medium, attained very high strength levels, with failure values typically above 2100 mm Hg.

A variety of techniques were used to manufacture three-dimensional polymeric scaffolds [55], including solvent casting/solvent leaching [56], gas foaming [57-59], phase separation [60, 61], three-dimensional printing [62] and the use of various types of preformed constructs [63, 64]. The major drawbacks of these techniques stem from their limited control of the pore size and from the partial degree of interconnectivity they achieve. Furthermore, these ill-defined porous structures are essentially isotropic, while the tissue to be generated exhibits significant levels of anisotropy.

The work done by Hutmacher [65-67] and his group represents a significant step forward in scaffold design, since it allows the controlled fabrication of fully interconnected porous templates. The authors capitalized on the unique features of the rapid prototyping Fused Deposition Modeling (FDM) technology, to create this novel type of scaffolds. The computer-programmed FDM technique entails the extrusion of the molten polymer and its deposition following specific predetermined lay-down patterns, aiming at building precise three-dimensional architectures. Even though most of this group's work has concentrated on osseous tissues, the potential of this method in the arterial field is apparent.

In a recent publication, Matsuda and his group [68] described small diameter arterial prostheses engineered so that they displayed a J-shaped stress-strain curve, under a physiological loading regime. This was achieved by fabricating coaxial tubular structures made of polyurethane elastomers, where the mechanical properties of the different concentric tubes (two or three), were tailored so that their compliance decreased from the center outwards. While the distensibility of the tubes was controlled by varying the mechanical properties of the polymer as well as the porosity and thickness of the tube wall, the transition from one region to the other was determined by the inter-tubular distance.

5.   Composite Materials in tissue engineered vascular grafts.

Since the composition and architecture of natural tissues are greatly affected by the stress field induced by the implant, a clear dependency exists between the mechanical behavior of the template and that of the regenerated tissue. In contrast to requirements, available vascular grafts are rigid structures, lacking anisotropy and non-linear compliance. It is, therefore, apparent that a new generation of arterial prostheses, characterized by an isocompliant and anisotropic mechanical behavior, is called for.

Fiber-reinforced polymeric composites have attracted much attention as the basis for improved biomedical devices, especially in the orthopedic field [69-72]. Advantages derived from the heterogeneity and, most importantly, from the anisotropy of synthetic composite biomaterials, can be readily anticipated.

Anisotropy is a prerequisite for biomedical structures that require specific stress and stiffness directionalities, such as the different circumferential and axial moduli and strength of native blood vessels.

From a Materials Science viewpoint, the remaining of this article will underscore the advantages of using composite design tools for the engineering of filament wound scaffolds for arterial tissue regeneration. From a biological perspective, and in contrast to the *ex vivo* work described so far, the study described below capitalized on the advantages of *in vivo* Tissue Engineering. In this case, we view the implanted prosthesis as performing initially as a transient vascular graft and, later on, as a scaffold for arterial tissue regeneration. The programmed degradation of the prosthesis allowed the incorporation of the implant and promoted the regeneration of arterial tissue, *in situ*. That, due to both the action of autologous biomolecules, cells and tissue components, and the dynamic physiological loading regime operative at the site of implantation.

Filament wound vascular grafts displaying various levels of biodegradability can be manufactured, from selectively to totally biodegradable prostheses. While the former comprise non-degradable fibers and a biodegradable sealing component, all the constituents of the latter degrade over time, following different degradation kinetics.

The two-phase design philosophy on which these selectively biodegradable filament wound arterial prostheses are founded, differentiates conceptually between the fiber and matrix roles both mechanically as well as biologically. Typically, the reinforcing fibers are responsible for the mechanical behavior of the vascular grafts, while the matrix governs their hemocompatibility and built-in, selective biodegradability. The elastic properties of the graft determine its compliance and pulsatility, while its tendency to develop aneurysms, its burst strength and its suture holding strength, will depend on the ultimate strength of the prosthesis. These important clinical characteristics are, therefore, determined by the properties of the constituent materials of the graft and by their relative volume fractions. In addition, the orientation of the fibers, dictated by the winding angle, adds yet another dimension to the design options, allowing for larger latitude in compliance and strength tailoring.

The production of these filament wound prostheses comprised the following steps: [a] in the first stage, the continuous fiber was dipped in a solution of the binder and, guided by a traverse head that moved back and forth, was wound on a rotating mandrel, to form a filament wound scaffold. [b] the filamentous construct was coated with a thin layer of a highly flexible biodegradable polymer developed in our laboratory [73-77], to generate a initially impervious prosthesis. Since the winding angle is determined by the ratio of the circumferential velocity of the rotating mandrel and the longitudinal velocity of the carriage, by varying them independently, a broad selection of winding angles was generated (30-75°).

Fig. 2 presents an orthotropic lamina, showing its longitudinal and transverse (principal) axes oriented at an angle θ with respect to the coordinate x,y axes.

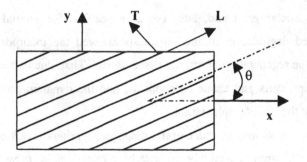

Figure 2. A schematic presentation of an orthotropic lamina.

In general, the compliance along the x axis, $1/E_x$, of such a lamina, is given by [78]:

$$\frac{1}{E_x} = \frac{\cos^4\theta}{E_L} + \frac{\sin^4\theta}{E_T} + \frac{1}{4}\left(\frac{1}{G_{LT}} - \frac{2\gamma_{LT}}{E_L}\right) \cdot \sin^2 2\theta \qquad [1]$$

where E_x is the Young's modulus in the x direction, E_L, E_T, G_{LT} and v_{LT} are the Young's moduli, the shear modulus and Poisson's ratio, respectively, of the composite lamina and L and T denote the principal material axes (longitudinal and transverse). While the material properties and relative content of the fibers and matrix dictate the basic lamina behavior, the orientation angle θ adds yet another dimension to the design options, as expressed by *Equation 1*. The much higher level of design versatility of composite systems is an extremely important feature, when engineering biomedical systems that attempt to mimic the behavior of natural tissues.

As readily derived from *Equation 1*, it is obvious that when $\theta=0°$, E_x is equal to E_L, and at $\theta=90°$, E_x is equal to E_T [79, 80]. While the values of E_x at these two extreme positions are readily obtained, E_x changes in the interval between $0°$ and $90°$, in a manner dictated by the values of the elastic properties of the lamina. The $E_x = f(\theta)$ function can then be determined by calculating the first and second derivatives of *Equation 1*. Furthermore, it can be demonstrated that, depending on the values of the four principal elastic constants of the material, E_x can either be larger then E_L or smaller then E_T at some in-between value of θ, as dictated by the conditions defined by *Equations 2* and *3*, respectively.

$$G_{LT} > \frac{E_L}{2(1 + \gamma_{LT})} \qquad [2]$$

$$G_{LT} < \frac{E_L}{2(E_L/E_T + \gamma_{LT})} \qquad [3]$$

These equations define the conditions for the occurrence of a maximum and minimum, respectively, for the $E_x = f(\theta)$ function. *Equations 2* and *3* also define the conditions under which the value of E_x will be bound by those of E_L and E_T (i.e. without maximum or minimum) shown in *Equation 4*.

$$\frac{E_L}{2(1 + \gamma_{LT})} > G_{LT} > \frac{E_L}{2(E_L/E_T + \gamma_{LT})} \qquad [4]$$

When the first derivative of *Equation 1* is equated to zero, the position of the maximum or minimum of the function, is easily obtained from *Equation 5*.

$$\tan^2\theta = \frac{(2/E_L)+(2\gamma_{LT}/E_L)-(1/G_{LT})}{(2/E_T)+(2\gamma_{LT}/E_L)-(1/G_{LT})}$$ [5]

The fact that the $E_x(\theta)$ function can display a minimum or a maximum for a specific reinforcement angle, has significant implications when designing a filament wound arterial prosthesis, namely, that a specific angle exists, where the compliance attains a maximum (or a minimum), as dictated by both the elastic constants of the system and their relative values. This can be illustrated by the following highly compliant, $\pm\theta$ angle-ply composite structure, the main elastic constants of which are $E_L = 11.0$ MPa, $E_T = 8.5$ MPa, $G_{LT} = 2.0$ MPa and $v_{LT} = 0.4$. In accordance with the condition derived from *Equation 3*, this system is expected to generate a minimum for E_x. Then, from *Equation 5*, it can be calculated that the $E_x(\theta)$ function reaches a minimum at $\theta_{min} = 48.5°$. It is worth stressing that, given the elastic constants combination of the system, the tensile loading along the x axis causes sizeable longitudinal strains and concomitant transverse Poisson shrinkages, which are accommodated by a decrease in θ.

The occurrence of a minimum (or a maximum) in the $E_x = f(\theta)$ function and the fact that θ drops with stress, define the elastic behavior of the system, depending on the initial value of the reinforcement angle, θ_{init}, relative to the minimum (or maximum) in the function. Because of its relevance to the biological performance of filament wound arterial prostheses, the present study focused on the instance where the function exhibited a minimum, i.e maximum compliance. When the $E_x = f(\theta)$ function displays a minimum, three distinct scenarios exist, depending on the relative positions of the initial winding angle (θ_{init}) and the minimum in the function. Fig. 3 presents a graphic representation of E_x as a function of θ, based on the hypothetical values set above for the elastic constants of the material ($E_L = 11.0$ MPa, $E_T = 8.5$ MPa, $G_{LT} = 2.0$ MPa and $v_{LT} = 0.4$). It is apparent from both Fig. 3 *and Equation 1*, that when the initial reinforcement angle (θ_{init}) is smaller than θ_{min}, a gradual modulus increase is obtained, as the reinforcement angle decreases due to the applied stress. Directly derived from this behavior, a convex, J-shaped, stress-strain curve was produced, displaying a decrease in compliance as stress (pressure) increased. A similar line of reasoning predicts that when θ_{init} is larger than θ_{min}, a decrease in the reinforcement angle will be measured, with a concave stress-strain curve being generated. Lastly, when θ_{init} and θ_{min} are close, E_x is only marginally affected by changes in θ, hence a linear stress-strain curve with a constant modulus value is anticipated for the soft angle-ply lamina.

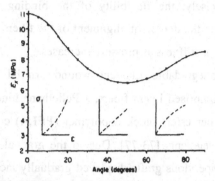

Figure 3. A theoretical plot of the modulus in the x direction, as a function of the angle θ.

This behavior can be illustrated for a elastomeric scaffold comprising Lycra fibers and a Pellethane binding matrix, where a significant angle decrease was witnessed, as a function of strain (see Fig. 4). It is apparent from the data presented that, within that strain range investigated (up to 40%), the decrease in the reinforcement angle was nearly linear, the angle decreasing by 38% from its initial value (33.5°). Since the stress field operating on the artery produces an essentially pure hoop loading mode, the required J-shaped hoop stress-strain response can be generated by an angle-ply design where the initial reinforcement angle is larger than θ_{min}.

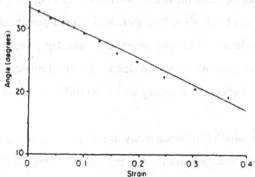

Figure 4. The decrease in θ with strain, for a Lycra-Pellethane material.

The basic performance of these elastomeric filament wound vascular prostheses allows to engineer compliant tubular structures that exhibit an increase in compliance as pressure increases. It is also worth stressing that this behavior is mainly due to the cyclic scissoring of the fibers and the concomitant decrease in the reinforcement angle, upon the application of stress. Hence, it is apparent that the mechanical properties of the matrix binding the fibers, play a fundamental role in determining the dynamic behavior of the prosthesis. More precisely, the flexibility of the binding polymer and its resistance to deformation will dictate the degree of alignment of the fibers under stress and, consequently, the degree of increase in stiffness as pressure increases.

Selectively biodegradable filament wound arterial prostheses engineered in our laboratory typically comprised Lycra fibers, a Pellethane binding material and a elastomeric biodegradable poly(ether ester) block copolymer (PELA) comprising poly(lactic acid) and poly(ethylene oxide) segments [73-77]. Due to the gradual degradation and dissolution of PELA, the initially impervious grafts displayed gradually increasing mural porosity. *In vitro* degradation studies showed that PELA degraded substantially over a 2-4 week period, the porosity increasing accordingly. Concomitant with the degradation and dissolution of PELA, the modulus of the graft decreased steadily by approximately one decade over a short period of time. Measurements of the axial and hoop compliances, under both static and pulsatile internal pressure, demonstrated that the filament wound structures exhibited an anisotropic dimensional response, resembling that of natural arteries.

Aiming at closely matching the mechanical behavior of the prosthesis and the host artery, the grafts were fabricated using a $50°$ winding angle and a wall thickness of 440 μm. The initial compliance of these prostheses was 2.5% mm Hg^{-1} 10^{-2}, which is equivalent to an initial tensile modulus of 7.1 MPa. The as-manufactured prosthesis exhibited extremely low water permeability values (1-2 ml water/min. cm^{-2} mm Hg), while a progressively increasing leakage developed, as porosity increased due to the programmed degradation of PELA. The wound graft exhibited excellent handling and suturability characteristics as well as enhanced burst strength.

In a three month short-term study, 6 mm internal diameter vascular grafts were implanted in the canine carotid artery. At implantation time, the filament wound prostheses were impermeable to blood, no oozing being detectable. Due to their biodegradable sealing component, the implants combined minimal intraoperative blood loss and high healing porosity. The smooth, thin layer of pseudointima, strongly adhered to the luminal surface of the graft, permitted clear visualization of the underlying filamentous structure (see Fig. 5).

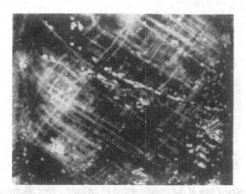

Figure 5. Gross observation of a filament wound graft after 90 days implantation.

Fig. 6 shows the luminal surface of the filament wound graft. It is apparent from the micrograph that the polymeric surface was covered with endothelial cells, displaying cobblestone pattern of confluent cells with somewhat bulging nuclei, oriented mainly parallel to the direction of blood flow.

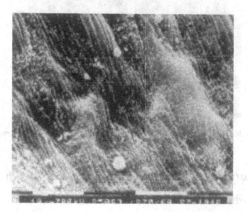

Figure 6. SEM micrograph of the luminal surface of an explanted filament wound graft.

This behavior was attributed to the presence of PEG chains on the surface, known to enhance blood compatibility by minimizing cell and blood proteins adsorption to the surface [81-84].

The pseudointima blended intimately with the periprosthetic tissue which invaded the increasingly porous wall of the prosthesis.

384

Fig. 7 presents SEM photographs showing that the subendothelial layer protruding among the elastomer fiber bundles and, by integrating with the adventitial tissue, forming a single organic tissular structure.

Figure 7. SEM micrograph of the cross-section of an exaplanted filament wound graft.

The presence of numerous vasa vasora, as seen in Fig. 8, demonstrated the vascularized nature of the regenerating tissue and the red blood cells within these new blood vessels.

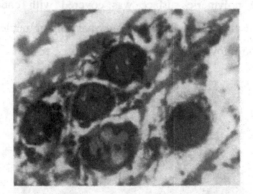

Figure 8. A Mason Trichrome stained section of an explanted filament wound graft.

The capillaries were surrounded by collagenous tissue and smooth muscle cells, stained blue and red, respectively, when stained with Masson stain.

Since the pulsatility of the arterial wall is directly related to its elastin content, sections of the different prostheses were stained with Van Giessen stain (see Fig. 9).

Figure 9. A Van Giesen stained section of an explanted filament wound graft.

The stained samples revealed the presence of wavy elastin fibers intermingled with the residual scaffold of the progressively degrading prosthesis.

6. Conclusions

The remarkable progress made in Biomaterials Science, witnessed during the last two decades, has made possible the development of a diversity of biomedical systems, which have contributed significantly to modern medicine. Nevertheless, only a new generation of devices, based on novel tailor made materials and new engineering and biological concepts, will permit further progress.

Developing a new vascular graft is a multifaceted challenge where hemocompatibility, porosity and compliance requirements must be organically integrated into the prosthesis design. By capitalizing on the advantages of Composite Materials and by choosing the appropriate material combination, a variety of new vascular grafts were produced, exhibiting gradually increasing porosity and mechanical properties akin to those of natural arteries. Due to its enhanced hydrophilicity, the PELA-rich mural composition attained high water uptake levels, resulting in a prosthesis which combined tissue-like soft consistency and the high compliance with the enhanced strength provided by the elastomeric scaffold. The selectively biodegradable graft resulted in excellent healing and incorporation processes, this being fully corroborated by histological studies. It was also surmised that PELA played an important role, not only as the gradually degrading sealant of the graft, but also by minimizing its surface thrombogenicity.

The selective biodegardable arterial prosthesis described above is seen as the first step in the development of tissue engineered arterial implants able to induce regeneration of functional tissue, by providing the appropriate scaffolds on which cells will be able to grow, differentiate and generate new tissue.

7. References

1. Carrell A. (1908) Results of the transplantation of blood vessels, organs and limbs, *JAMA* **51**, 1662-1667.

2. Tuffier, T. (1915) L'Intubation dans les plaies de grose arteres. *Bul. Acad. Nat. Med. (Paris)* **74**, 455.

3. Blakemore, A. H., Lord, J. V. and Stefko, P. L. (1942) The severed primary artery in the war wounded: A non-suture method of bridging arterial defects. *Surgery* **12**, 488-508.

4. Blakemore, A. H. and Voorhees, A. B. (1954) The use of tubes constructed from Vinyon N cloth in bridging arterial defect. Experimental and clinical. *Ann. Surg.* **140**, 324-338.

5. Zacharias, R. K., Kirkman, T. R. and Clowes, A. W. (1987) Mechanisms of healing in synthetic grafts, *J. Vasc Surg.* **6**, 429-436.

6. Callow, D. (1982) Current status of vascular grafts, *Surg. Clin. N. Am.*, **62**, 501-506.

7. Klement, P., Du, Y. J., Berry, L. Andrew, M. and Chan, A. K. C. (2002) Blood-compatible biomaterials by surface coating with a novel antithrombin-heparin covalent complex, *Biomaterials* **23(2)**, 527-535.

8. Kouvroukoglou, S., Dee, K. C., Bizios, R. McIntire, L. V. and Zygourakis, K. (2000) Endothelial cell migration on surfaces modified with immobilized adhesive peptides, *Biomaterials* **21(17)**, 1725-1733.

9. Lu, A. and Sipehia, R. (2001) Antithrombotic and fibrinolytic system of human endothelial cells seeded on PTFE: the effects of surface modification of PTFE by ammonia plasma treatment and ECM protein coatings, *Biomaterials* **22(11)**, 1439-1446.

10. Wyers, M. C., Phaneuf, M. D., Rzucidlo, E. M., Contreras, M. A., LoGerfo, F. W. and Quist, W. C. (1999) *In Vivo* assessment of a novel Dacron surface with covalently bound recombinant hirudin, *Cardiovascular Pathology* **8(3)**, 153-159.

11. Yoneyama, T., Sugihara, K., Ishihara, K., Iwasaki, Y. and Nakabayashi, N. (2002) The vascular prosthesis without pseudointima prepared by antithrombogenic phospholipid polymer, *Biomaterials* **23**, 1455-1459.

12. Baquey, Ch., Palumbo, F., Porte-Durrieu, M., Legeay, C. G., Tressaud, A. and d'Agostino, R. Plasma treatment of expanded PTFE offers a way to a biofunctionalization of its surface, *Nuclear Instruments and Methods in Physics Research Section B: Beam Interactions with Materials and Atoms* **151**, 255-262.

13. Chandy, T., Das, G. S., Wilson, R. F. and Rao, G. H. R. (2000) Use of plasma glow for surface-engineering biomolecules to enhance blood compatibility of Dacron and PTFE vascular prostheses, *Biomaterials* **21**, 699-712.

14. Korematsu, A., Takemoto, Y., Nakaya, T. and Inoue, H. (2002) Synthesis, characterization and platelet adhesion of segmented polyurethanes grafted phospholipid analogous vinyl monomer on surface, *Biomaterials* **23**, 263-271.

15. Lee, J. H., Ju, Y. M. and Kim, D. M. (2000) Platelet adhesion onto segmented polyurethane film surfaces modified by addition and crosslinking of PEO-containing block copolymers, *Biomaterials* 21, 683-691.

16. Hermansen, C., Kraglund, K., Ludwigsen, E., and Mouritzen, C. (1980) Influence of porosity on the viability of the neointima," *Eur. Surg. Res.,* 12, 349-362.

17. Guidoin, R., Marceau, D., Rao, T. J., King, M., Merhi, Y., Roy, P. E., Martin, L. and Duval, M. (1987) *In vitro* and *in vivo* characterization of an impervious polyester arterial prosthesis: the Gelseal Triaxial graft," *Biomaterials* 8, 433-441.

18. Uretzky, G., Appelbaum, Y., Younes, H., Udasin, R., Nataf, P., Baccioglu, E., Pizof, G., Borman, J. B. and Cohn, D. (1990) Long-term evaluation of a new selectively biodegradable vascular graft for right ventricular conduit. An experimental study, *J. Thorac. Cardiovasc. Surg.,* 100, 769-780.

19. Abbott, W.M. and Bouchier-Hayes, D.J. (1978) The role of mechanical properties in graft design, in H. Dardik (ed.) *Graft Materials in Vascular Surgery*, Chicago, USA

20. Seifert, K.B., Albo, D. and Knowlton, H. Effect of elasticity of prosthetic wall on patency of small diameter arterial prostheses (1979) *Surg. Forum* 30, 206-208.

21. Tu R, McIntyre J, Hata C, Lu CL, Wang E, Quijano RC. (1991) Dynamic internal compliance of a vascular prosthesis. *Trans Am Soc Intern Organs* 37, M470-M472.

22. Abbott W. and Cambria R. (1982) Control of physical characteristics of vascular grafts, in J.C. Stanley (ed.) *Biologic and Synthetic Vascular Prostheses,* Grune & Stratton, New York, pp. 189-220.

23. Pevec W.C., Darling D, L'Italien D.G. and Abbott W.M. (1992) Femoropopliteal reconstruction with knitted, non-velour Dacron *versus* expanded poly-tetrafluorethylene, *J Vasc Surg* 16, 60-65.

24. Stewart, S.F.C. and Lyman D.J. (1992) Effects of a vascular graft/natural artery compliance mismatch on pulsatile flow. *J Biomech* 25, 297-310.

25. Weston M.W, Rhee K. and Tarbell J.M. (1996) Compliance and diameter mismatch affect the wall shear rate distribution near an end-to-end anastomosis, *J Biomech* 29, 187-198.

26. Clark, R. E., Apostolou, S. and Kardos, J. L. (1976) Mismatch of mechanical properties as a cause of arterial prostheses thrombosis, *Surg. Forum,* 27, 208-210.

27. Baird, R.N., Kidson, J. L., Italien I.G., and Abbot, W.M. (1977) Dynamic compliance of arterial grafts, *Am. J. Physiol.,* 233, H568-H572.

28. Bos, G.W., Poot A.A., Beugeling, T., Van Aken, W.G. and Feijen J. (1998) Small-diameter vascular graft prostheses: Current status, *Arch Physiol Biochem* 106, 100-115.

29. Abott W.M., (1997) Prosthetic above-knee femoral-popliteal bypass: Indications and choice of graft, *Semin Vasc Surg* 10, 3-7.

30. Annis, D., Bornat A., Edwards, R.O., Higham, A., Loveday, B. and Wilson, J., (1978) An elastomeric vascular prosthesis *Trans. Am. Soc. Artif. Intern. Organs* 24, 209-214.

31. Van der Lei, B., Nieuwenhuis, B., Molennar, I. and Wildevuur, Ch. R. H. (1987) Long-term biological fate of neoarteries regenerated in microporous, compliant, biodegradable small-caliber vascular grafts in rats, *Surgery* 101, 459-467.

388

32. Cohn, D., Elchai, Z., Gershon, B., Karck, M., Lazarovici, G., Sela, J., Chandra, M., Marom, G. and Uretzky, G. (1992) Introducing a selectively biodegradable filament wound arterial prosthesis: a short-term implantation study, *J Biomed Mater Res* **26**(9) 1184-1204.

33. Doi, K., Nakayama, Y. and Matsuda, T. (1996) Novel compliant and tissue-permeable microporous polyurethane vascular prostheses fabricated using an excimer laser ablation technique, *J Biomed Mater Res* **31**, 27-33.

34. Edwards, A., Carson, R.J., Bowald, S. and Quist, W.C. (1995) Development of a microporous compliant small bore vascular graft, *J Biomater Appl* **10**, 171-187.

35. Tai, N.R., Salacinski, H., Edwards, A., Hamilton, G. and Seifalian, A.M. (2000) Compliance properties of conduits used in vascular reconstruction. *Br J Surg* **87**, 1480-1488.

36. Hayashi, K., Takamizawa, K., Saito, T., Kira, K., Hiramatsu, K. and Kondo, K. (1989) Elastic properties and strength of a novel small-diameter, compliant polyurethane vascular graft, *J Biomed Mater Res* **23**(Suppl), 229-244.

37. Edwards, A., Carson, R.J., Szycher, M. and Bowald, S. *In vitro* and *in vivo* biodurability of a compliant microporous vascular graft. *J Biomater Appl* **13**, 23-45.

38. L'Heureux, N., Pâquet, S., Labbé, R., Germain, L. and Auger, F. A. (1998) A completely biological tissue-engineered human blood vesse,l *FASEB J.* **12**, 47-56.

39. Weinberg, C. B. and Bell, E. (1986) A blood vessel model constructed from collagen and cultured vascular cells, *Science* **231**, 397–400

40. Hirai, J. and Matsuda, T. (1996) Venous reconstruction using hybrid vascular tissue composed of vascular cells and collagen-tissue regeneration process, *Cell Transplant.* **5**, 93–105

41. L'Heureux, N., Germain, L., Labbe, R. and Auger, F. A. (1993) *In vitro* construction of a human blood vessel from cultured vascular cells: a morphologic study, *J. Vasc. Surg.* **17**, 499–509

42. Zilla, P., Deutsch, M., Meinhert, J., *et al.* (1994) Clinical *in vitro* endothelialization of femoropopliteal bypass grafts: an actuarial follow-up over three years. J. Vascu Surg. **19**, 540.

43. Stanley, J.C., Burkel, W.E., Ford, J.W., *et al.* (1982) Enhanced patency of small-diameter, externally supported Dacron iliofemoral grafts seeded with endothelial cells, *Surgery* **92**, 994.

44. Kobashi, T. and Matsuda, T. (1999) Fabrication of branched hybrid vascular prostheses, *Tissue Engineering* **5**(6), 515-523.

45. Langer, R. and Vacanti, J. P. (1993) Tissue engineering, *Science* **260**, 920-926.

46. Guthrie, C. C. (1919) *JAMA* **73**, 188.

47. Niklason, L. E., Gao, J., Abbott, W. M., Hirschi, K. K., Houser, S., Marini, R. and Langer, R. (1999) Functional arteries grown *in vitro*, *Science* **284**, 489-493.

48. Risau, R. and Flamme, I. (1995) *Annu. Rev. Cell Dev. Biol.* **11**, 73.

49. Fernandez, P., Bareille, R., Conrad, V., Midy, D. and Bordenave, L. (2001) Evaluation of an *in vitro* endothelialized vascular graft under pulsatile shear stress with a novel radiolabeling procedure, *Biomaterials* **22**(7), 649-658.

50. Birchall, I. E., Lee, V. W. K. and Ketharanathan, V. (2001) Retention of endothelium on ovine collagen biomatrix vascular conduits under physiological shear stress, *Biomaterials* **22**(23), 3139-3144.

51. Papadaki, M, and Eskin, S.G. (1997) Effects of fluid shear stress on gene regulation of vascular cells, *Biotechnology Progress* **13**(3), 209-221.

52. Nerem, R. M. and Seliktar, D. (2001) Vascular tissue engineering, *Annual Review of Biomedical Engineering* **3**, 225-243.

53. Grottkau, B.E., Noordin, S., Shortkroff, S., Schaffer, J.L., Thornhill, T.S. and Spector, M. (2002) Effect of mechanical perturbation on the release of PGE$_2$ by macrophages *in vitro*, *J Biomed Mater Res*, **59**(2), 288-293.

54. Fisher, A.B., Chien, S., Barakat, A. I. and Nerem, R.M. (2001) Endothelial cellular response to altered shear stress, *Am J Physiol Lung Cell Mol Physiol*, **281**(3), 529-533.

55. Widmer, M.S. and Mikos A.G. (1998) Fabrication of biodegradable polymer scaffolds for tissue engineering, in: C.W. Patrick Jr, A.G. Mikos and L.V. McIntire (eds.), *Frontiers in Tissue Engineering*. Elsevier Science, New York, pp. 107-120.

56. Mikos, A.G., Sarakinos, G., Leite, S.M., Vacanti, J.P. and Langer, R. (1993) Laminated three-dimensional biodegradable foams for use in tissue engineering, *Biomaterials* **14**, 323-330.

57. Mooney, D.J., Baldwin, D.F., Suh, N.P., Vacanti, J.P. and Langer, R. (1996) Novel approach to fabricate porous sponges of poly(D,L-lactic-co-glycolic acid) without the use of organic solvents, *Biomaterials* **17**, 1417-1422.

58. Nam, Y.S., Yoon, J.J. and Park, T.G. (2000) A novel fabrication method for macroporous scaffolds using gas foaming salt as porogen additive, *J Biomed Mater Res, Appl Biomater* **53**, 1-7.

59. Yoon, J.J. and Park, T.G. (2001) Degradation behavior of biodegradable macroporous scaffolds prepared by gas foaming of effervescent salts, *J Biomed Mater Res* **55**(3), 401-408.

60. Nam, Y.S. and Park, T.G. (1999) Porous biodegradable polymeric scaffolds prepared by thermally induced phase separation. *J Biomed Mater Res* **47**, 8-17.

61. Lo, H., Kadiyala, S., Guggino, S.E. and Leong, K.W. (1996) Poly(L-lactic acid) foams with cell seeding and controlled-release capacity, *J Biomed Mater Res* **30**, 475-484.

62. Park, A., Wu, B. and Griffith, L.G. (1998) Integration of surface modification and 3D fabrication techniques to prepare patterned poly(L-lactide) substrates allowing regionally selective cell adhesion, *J Biomater Sci, Polym Edn* **9**, 89-110.

63. Mooney, D.J., Mazzoni, C.L., Breuer, C., McNamara, K., Hern, D., Vacanti, J.P. and Langer, R. (1996) Stabilized polyglycolic acid fiber-based tubes for tissue engineering, *Biomaterials* **17**, 115-124.

64. Freed, L.E., Marquis, J.C., Nohria, A., Emmanual, J., Mikos, A.G. and Langer, R. (1993) Neocartilage formation *in vitro* and *in vivo* using cells cultured on synthetic biodegradable polymers, *J Biomed Mater Res* **27**, 11-23.

65. Hutmacher, D. W., Schantz, T., Zein, I., Ng, K. W., Teoh, S. H. and Tan, K. C. (2001) Mechanical properties and cell cultural response of polycaprolactone scaffolds designed and fabricated *via* fused deposition modeling, *J Biomed Mater Res* **55**(2), 203-216.

66. Hutmacher, D.W. (2000) Scaffolds in tissue engineering bone and cartilage, *Biomaterials* **21**(24), 2529-2543.

67. Zein, I., Hutmacher, D.W., Tan, K.C. and Teoh, S.H. (2002) Fused deposition modeling of novel scaffold architectures for tissue engineering applications, *Biomaterials* **23**(4), 1169-1185.

68. Sonoda, H., Takamizawa, K., Nakayama, Y., Yasui, H. and Matsuda, T. (2001) Small-diameter compliant arterial graft prosthesis: Design concept of coaxial double tubular graft and its fabrication, *J Biomed Mater Res* **55**(3), 266-276.

69. Ramakrishna, S., Mayer, J., Wintermantel, E. and Leong, K.W. (2001) Biomedical applications of polymer-composite materials: a review, *Composites Science and Technology* **61**(9), 1189-1224.

70. Espigares, I., Elvira, C., Mano, J.F., Vázquez, B., San Román, J. and Reis, R.L. (2002) New partially degradable and bioactive acrylic bone cements based on starch blends and ceramic fillers, *Biomaterials* **23**(8), 1883-1895.

71. Seal, B. L., Otero, T. C. and Panitch, A. (2001) Polymeric biomaterials for tissue and organ regeneration, *Materials Science and Engineering: Reports,* **34**(4-5), 147-230.

72. Burg K.J.L., Porter, S. and Kellam, J.F. (2000) Biomaterial developments for bone tissue engineering, *Biomaterials,* **21**(23), 2347-2359.

73. Cohn, D. and Younes, H. (1988) Biodegradable PEO/PLA block copolymers. *J Biomed Mater Res* **22**, 993-1009.

74. Younes, H., Nataf, P.R., Cohn, D., Appelbaum, Y.J., Pizov, G. and Uretzky, G. (1988) Biodegradable PELA block copolymers: *in vitro* degradation and tissue reaction, *Biomaterials, Artificial Cells, and Artificial Organs* **16**(4), 705-719.

75. Cohn, D. and Younes,H. (1989) Compositional and structural analysis of PELA biodegradable block copolymers degrading under *in vitro* conditions. *Biomaterials* **10**, 466-474.

76. Hotovely-Salomon, A. (1999) Biodegradable polymeric scaffolds for vascular tissue engineering [Ph.D. thesis]. Jerusalem, Hebrew University.

77. Cohn, D., Stern, T., González, M.F. and Epstein, J. (2002) Biodegradable poly(ethylene oxide)/poly(epsilon-caprolactone) multiblock copolymers, *J Biomed Mater Res* **59**(2), 273-281.

78. Hull, D., (1981) *An Introduction* to *Composite Materials,* Cambridge University Press, UK.

79. Agarwal, B.D. and Broutman, L.J. (1980) *Analysis and Performance of Fiber Composites,* John Wiley & Sons, USA.

80. Mallick, P.K., (1988) *Fiber-Reinforced Composites,* Marcel Dekker, USA.

81. Merrill, E.W. and Salzman, E.W. (1983) Polyethylene oxide as a biomaterial, *ASAIO J.* **6**(2), 60-64.

82. Drumheller, P.D. and Hubbell, J.A. (1995) Densely crosslinked polymer networks of poly(ethylene glycol) in trimethylolpropane triacrylate for cell-adhesion-resistant surfaces, *JBiomed Mater Res* **29**(2), 207-215.

83. Kidane, A., McPherson, T., Shim, H.S. and Park, K. (2000) Surface modification of polyethylene terephthalate using PEO-polybutadiene-PEO triblock copolymers, *Colloids and Surfaces B: Biointerfaces,* **18**(3-4) 347-353.

84. Zou, X. P., Kang, E. T. and Neoh, K. G. (2002) Plasma-induced graft polymerization of poly(ethylene glycol) methyl ether methacrylate on poly(tetrafluoroethylene) films for reduction in protein adsorption, *Surface and Coatings Technology* **149**(2-3), 119-128.

TISSUE-ENGINEERED VASCULAR GRAFTS FOR SMALL-DIAMETER ARTERIAL REPLACEMENT

M.J.B. Wissink, J. Feijen

Institute for Biomedical Technology, Polymer Chemistry and Biomaterials Group, Department of Chemical Technology, University of Twente, P.O. Box 217, 7500 AE Enschede, The Netherlands.

1. Introduction

The human blood circulation is a closed system of conduits that carries blood from the heart to all tissues, and from these tissues back to the heart. In this system, the aorta and other, smaller arteries distribute blood from the heart to the various parts of the body.

A normal artery has a tube-like structure comprising of three distinctive layers: the intima, the media and the adventitia (figure 1). The intima, the inner layer of the artery that is in contact with blood, consists of a monolayer of endothelial cells supported by an underlying connective tissue membrane. The media comprises of e.g. collagens and elastin, with a dense population of smooth muscle cells (SMCs) that are oriented perpendicular to the blood flow. The adventitia, the outer layer of the artery, is a collagenous matrix containing fibroblasts, small blood vessels and nerves.

Normal vascular endothelium maintains a delicate balance in the vascular system, between growth promotion and inhibition of the underlying SMCs and fibroblasts, vasoconstriction and vasodilatation, blood cell adherence, and anticoagulation and procoagulation. In this process, endothelial cells actively secrete various factors affecting platelet adhesion and aggregation (prostacyclin, PGI_2), blood coagulation (von Willebrand factor, vWF) and fibrinolysis (tissue plasminogen activator, tPA, and tissue plasminogen activator inhibitor, PAI) [1]. Contraction and relaxation of the SMCs in the adventitia regulates blood flow, whereas the fibroblasts in the adventitia add to the rigidity and shape of a blood vessel.

R.L. Reis and D. Cohn (eds.),
Polymer Based Systems on Tissue Engineering, Replacement and Regeneration, 391–405.

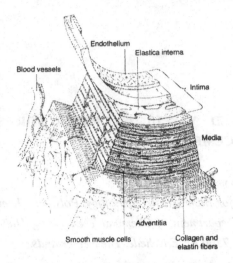

Figure 1: Anatomy of an artery (courtesy: Archief Natuur & Techniek, Diemen, the Netherlands).

The viscoelastic (mechanical) properties of a blood vessel are largely determined by the extracellular matrices in the media and the adventitia. Collagens provide stiffness, whereas elastin contributes to the elastic properties of a vessel. Glycosaminoglycans in the extracellular matrices, like heparan sulphate, determine compressibility [2].

Diseases of the circulatory system are the main cause of mortality in the western world. Degenerative arterial diseases like atherosclerosis and aneurysm formation are prominent circulatory disorders. Atherosclerosis is a slowly progressive process starting at relatively young age, which leads to gradual narrowing of arteries due to deposition of atherosclerotic plaque. Eventually, arterial occlusion may result in complications such as myocardial infarction, stroke, and gangrene of the extremities [3]. Aneurysm formation results in a local increase in arterial diameter, accompanied by a decreased thickness of the arterial wall (e.g. the aorta) and thrombosis. When an aneurysm ruptures, acute and severe blood loss occurs which may be life threatening.

Synthetic vascular grafts, e.g. made of Dacron (polyethylene terephthalate, figure 2a) or expanded Teflon (polytetrafluoroethylene, ePTFE, figure 2b) have been used successfully for many years in large-diameter applications like the aorta or the aortoiliac region. The long-term patency rates of small-diameter synthetic vascular grafts (diameter 5 mm or less), however, are disappointing, primarily due to (re-) stenosis and thrombus formation [4-7]. Autologous saphenous veins and iliac arteries have been used to repair small- and medium-diameter arteries [8]. In the lower extremities, long-term results of autologous saphenous vein grafts, used either as *in situ* or as reversed grafts, have been very good [9]. However, suitable autologous veins are often absent, for example due to previous utilization or disease [10]. Furthermore, there are arguments for the preferential use of synthetic graft materials in the upper legs, to preserve autologous veins for a later date [11]. Therefore, there is a need for alternative small-diameter vascular prostheses with improved patency rates.

Figure 2: Scanning electron micrographs of (A) Dacron (Cooley knitted Dacron, Meadox Medicals, Oakland, NJ, 100 ×) and (B) expanded Teflon (Gore-Tex, Gore & Associates, Flagstaff, AZ, 500 ×).

2. Tissue-engineered vascular grafts for small diameter arterial replacement

This review is focussed on developments in tissue engineering of (small diameter) vascular grafts. For an overview of other techniques that have been explored in the development of blood compatible vascular prostheses we refer to Engbers *et al.* [12].

The goal for tissue engineering of vascular grafts is the replacement of a diseased blood vessel with a biologically functional and structurally stable graft. Two major approaches can be identified that will be discussed in detail below:

1. Endothelial cell seeding. By applying a monolayer of endothelial cells in the lumen of a (synthetic) vascular graft, the performance of the graft can be significantly improved.

2. Completely tissue-engineered vascular grafts. These grafts have a close resemblance to native blood vessels, and are produced by endothelial cell seeding of a pre-formed tubular structure containing autologous vessel-wall cells and optionally a polymeric scaffold for improved mechanical strength.

For successful application, tissue-engineered vascular grafts should meet several criteria. The mechanical (viscoelastic) properties and compliance of the graft should be comparable to those of the host vessel. Furthermore, the tissue-engineered graft should have good suture retention, the graft should be able to withstand long-term cyclic loading due to the pulsatile flow of blood, and should have a sufficiently high burst strength. Regarding biological function, the lumen of the graft should be lined with a non-thrombogenic layer of e.g. endothelial cells displaying the same properties as endothelial cells in a normal vessel. When SMCs and/or fibroblasts are incorporated in the tissue-engineered vessel, uncontrolled proliferation of these cells after implantation resulting in narrowing of the graft lumen should be prevented.

3. Endothelial cell seeding of synthetic vascular grafts

Endothelial cell seeding is a recognized method to improve blood compatibility of small-diameter vascular grafts [13-15]. Endothelium is often described as the perfect natural blood-compatible surface [12, 16] which secretes various substances affecting platelet adhesion and aggregation, blood coagulation and fibrinolysis. Furthermore, expression of cell-surface components like heparan sulphate or tissue thromboplastin (tissue factor) regulate interactions with blood components. Endothelial cell function thus directly affects the balance of haemostasis and thrombosis in the cardiovascular system [16, 17]. Upon creating a functional inner lining of endothelial cells in (small diameter) vascular grafts, implants with similar non-thrombogenic surface characteristics as normal blood vessels can be obtained.

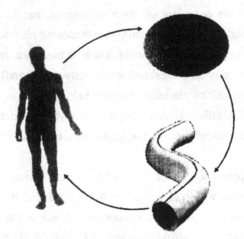

Figure 3: Endothelial cell seeding: Endothelial cells isolated from e.g. a blood vessel of a patient are seeded in the lumen of a synthetic vascular graft, which is subsequently used to replace a diseased artery of the patient.

In most animal models, endothelialization of vascular grafts will occur spontaneously [18, 19]. In contrast, reports of spontaneous endothelialization of vascular grafts in humans are anecdotal [20]. In humans, endothelialization is in general confined to a few millimetres pannus-ingrowth from both anastomoses [21]. To obtain continuously endothelium-lined vascular grafts in humans, endothelial cell seeding has to be applied. The potential of this strategy was demonstrated in a number of animal studies. Significantly increased patency rates [22-26] and a strong reduction in platelet adhesion and thrombus formation [27-29] were observed after endothelial cell seeding of vascular grafts.

3.1. SUBSTRATE COMPATIBILITY

Commercially available synthetic vascular graft materials are generally poor substrates for endothelial cell seeding [30-32]. Modification of existing graft materials by chemical surface treatment [33], glow discharge modification [30, 34-36] or application of a suitable protein coating may improve the potential of currently available graft materials for endothelial cell adhesion and proliferation. Regarding protein coating, substances like fibrin glue [37, 38], non-crosslinked collagen [28, 39] and fibronectin [40, 41] have been shown to result in improved *in vitro* endothelialization of vascular graft materials. Furthermore, immobilization of RGD-peptides [42] or antibodies against endothelial cell membrane antigens [43] may improve adherence and growth of seeded endothelial cells.

Pre-clotting [23, 25, 44, 45], fibrin glue [27] or coating with fibronectin is often used before endothelial cell seeding in animal models [26, 28, 46]. In patients, the use of either one of the above techniques may face certain disadvantages. Pre-clotting of vascular grafts (prior to cell seeding) is elaborate, and problematic notably in patients using anticoagulant treatment. The use of human fibrin glue or human fibronectin introduces a risk of transmitting viral diseases.

3.2. DIRECT SEEDING VERSUS TWO-STAGED PROCEDURES

To prevent graft rejection due to differences in histocompatibility and blood group antigens, and to prevent transmission of (viral) diseases between donor (of endothelial cells) and recipient, there is a strong preference for the use of autologous cells for seeding of vascular grafts. As a consequence, the supply of cells directly after isolation is limited because cells have to be harvested from pieces of vascular tissue from the patient undergoing vascular surgery. The number of cells harvested is usually not sufficient to obtain a confluent covering at the inside of the graft immediately after seeding.

Three basic methods for endothelialization of vascular grafts can be recognized: immediate seeding, delayed seeding or implantation of pre-lined grafts [15]. Delayed seeding is a two-stage procedure. After source tissue removal and endothelial cell harvesting during a first operation, standard cell culture techniques are applied to expand cell numbers. Subsequently, the vascular graft is seeded and implanted during a second operation. The second approach, i.e. implantation of pre-lined grafts, also involves a two-staged operation. Successful implantation of grafts pre-lined with endothelial cells has been reported by Zilla *et al.* [47, 48]. After endothelial cell isolation, cells were seeded directly into fibrin-glue coated ePTFE grafts. Thereafter, the cells were cultured *in situ* for an average time of 25 days, to obtain vascular grafts lined with a monolayer of endothelial cells. This approach has been successfully used clinically for eight years [13]. An excellent primary patency rate of 74% after 7 years has been reported for pre-lined grafts used for femoropopliteal reconstruction, which is comparable to the patency rate of autologous vein grafts [7, 49].

Drawbacks of delayed seeding or the use of pre-lined grafts are the long interval between the need of an endothelialized vascular graft and its availability, limiting its use to non-emergency situations, and the increased risk of bacterial infection due to prolonged *in vitro* endothelial cell culture [38]. For those reasons, immediate seeding is the preferred method for endothelialization of synthetic vascular grafts, although low seeding densities have to be accepted and dealt with. Studies with immediate cell seeding in humans, however, have mostly been disappointing. Most likely, seeding densities were insufficient to lead to the formation of a confluent endothelial lining in situ [24, 47, 50, 51].

3.3. COLLAGEN-COATED VASCULAR GRAFTS

Collagen, gelatin and albumin impregnated synthetic vascular grafts originally have been developed to eliminate the procedure of pre-clotting of the porous Dacron prior to implantation of the graft. Since non-crosslinked collagen is a suitable matrix for the growth of endothelial cells *in vitro* [52-54], application of a collagen coating on synthetic vascular graft materials may result in a matrix suitable for endothelial cell seeding. Furthermore, despite collagen thrombogenicity, patency rates of small-diameter collagen-coated grafts were comparable to or somewhat higher than patency rates of non-coated synthetic grafts [4, 55].

In order to prevent rapid *in vivo* resorption of the protein coating, in *commercially available* collagen-coated vascular grafts, glutaraldehyde or formaldehyde are commonly used to crosslink the matrix [8]. These agents are incorporated in the collagen matrix during crosslinking. During *in vitro* and *in vivo* degradation, notably glutaraldehyde-crosslinked collagen evokes cytotoxic reactions due to release of (unreacted) glutaraldehyde and glutaraldehyde derivatives [56-58], which hamper endothelialization of the luminal graft surface of commercially available vascular grafts [58, 59].

A collagen matrix crosslinked using a water-soluble carbodiimide was developed as an alternative coating for endothelial cell seeding of (small-diameter) vascular grafts [60]. In contrast to glutaraldehyde- or formaldehyde crosslinked collagen, this matrix demonstrated very good biocompatibility [61]. Compared to non-crosslinked type I collagen, *in vitro* proliferation of seeded endothelial cells was significantly increased on carbodiimide-crosslinked collagen. Neither the morphology of the cells nor the secretion of PGI_2, vWF, t-PA and PAI was affected by the crosslinking of the collagen substrate [62].

3.4. PEPTIDE GROWTH FACTORS

Peptide growth factors can generally be defined as proteins that promote proliferation and migration of cells [63]. For endothelial cells several peptide growth factors were identified, including acidic fibroblast growth factor (aFGF), basic fibroblast growth factor (bFGF) and vascular endothelial growth factor (VEGF) [64].

Local sustained release of endothelial growth factors from vascular grafts is expected to improve the proliferation of seeded endothelial cells. Furthermore, incorporation of peptide growth factors is likely to allow lower seeding densities than grafts without growth factor pre-loading, thus increasing the chance of successful endothelialization of vascular grafts after immediate, low density endothelial cell seeding.

Local sustained bFGF release has been shown to improve spontaneous endothelialization of vascular grafts *in vitro* as well as *in vivo*. A fibrin glue sealant containing heparin and aFGF was developed to induce spontaneous endothelialization when applied as coating of synthetic vascular graft materials [65, 66]. By adjusting the ratio of heparin and growth factor in the fibrin glue, proliferation of seeded endothelial cells was accelerated while proliferation of smooth muscle cells could be inhibited *in vitro* [67]. Spontaneous endothelialization of these vascular grafts was found to be increased when implanted in rats, dogs and rabbits [65, 68, 69]. Polyurethane grafts coated with a mixture of photoreactive gelatin, heparin and bFGF, demonstrated spontaneous endothelialization as a result of transmural in-growth of endothelial cells, when implanted in rat aortas [70].

For immediate endothelial cell seeding procedures a novel collagen matrix for synthetic vascular grafts was developed. Collagen was crosslinked using a water-soluble carbodiimide, after which heparin was immobilized onto the crosslinked collagen matrix. Pre-loading of these matrices with increasing quantities of bFGF resulted in progressively improved endothelial cell proliferation *in vitro*. Furthermore, after loading these matrices with bFGF, a confluent layer of endothelial cell was formed after seeding at very low densities (e.g. 250 cells/cm^2) [71].

Until now, local sustained release of bFGF or other growth factors from vascular grafts in combination with endothelial cell seeding *in vivo* has not been described.

4. Completely tissue-engineered vascular grafts

Completely tissue-engineered vascular grafts not only contain an intimal layer of endothelial cells, but also a medial layer of SMCs and/or an adventitial layer of fibroblasts. A (polymeric) scaffold may be incorporated in the construct to provide mechanical strength. Regarding this emerging field of research there is still a limited number of publications. Because of the biomedical potential of these completely engineered constructs these publications will be discussed separately in detail below.

The Sparks mandrel system, patented in 1972, was the first attempt to obtain autologous, tissue-engineered vascular grafts [72]. This system consisted of a solid tubular silicon mandrel surrounded by two porous, coarse-knitted Dacron tubes, and had to be inserted under a layer of muscles overlying the rib cage. During implantation, fibrous tissue was formed around the mandrel and Dacron mesh. After graft harvesting and removal of the silicon mandrel, a vessel of Dacron, connective tissue and fibroblasts was obtained. Both in animals and humans the performance of these grafts was disappointing. Because of the absence of endothelial cells the graft proved to be very thrombogenic. Furthermore, after implantation progressive fibroblast proliferation and connective tissue deposition resulted in graft occlusion or collapse of the vessel wall structure, which can also be contributed at least in part to the absence of an endothelial lining.

Endothelial cell seeding seems essential to maintain graft geometry when SMCs and/or fibroblasts are incorporated in a tissue-engineered vascular graft. In *in vitro* co-culture, endothelial cells inhibited the proliferation of both SMCs and fibroblasts. Furthermore, *in vivo* neo-intimal thickening (hyperplasia) of vascular grafts by SMC proliferation is inhibited by endothelial cell seeding [73]. Incorporation of SMCs seems to be beneficial for maintaining the integrity of the endothelial lining of an engineered graft directly after implantation [74].

In 1986 Weinberg and Bell constructed a complete tissue-engineered artery, with a Dacron mesh to provide mechanical strength [75]. A mixture of collagen and bovine SMCs were cast into an annular mould. Contraction of the matrix by the SMCs during culture resulted in the formation of a tubular structure. A Dacron mesh sleeve was placed around the SMC/collagen tube whereafter an adventitial layer of bovine fibroblasts was cultured around this structure. After removal of the mould, the lumen of the graft was lined with bovine endothelial cells (fig. 4A). Except for the Dacron mesh, the graft closely resembled a native artery. Orientation of SMCs was not observed. The lining of endothelial cells functioned as a permeability barrier, and produced vWF and PGI_2. Although the maximal burst strength of the graft was still rather low, some important factors influencing the burst strength of the graft were determined indicating the possibility of further optimization. These included culture parameters, cell seeding densities and graft collagen content.

Analogous to Weinberg *et al.*, Hirai cultured a collagenous tube containing SMCs, which was seeded with endothelial cells prior to use [76]. After implantation of autologous grafts in dogs, after 6 months a cumulative patency rate of 65% was observed. Upon implantation however the vessels had to be wrapped in a Dacron mesh to prevent early graft rupture. Six months after implantation, remodelling of the implants had resulted in a structure resembling native blood vessels.

Adding sodium ascorbate to cultures of SMCs and fibroblasts, and culturing of the tissue-engineered vessel under flow conditions greatly improved the mechanical strength of a vessel, as demonstrated by L'Heureux *et al* [77]. Human vascular SMC were cultured to form a cohesive cellular sheet. This sheet was placed around a tubular support to produce the media of the vessel, and a sheet of human fibroblasts was wrapped around the media to provide the adventitia. After a maturation period of 8 weeks the mandrel was removed and endothelial cells were seeded in the lumen of the graft (fig. 4B). The vessel obtained had a well-defined intima, media and adventitia, and contained extracellular matrix components including collagens, elastin, laminin and glycosaminoglycans. The monolayer of endothelial cells secreted vWF and PGI_2 and inhibited platelet adhesion *in vitro*. The vessel demonstrated good handling and suturability characteristics, and had a burst strength comparable to human vessels, 2000 mm Hg.

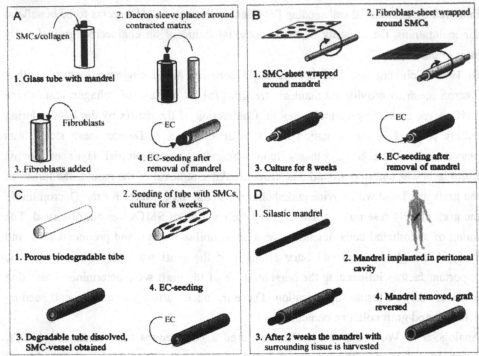

Figure 4: Different approaches to tissue-engineered vascular grafts. Fig 4A: Complete tissue-engineered graft with Dacron mesh (Weinberg, *et al.* [75]). Fig 4B: Graft constructed of sheets of SMCs and fibroblasts, seeded with EC (L'Heureux, *et al.* [77]). Fig 4C: Adventitia cultured on a degradable scaffold, seeded with EC (Niklason, *et al.* [78]). Fig 4D: Mandrel-grown graft (Campbell, *et al.* [79]). For a detailed description of procedures is referred to the text. EC: endothelial cells.

Niklason *et al.* cultured bovine SMCs on a tubular biodegradable polyglycolic acid mesh. After 8 weeks of culture under pulsatile flow conditions the polymeric mesh was largely degraded, and a structure resembling the media of an artery was obtained [78]. This vessel was seeded with bovine endothelial cells and cultured for another three days (fig. 4C). Addition of ascorbic acid, copper ions and amino acids to the culture medium proved to be essential for collagen synthesis and collagen crosslink formation, providing mechanical strength to the graft. When compared to vessels cultured under static conditions, vessels cultured under pulsatile flow conditions had a histologic appearance more similar to native vessels. Engineered vessels had a burst strength greater than 2000 mm Hg, a good suture retention and collagen contents of up to 50 percent. Cultured vessels also showed SMC contraction in response to pharmacological stimuli. When implanted as xenografts in swine, these grafts remained open during the whole implantation period of four weeks.

Recently, a new approach to *in vivo* mandrel grown vascular substitutes was developed [79]. A solid silastic tube was placed inside the peritoneal cavity of rats and rabbits. After two weeks the graft with surrounding granulation tissue was harvested, the silastic tube was removed, and the tube of tissue was reversed (fig. 4D). The reversed vessel had an inner lining of non-thrombogenic mesothelial cells, a media of myofibroblasts, collagen and elastin, and a collagenous adventitia. When used to partially replace a carotid artery, the overall patency rate of these grafts was approximately 70% in both rats and rabbits. The grafts became responsive to contractile agonists, similar to the vessel into which they had been grafted. Burst strength was not determined.

Bader *et al.* developed a mild procedure for de-cellularization of pig aortas, leaving the structural organization and different components of the extracellular matrix intact [80]. This matrix was seeded with pre-expanded human endothelial cells and myofibroblasts, and was subsequently cultured under flow conditions. A graft was obtained with a confluent layer of endothelial cells covering the luminal surface, which functionality was assessed by their NO secretion. Myofibroblasts had migrated into positions formerly occupied by the xenogenic cells. The total preparation time of the graft was one to two weeks, including pre-expansion of the endothelial cells.

5. Conclusions and outlook

Implantation of completely endothelialized small diameter vascular grafts is the only successful approach to tissue-engineered vascular grafts reported until now. Fibrin-glue coated ePTFE grafts lined with a monolayer of endothelial cells have been successfully used in the clinic for eight years, with patency rates comparable to the patency rate of autologous vein grafts. A drawback is the considerable time needed to obtain a completely endothelialized graft after initial endothelial cell harvest, limiting its use to non-emergency procedures, and the increased risk of bacterial infection associated with prolonged culture.

New optimized substrates for endothelial cell seeding in combination with peptide growth factors increase the potential of direct endothelial cell seeding techniques.

Directly seeded grafts could combine a simple and quick procedure for graft preparation with the successful patency rates of completely lined grafts.

Completely tissue-engineered vascular grafts have a great potential in autologous small diameter blood vessel replacement. Tissue-engineered grafts with a structure and burst strength similar to native arteries already have been developed. At this time however studies describing these grafts are mainly focussed on *in vitro* graft culture. There is still a lack of convincing data regarding the performance of these grafts in experimental animals, and no (pre-) clinical data is to be expected on short terms. Ongoing research, improved graft culture techniques, SMC-orientation, stimulation of elastin formation, incorporation of peptide growth factors and perhaps the use of progenitor cells may contribute to the successful development of completely tissue-engineered vascular grafts.

References

1   Davies, M.G., and Hagen, P.O. (1993) The vascular endothelium, *Ann.Surg.* **218**, 593-609.
2   Ratcliffe, A. (2000) Tissue engineering of vascular grafts, *Mat. Biol.* **19**, 353-357.
3   Ross, R. (1993) The pathogenesis of atherosclerosis: A perspective for the 1990s, *Nature* **362**, 801-809.
4   Koch, G., Gutschi, S., Pascher, O., Fruhwirth, J., and Hauser, H. (1996) Femoropopliteal vascular replacement: Vein, ePTFE or ovine collagen?, *Zentralbl.Chir.* **121**, 761-767.
5   Pevec, W.C., Darling, R.C., L'Italien, G.J., and Abbot, W.M. (1992) Femoropopliteal reconstruction with knitted nonvelour Dacron versus expanded polytetrafluoroethylene, *J.Vasc.Surg.* **16**, 60-65.
6   Wilson, Y.G., Wyatt, M.G., Currie, I.C., Baird, R.N., and Lamont, P.M. (1995) Preferential use of vein for above-knee femoropopliteal grafts, *Eur.J.Vasc.Endovasc.Surg.* **10**, 220-225.
7   Watelet, J., Soury, P., Menard, J.F., Plissonnier, D., Peillon, C., Lestrat, J.P., and Testart, J. (1997) Femoropopliteal bypass: In situ or reversed vein grafts? Ten-year results of a randomized prospective study., *Ann.Vasc.Surg.* **11**, 510-519.
8   Weadock, K.S., and Goggins, J.A. (1993) Vascular graft sealants, *J. of Long Term Effects of Med. Impl.* **3**, 207-222.
9   Conte, M.S. (1998) The ideal small arterial substitute: A search for the holy grail?, *FASEB J.* **12**, 43-45.
10 Senatore, F., Shankar, H., and Ventaramani, E.S., Development of biocompatible vascular prostheses. In: CRC Critical reviews in biocompatibility, S.K. Williams (Ed.), CRC Press, Ann Arbor, 1988.
11 Berlakovich, G.A., Herbst, F., Mittlbock, M., and Kretschmer, G. (1994) The choice of material for above knee femoropopliteal bypass, *Arch.Surg.* **129**, 297-302.
12 Engbers, G.H.M., and Feijen, J. (1991) Current techniques to improve the blood compatibility of biomaterial surfaces., *Int.J.Art.Org.* **14**, 199-215.
13 Meinhart, J., Deutsch, M., and Zilla, P. (1997) Eight years of clinical endothelial cell transplantation, *ASAIO J.* **43**, M515-M521.
14 Schmidt, S.P., Sharp, W.V., Evancho, M.M., and Meerbaum, S.O., Endothelial cell seeding of prosthetic vascular grafts - Current status. In: High performance biomaterials, M. Scyzer (Ed.), Technomic, Lancaster, 1991.
15 Herring, M.B., Endothelial seeding of blood flow surfaces. In: Vascular grafting. Clinical applications and techniques, C.B. Wright, *et al.* (Eds.), John Wright, Boston, 1983.
16 Weksler, B.B., Platelet interactions with the vessel wall. In: Hemostasis and thrombosis. Basic principles and clinical practice, R.W. Colman, *et al.* (Eds.), J.B. Lippincott, Philadelphia, 1987.
17 Kefalides, N.A., Biochemical aspects of the vessel wall. In: Hemostasis and thrombosis. Basic principles and clinical practice, R.W. Colman, *et al.* (Eds.), J.B. Lippincott, Philadelphia, 1987.
18 Parsson, H., Jundzill, W., Johansson, K., Jonung, T., and Norgren, L. (1994) Healing characteristics of polymer-coated or collagen-treated Dacron grafts: An experimental porcine study, *Cardiovasc.Surg.* **2**, 243-248.

19 Bull, D.A., Hunter, G.C., Holubec, H., Aguirre, M.L., Rappaport, W.D., and Putnam, C.W. (1995) Cellular origin and rate of endothelial cell coverage of PTFE grafts, *J.Surg.Res.* **58**, 58-68.

20 Wu, M.H.D., Shi, Q., Wechezak, A.R., Clowes, A.W., Gordon, I.L., and Sauvage, L.R. (1995) Definitive proof of endothelialization of a Dacron arterial prosthesis in a human being, *J.Vasc.Surg.* **21**, 862-867.

21 Russel, S.D., and Herring, M.B., Endothelial cell seeding of vascular prostheses. In: Biology of endothelial cells, E.A. Jaffe (Ed.), Martinus Nijhof Publishers, Boston, 1984.

22 Shindo, S., Takagi, A., and Wittermore, A.D. (1987) Improved patency of collagen-impregnated grafts after in vitro autogenous endothelial cell seeding, *J.Vasc.Surg.* **6**, 325-332.

23 Pasic, M., Mullerglauser, W., Vonsegesser, L.K., Lachat, M., Mihaljevic, T., and Turina, M.I. (1994) Superior late patency of small-diameter Dacron grafts seeded with omental microvascular cells: An experimental study, *Ann.Thorac.Surg.* **58**, 677-684.

24 Williams, S.K., Rose, D.G., and Jarrel, B.E. (1994) Microvascular endothelial cell sodding of ePTFE vascular grafts: Improved patency and stability of the cellular lining, *J.Biomed.Mat.Res.* **28**, 203-212.

25 Campbell, J.B., Glover, J.L., and Herring, B. (1988) The influence of endothelial seeding and platelet inhibition on the patency of ePTFE grafts used to replace small arteries - An experimental study, *Eur.J.Vasc.Surg.* **2**, 365-370.

26 Budd, J.S., Allen, K.E., Hartley, G., and Bell, P.R.F. (1991) The effect of preformed confluent endothelial cell monolayers on the patency an thrombogenicity of small calibre vascular grafts, *Eur.J.Vasc.Surg.* **5**, 397-405.

27 Zilla, P., Preiss, P., Groscurth, P., Rosemeier, F., Deutsch, M., Odell, J., Heidinger, C., Fasol, R., and Vonoppell, U. (1994) In vitro lined endothelium: Initial integrity and ultrastructural events, *Surgery* **116**, 524-534.

28 Schneider, P.A., Hanson, S.R., Price, T.M., and Harker, L.A. (1988) Preformed confluent endothelial cell monolayers prevent early platelet deposition in vascular prostheses in baboons, *J.Vasc.Surg.* **8**, 229-235.

29 Ahlswede, K.M., and Williams, S.K. (1994) Microvascular endothelial cell sodding of 1-mm expanded polytetrafluoroethylene vascular grafts, *Arterioscl.Thromb.* **14**, 25-31.

30 Dekker, A., Reitsema, K., Beugeling, T., Bantjes, A., and Feijen, J. (1991) Adhesion of endothelial cells and adsorption of serum proteins on gas plasma treated polytetrafluoroethylene, *Biomaterials* **12**, 130-137.

31 Absolom, D.R., Hawthorne, L.A., and Chang, G. (1988) Endothelialisation of polymer surfaces, *J.Biomed.Mat.Res.* **22**, 271-280.

32 Foxall, T.L., Auger, K.R., Callow, A.D., and Libby, P. (1986) Adult human endothelial cell coverage of small calibre Dacron and polytetrafluoroethylene vascular prostheses in vitro, *J.Surg.Res.* **41**, 158-172.

33 McKeown, N.B., Chang, G., Niven, J., Romaschin, A.D., Wilson, G.J., Thompson, M., and Kalman, P.G. (1991) Facilitation of endothelial cell growth on hydroxylated ePTFE vascular grafts, *ASAIO Trans.* **37**, M477-478.

34 Ertel, S.I., Ratner, B.D., and Horbett, T.A. (1990) Radiofrequency plasma deposition of oxygen-containing films on polystyrene and poly(ethylene terephthalate) substrates improves endothelial cell growth, *J.Biomed.Mat.Res.* **24**, 1637-1659.

35 Ertel, S.I., Chilkoti, A., Horbet, T.A., and Ratner, B.D. (1991) Endothelial cell growth on oxygen-containing films deposited by radio-frequency plasmas: The role of surface carbonyl groups, *J.Biomat.Sci.Pol.Ed.* **3**, 163-183.

36 Pratt, K.J., Williams, S.K., and Jarrell, B.E. (1989) Enhanced adherence of human adult endothelial cells to plasma discharge modified polyethylene terephthalate, *J.Biomed.Mat.Res.* **23**, 1131-1147.

37 Mazzucotelli, J.P., Klein-Soyer, C., Beretz, A., Brisson, C., Archipoff, G., and Cazenave, J.P. (1991) Endothelial cell seeding: Coating Dacron and expanded polytetrafluoroethylene vascular grafts with a biological glue allows adhesion and growth of human saphenous vein endothelial cells, *Int.J.Art.Org* **14**, 482-490.

38 Zilla, P., Fasol, R., Preiss, P., Kadletz, M., Deutsch, M., Schima, H., and Tsangaris, S. (1989) Use of fibrin glue as a substrate for in vitro endothelialisation of PTFE vascular grafts, *Surg.* **105**, 515-522.

39 Kaehler, J., Zilla, P., Fasol, R., Deutsch, M., and Kadletz, M. (1989) Precoating substrate and surface configuration determine adherence and spreading of seeded endothelial cells on polytetrafluoroethylene grafts, *J.Vasc.Surg.* **9**, 535-541.

40 van Wachem, P.B., Vreriks, C.M., Beugeling, T., Feijen, J., Bantjes, A., Detmers, J.P., and van Aken, W.G. (1987) The influence of protein adsorption on the interactions of cultured human endothelial cells with polymers., *J.Biomed.Mat.Res.* **21**, 701-718.

404

41   James, N.L., Schindhelm, K., Slowiaczek, P., Milthorpe, B.K., Dudman, N.P.B., Johnson, G., and Steele, J.G. (1990) Endothelial cell seeding of small diameter vascular grafts, *Art.Org.* **14**, 355-360.

42   Massia, S.P., and Hubbel, J.A. (1991) Human endothelial cell interactions with surface coupled adhesion peptides on a nonadhesive glass substrate and two polymeric materials, *J.Biomed.Mat.Res.* **25**, 223-242.

43   Dekker, A., Poot, A.A., van Mourik, J.A., Workel, M.P.A., Beugeling, T., Bantjes, A., Feijen, J., van and Aken, W.G. (1991) Improved adhesion and proliferation of human endothelial cells on polyethylene precoated with monoclonal antibodies directed against cell membrane antigens and extracellular matrix proteins, *Thromb.Haemos.* **66**, 715-724.

44   Wang, Z., Du, W., Li, G., Pu, L., and Sharefkin, J.B. (1990) Rapid cellular luminal coverage of Dacron interior vena cava prostheses in dogs by immediate seeding of autogenous endothelial cells derived from omental tissue: Results of a preliminary trail, *J.Vasc.Surg.* **12**, 168-179.

45   Baitella-Eberle, G., Groscurth, P., Zilla, P., Lachat, M., Mullerglauser, W., Schneider, J., Neudecker, A., Vonsegesser, L.K., Dardel, E., and Turina, M. (1993) Long-term results of tissue development and cell differentiation on Dacron prostheses seeded with microvascular cells in dogs, *J.Vasc.Surg.* **18**, 1019-1028.

46   Poole-Warren, L.A., Schindhelm, K., Graham, A.R., Slowiaczek, P.R., and Noble, K.R. (1996) Performance of small diameter synthetic vascular prostheses with confluent autologous endothelial cell linings, *J.Biomed.Mat.Res.* **30**, 221-229.

47   Zilla, P., Deutsch, M., Meinhart, J., Puschmann, R., Eberl, T., Minar, E., Dudchak, R., Lugmaier, H., Schmidt, P., Noszian, I., and Fischlein, T. (1994) Clinical in vitro endothelialisation of femoropopliteal bypass grafts: An actuarial follow up over three years, *J.Vasc.Surg.* **19**, 540-548.

48   Fischlein, T., Zilla, P., Deutsch, M., Meinhart, J., Puschmann, R., Vesely, M., Eberl, T., Balon, R., and Deutsch, M. (1994) In vivo endothelialisation of a mesosystemic shunt: A clinical case report, *J.Vasc.Surg.* **19**, 549-554.

49   Woratyla, S.P., Darling, R.C., Chang, B.B., Paty, P.S., Kreienberg, P.B., Leather, R.P., and Shah, D.M. (1997) The performance of femoropopliteal bypasses using polytetrafluoroethylene above the knee versus autogenous vein below the knee, *Am.J.Surg.* **174**, 169-172.

50   Fasol, R., Zilla, P., Deutsch, M., Grimm, M., Fishlein, T., and Laufer, G. (1989) Human endothelial cell seeding: Evaluation of its effectiveness by platelet parameters after one year, *J.Vasc.Surg.* **9**, 432-436.

51   Herring, M., Smith, J., Dalsing, M., Glover, J., Compton, R., Etchberger, K., and Zollinger, T. (1994) Endothelial seeding of polytetrafluoroethylene femoral popliteal bypasses: The failure of low-density seeding to improve patency, *J.Vasc.Surg.* **20**, 650-655.

52   Delvos, U., Gajdusek, C., Sage, H., Harker, L.A., and Swartz, S.M. (1982) Interactions of vascular wall cells with collagen gels, *Lab.Invest.* **46**, 61-72.

53   Sato, K., Taira, T., Takayama, R., Ohtsuki, K., and Kawabata, M. (1995) Improved chromatographic purification of human and bovine type V collagen sub-molecular species and their subunit chains from conventional crude preparations. Application to cell-substratum assays for human umbilical vein endothelial cell, *J.Chromatogr.* **663**, 25-33.

54   Schor, A.M., Schor, S.L., and Kumar, S. (1979) Importance of a collagen substratum for stimulation of capillary endothelial cell proliferation by tumor angiogenesis factor, *Int.J.Cancer* **24**, 225-234.

55   Brewster, D.C., LaSalle, A., Robinson, J.G., Strayhorn, E.C., and Darling, R.C. (1983) Factors affecting patency of femoropopliteal bypass graft, *Surg.Gyn.Obst.* **157**, 437-442.

56   van Luyn, M.J.A., van Wachem, P.B., Olde Damink, L.H.H., Dijkstra, P.J., and Feijen, J. (1992) Relations between in vitro cytotoxicity and crosslinked dermal sheep collagens, *J.Biomed.Mat.Res.* **26**, 1091-1110.

57   Huang-Lee, L.L.H., Cheung, D.T., and Nimni, M.E. (1990) Biochemical changes and cytotoxicity associated with the degradation of polymeric glutaraldehyde derived crosslinks, *J.Biomed.Mat.Res.* **24**, 1185-1201.

58   Eyble, E., Griesmacher, A., Grimm, M., and Wolner, E. (1989) Toxic effects of aldehydes released from fixed pericardium on bovine aortic endothelial cells, *J.Biomed.Mat.Res.* **23**, 1355-1369.

59   Bengtsson, L., Ragnarson, B., and Haegerstrand, A. (1993) Lining of viable and nonviable allogenic and xenogenic cardiovascular tissue with cultured adult human venous endothelium, *J.Thorac.Cardiovasc.Surg.* **106**, 434-443.

60   Wissink, M.J.B., Beernink, R., Poot, A.A., Engbers, G.H.M., Beugeling, T., van Aken, W.G., and Feijen, J. (2001) Relation between cell density and the secretion of von Willebrand factor and prostacyclin by human umbilical vein endothelial cells, *Biomaterials* **22**, 2283-2290.

61   van Wachem, P.B., Plantinga, J.A., Wissink, M.J.B., Beernink, R., Poot, A.A., Engbers, G.H.M., Beugeling, T., van Aken, W.G., Feijen, J., and van Luyn, M.J.A. (2001) In vivo biocompatibility of

carbodiimide-crosslinked collagen matrices: Effects of crosslink density, heparin immobilization, and bFGF loading, *J.Biomed.Mat.Res.* **55**, 368-378.

62 Wissink, M.J.B., van Luyn, M.J.A., Beernink, R., Dijk, F., Poot, A.A., Engbers, G.H.M., Beugeling, T., van Aken, W.G., and Feijen, J. (2000) Endothelial cell seeding on crosslinked collagen: Effect of crosslinking on endothelial cell proliferation and functional parameters, *Thromb. Haemos.* **84**, 325-331.

63 Sporn, M.B., and Roberts, A.B., The multifunctional nature of peptide growth factors. In: Peptide growth factors and their receptors I, M.B. Sporn and A.B. Roberts (Eds.), Springer Verlag, Berlin, 1990.

64 Bradshaw, R.A., and Cavanaugh, K.P., Isolation and characterisation of growth factors. In: Peptide growth factors and their receptors I, M.B. Sporn and A.B. Roberts (Eds.), Springer Verlag, Berlin, 1990.

65 Greisler, H.P., Nunez, H.A., and Drohan, W.N., Tissue sealant and growth factor containing composites that promote accelerated wound healing, WO 92 920 9301 (1992 patent, 44 pages).

66 Nunez, H.A., Drohan, W.N., Burgess, W.H., Greisler, H.P., Hollinger, J.O., Lasa, C.I., Maciag, T., and Macphee, M.J., Supplemented and unsupplemented fibrin tissue sealants, methods of their production and use, WO 9617633 (1996 patent, 158 pages).

67 Greisler, H.P., Gosselin, C., Reen, D., Kang, S.S., and Kim, D.U. (1996) Biointeractive polymers and tissue engineered bloodvessels, *Biomaterials* **17**, 329-336.

68 Greisler, H.P. (1996) Growth factor release from vascular grafts, *J.Contr.Rel.* **39**, 267-280.

69 Gray, J.L., Kang, S.S., Zenni, G.C., Kim, D.U., Kim, P.I., Burgess, W.H., Drohan, W., Winkles, J.A., Haudenschild, C.C., and Greisler, H.P. (1994) FGF-1 affixation stimulates ePTFE endothelialisation without intimal hyperplasia, *J.Surg.Res.* **57**, 596-612.

70 Doi, K., and Matsuda, T. (1997) Enhanced vascularisation in a microporous polyurethane graft impregnated with basic fibroblast growth factor and heparin, *J.Biomed.Mat.Res.* **34**, 361-370.

71 Wissink, M.J.B., Beernink, R., Scharenborg, N.M., Poot, A.A., Engbers, G.H.M., Beugeling, T., van Aken, W.G., and Feijen, J. (2000) Endothelial cell seeding of (heparinized) collagen matrices: Effect of bFGF pre-loading on proliferation (after low density seeding) and pro-coagulant factors, *J. Contr. Rel.* **67**, 141-155.

72 Guidoin, R., Noel, H.P., Marois, M., Martin, L., Laroche, F., Beland, L., Cote, R., and Gosselin, C. (1980) Another look at the Sparks-Mandril arterial graft precursor for vascular repair. Pathology by scanning electron microscopy, *Biomat.Med.Dev.Art.Org* **8**, 145-167.

73 Miwa, H., Matsuda, T., Tani, N., Kondo, K., and Iida, F. (1993) An in vitro endothelialised compliant vascular graft minimizes anastomotic hyperplasia, *ASAIO J.* **39**, M501-505.

74 Matsuda, T., and Miwa, H. (1995) A hybrid vascular model biomimicking the hierarchic structure of arterial wall: neointimal stability and neoarterial regeneration process under arterial, *J. Thorac. Cardiovasc. Surg. 1995, 110 (4/1), 988-997* **110**, 988-995.

75 Weinberg, C.B., and Bell, E. (1986) A blood vessel model constructed from collagen and cultured vascular cells, *Science* **231**, 397-400.

76 Hirai, J., and Matsuda, T. (1996) Venous reconstruction using hybrid vascular tissue composed of vascular cells and collagen tissue regeneration process, *Cell Transplant* **5**, 93-105.

77 L'Heureux, N., Paquet, S., Labbe, R., Germain, L., and Auger, F.A. (1988) A completely biological tissue-engineered human blood vessel, *FASEB J.* **12**, 47-56.

78 Niklason, L.E., Gao, J., and Abbott, W.M. (1999) Functional arteries grown in vitro, *Science* **284**, 489-493.

79 Campbell, J.H., Efendy, J.L., and Campbell, G.R. (1999) Novel vascular graft grown within recipient's own peritoneal cavity, *Circ. Res.* **85**, 1173-1178.

80 Bader, A., Steinhof, G., Strobl, K., Schilling, T., Brandes, G., Mertsching, T., Brandes, G., Mertsching, D., Froelich, J., and Haverich, A. (2000) Engineering of human vascular aortic tissue based on a xenogenic starter matrix, *Transplant.* **70**, 7-14.

Industrial and Innovation Opportunities

AN EXPLORATORY ANALYSIS OF THE DETERMINANTS OF INNOVATION IN THE EMERGING TISSUE ENGINEERING INDUSTRY

Nitin Pangarkar *, Dietmar W. Hutmacher #,&
*Faculty of Business Administration, National University of Singapore
FBA2 17, Law Link, Singapore 117591, # Division of Bioengineering,
Faculty of Mechanical Engineering, &Department of Orthopaedic Surgery,
Faculty of Medicine National University of Singapore, 10 Kent Ridge
Crescent, Singapore 119260

Abstract

Tissue engineering is a young and inter-disciplinary scientific discipline but it offers exciting opportunities to improve the quality of health care for hundreds of thousands of patients. Lured by its potential, several start-up companies, pharmaceutical corporations, and medical device enterprises alike are investing heavily in this sector. Innovation is a key driver of competition in this sector. In this study, we aim to explain the variation in innovation output across different firms in the sector. Our major premise is that firms that forge alliances will be able to tap into the expertise of their partners and thus improve their chances of coming out with an innovation. We further argue that alliances that enable technology acquisition or learning will enhance the innovation output of firms more than other kinds of alliances. We measure the innovation output of a company by the number of patents filed. Based on a preliminary analysis of eight companies, we find support for the hypotheses.

1.   Background

The last two decades have witnessed an escalating pace of technological change. Competitive environment in many industries is characterized by short product life cycles and rising R&D costs. In high technology industries such as biotechnology and tissue engineering, innovation and new products are becoming a key source of competitive advantage. [1,2]

This study aims to examine the drivers of innovation in the tissue engineering sector. Our choice of this sector is based on the belief that it is an important and growing sector. A recent study has concluded that at the beginning of 2001, tissue engineering R&D was being pursued by 3,300 scientists and support staff in more than 70 countries with a combined annual expenditure of over $600 million. [3] Furthermore, the aggregate investment in the sector since 1990 exceeds

409

R.L. Reis and D. Cohn (eds.),
Polymer Based Systems on Tissue Engineering, Replacement and Regeneration, 409–418.
© 2002 Kluwer Academic Publishers. Printed in the Netherlands.

$3.5 billion and the sector has witnessed the entry of many new startup firms. As many as 16 startup firms focusing on this sector have reached the milestone of Initial Public Offerings (IPOs) and have a combined market capitalization of $2.6 billion. [3]

Though much research has been undertaken on the drivers of innovation in firms, we believe that our study has two distinguishing characteristics, both related to the nature of the industry under consideration. First, tissue engineering is an emerging sector, the technology is still at an early stage of development. Secondly, tissue engineering is an interdisciplinary sector—it is based on knowledge from several different, yet related, disciplines. Both these factors imply that the drivers of innovation in this industry may be different from more technologically mature or less-interdisciplinary sectors.

A major premise of this study is that external linkages, in the form of alliances, are critical for innovation. Since the technology is interdisciplinary, few firms can hope to have or cultivate, in a speedy manner, all the skills necessary to compete, thus creating the need for collaboration.

The rest of this paper is organized as follows. We review the conceptual arguments and the main hypotheses of the paper in the next section. Methodological aspects of the study will be discussed next. The paper will conclude by summarizing the implications of the results, the limitations of the paper and identifying the directions for further research in the area.

2.   Conceptual arguments and hypotheses development

It is widely recognized that innovation or the ability to come out with new products is a key driver of competitive success in today's environment. Several successful corporations owe their profitability to their ability to come out with new products [4], especially in sectors such as pharmaceuticals. [5] In addition innovation bestows a number of advantages to its owner. It increases the visibility of firms, provides them with greater legitimacy, early market share (first mover advantage) and improves their chances of survival. [6] Greater legitimacy, in turn might also help a firm in recruiting allies, customers or sometimes even attracting financing.

Innovation, however, is not without risks. The costs of innovation can be substantial and there are no guarantees of success or returns. Some studies estimate that nearly 50% of new products, introduced to the marketplace during any given year, fail, causing financial loss and possibly embarrassment to promoters. [7,8]

A fundamental premise of this study is that formation of alliances with external parties will boost the innovation output of firms. In fact, Teece [9] concludes that "To be successful. Innovating firms must form linkages, upstream and downstream, lateral and horizontal." Several other studies [10,11] have also arrived at the same conclusion regarding strategic alliances and innovation. In high technology industries, boosting innovation is one of the key motivations behind forming alliances. Bidault and Cummings [12] remark that "among the various motivations for partnering, innovation is said to be a rationale of singular importance."

Strategic alliances include a broad array of cooperative agreements between two or more firms. Several types of alliances, which are particularly relevant for innovation, include the following: licensing agreements, joint (or collaborative) R&D projects and one way research agreements where only one party does the R&D work whereas the other provides funding, market access or other skills.

The above premise, regarding the critical role of strategic alliances in innovation, is grounded in several factors. Strategic alliances enable accessing complementary resources such as knowledge of a technical field where the particular firm is lacking. Such knowledge may be either difficult (e.g., might take a long time to develop) or expensive to acquire for the firm. Combining complementary resources might give the allying firms an advantage over others that are unable to do so. [13] Strategic alliances could also lead to a shorter time to market improving the chances of success. Glaxo's anti-ulcer drug, Zantac, is a good example in this regard. Though SmithKline's Tagamet preceded Zantac in the US market, Glaxo was able to be the first in several international markets, largely by signing-up allies with strong distribution networks in individual countries. As a result, Zantac was not only able to surpass Tagamet, but also become the world's largest selling drug for a significant length of time.

Strategic alliances also allow firms to share risks. In sectors where new product development is expensive, strategic alliances enable firms to spread the costs over several firms. [14] Risk reduction due to alliances goes well beyond simply dividing up the investment requirement—it is attributable to complementary resources (specialization), cross-fertilization of ideas and better speed to market. In addition, redundancies can be eliminated across partners, saving costs in the process (Gemuden, Heydebreck and Herden, 1992).

Strategic alliances might also enable partners to learn new skills, boosting future innovation. [115,16] Previous research has observed an increase in alliance formation is settings where cooperative learning is particularly important. [17]

The preceding arguments suggest the following hypothesis: **H1:** Firms that have forged a larger number of alliances will have greater learning opportunities due to their multiple alliances and hence exhibit greater innovation output than firms that forge fewer alliances.

Not all types of alliances, however, are likely to lead to increased innovation or at least not to the same extent. We divide alliances into two broad types: technology acquisition or technology learning alliances and other alliances. In the first category of alliances, the tissue engineering firm experiences a direct inflow of technical knowledge (say, through inward licensing) or at least has an opportunity to learn from it alliance partner (say through a joint R&D project or a joint venture). The latter type of alliance will enable a firm to directly observe, and possibly imbibe the skills of its partner. [18]

In the second category of alliances, the tissue engineering firm may not experience a direct inflow of technology skills or knowledge. This might include an arrangement where the tissue engineering firm receives funding from another firm (or a government agency such as NIH) for carrying out a research project. Other alliances in this category might include the following: marketing/ distribution agreements, outward licensing agreements where a firm provides another with technology in exchange for a fee. In both these cases, the tissue engineering firm is either a donor of technical knowledge or it is utilizing the expertise of its partner in a different functional area (distribution/ marketing). While these kinds of alliances might improve the financial viability of the allying firm, they may not add directly to the technological knowledge base of the tissue engineering firm and hence are unlikely to boost innovation to the same extent as the first category of alliances. This leads us to the second hypothesis: **H2**: Technology acquisition or technology learning alliances will have a greater impact on the innovation output of a tissue engineering firm than other types of alliances.

3.   Methodology

This study is based on the analysis of various secondary data sources. We adopted the following procedure for data collection. First, since one of the partners in the research project is actively involved in undertaking technical research in the tissue engineering sector, we were able to identify a comprehensive list of firms that are participating in this sector. Next, we identified the patents obtained by each of these firms. The key data source for this step was homepage of the US patent office (http://www.uspto.gov/). Our next task was to identify the strategic alliances for each of these firms. We purchased a database on strategic actions published by the BioAbility Inc (http://www.bioability.com/). From this database, we identified data regarding alliance formation by eight firms: Advanced Tissue Sciences, Organogenesis, Osiris Therapeutics, Atrix Laboraties, Creative Biomolecules, Integra Lifesciences, Lifecell Corporation and Stemcells (formely Cytotherapeutics). While the sample is relatively small, we believe that we can get some initial results regarding the impact of alliance formation on innovation output. In the next stage of research, we will expand the scope of analysis to include all the major firms in the sector.

To test the hypotheses informally, we tracked the alliance activity of the sample firms. Alliances were further divided into the two categories identified above: technology acquisition (or learning) alliances or other alliances. We then correlated the counts of these individual types of alliances with the number of patents filed.

4.   Results and discussion

Table 1 identifies the key players in the Tissue Engineering sector, their main lines of business, date of founding and the total number of tissue engineering patents obtained by them. It is apparent from the tables that, if we consider number of patents as a measure of success, companies focused on the tissue engineering sector have enjoyed much greater success. Out of the top five patent-holders, three companies (Advanced Tissue Sciences, Stemcells and Organogenesis) are purely

TABLE 1. Profile of key players in the Tissue Engineering sector

Company	Main Business	Date of founding	Date of initial public offering	No of patents in Tissue Engineering
Accorda Therapeutics	Tissue Engineering Company	NA	NA	1
Advanced Tissue Sciences Inc (formerly Marrow Technology)	Tissue Engineering Company	1986	1988	35
Atrix Laboratories	Drug delivery + tissue engineering	1987	N.A.	16
Biomatrix	Drug delivery + tissue engineering	1981	1991	17
Biora AB	Drug delivery + tissue engineering	NA	NA	1
Bionx Implants	Also in medical devices	NA	NA	1
Creative Biomolecules	Mainly a biotech company	1981	1993	4
Genetics Institute Inc	Mainly a biotech company	1980	Subsidiary of American Home Products	12
Genentech	Mainly a biotech company	1976	1980	
Guilford Pharmaceuticals	Drug Delivery Company	1993	1994	2
Integra Life Sciences	Drug Delivery & Scaffold Company	1985	1995	5
Lifecell Corporation	Biotechnology & Tissue Engineering	1986	1992	3
Organogenesis	Tissue Engineering Company	1985	1986	20
Orquest	Tissue Engineering Company	1994	Private	6
Ortec International	Tissue Engineering Company	NA	NA	1
Osiris Therapeutics	Tissue Engineering Company	1993	Private	17
Osteotech	Medical devices	1986	1991	4
Stemcells (formerly Cytotherapeutics)	Tissue Engineering Company	1988	1992	28

Notes:
Sources: www.yahoo.com and Individual company homepages
N.A.—Not Available

414

TABLE 2. Brief profile of companies included in the analysis

Company	Revenues1 (1)	Profits1 (2)	Patents² (3)	TE Patent Rank (4)	Total alliances (5)	Learning alliances (6)	Alliance Rank (7)
Advanced Tissue Sciences Inc	20.8	(19.3)	35 (35)	1	14	8	1
Organogenesis	7.6	(22.6)	123 (20)	3	13	2	3
Osiris Therapeutics	NA	NA	17 (17)	4	3	2	3
Atrix Laboratories	10.8	(22.5)	40 (16)	5	1	0	7
Creative Biomolecule			69 (4)	7	4	0	7
Integra Life Science	68.4	7.9	5 (5)	6	1	1	5
Lifecell	20.9	(2.6)	7 (3)	8	4	1	5
Stemcells (Cytotherapeutics)	0.677	.076	28 (28)	2	7	4	2

Notes:
1   Revenues and profits are for the most recent 9 months (ending 30/9/01) in US$ millions
2   Figures without parentheses indicate the total number of patents obtained by the firm.
      Figures inside parentheses indicate the patents obtained in the tissue engineering sector.
Sources: www.yahoo.com and individual company homepages.

focused on tissue engineering and two are focused on drug delivery and tissue engineering sectors (Atrix Laboratories and Biomatrix).

A brief profile of the eight companies, which were further analysed, is listed in Table 2. In terms of date of entry into the sector, Creative Biomolecules was the earliest (1981) entrant while Osiris is the latest (1993). The rest of the companies were founded between 1985 and 1988. Among the eight companies, Osiris is the only privately held company, possibly because it is also the youngest ion terms of age.

In terms of tissue engineering patents, Advanced Tissue Sciences is the clear leader (35 patents) followed by Stemcells (Cytotherapeutics) (28 patents) and Organogenesis (20 patents). Osiris (17 patents) and Atrix Laboratories (16 patents) fall in the middle band, while Creative Biomolecule, Integra Life Sciences and Lifecell are at the low end of the spectrum.

In terms total alliance formation (number of alliances of all kinds), Advanced Tissue Sciences and Organogensis are quite comparable, while most other companies have been far less active in forming partnerships. We will, however,

TABLE 3. Patterns of alliance formation and patent filing for selected Tissue Engineering Companies

		87-89	90-92	93	94	95	96	97	98	99	00	01	Tot.
Advanced Tissue Sciences Inc	Patents	0	0	1	0	3	9	2	9	6	4	1	35
	Alliances (1)	0	3	0	2	0	0	2	1	0	0	0	8
	Alliances (2)	0	0	2	0	1	0	0	0	1	1	0	5
	Total alliances	0	3	2	2	1	0	2	1	1	1	0	13
Organogenesis	Patents	1	0	1	1	2	2	1	4	4	0	2	18
	Alliances (1)	1	0	0	0	0	1	0	0	0	0	0	2
	Alliances (2)	7	0	0	1	1	1	0	1	0	0	0	11
	Total alliances	8	0	0	1	1	2	0	1	0	0	0	13
Osiris Therapeutics	Patents	0	0	0	0	0	1	1	3	3	4	2	14
	Alliances (1)	0	0	0	0	1	1	0	0	0	0	0	2
	Alliances (2)	0	0	0	0	0	1	0	0	0	0	0	1
	Total alliances	0	0	0	0	1	2	0	0	0	0	0	3
Atrix Labs	Patents	0	0	0	2	0	1	3	5	2	1	2	16
	Alliances (1)	0	0	0	0	0	0	0	0	0	0	0	0
	Alliances (2)	0	0	0	0	0	1	0	0	0	0	0	1
	Total alliances	0	0	0	0	0	1	0	0	0	0	0	1
Creative Biomolecule	Patents	0	2	0	1	0	0	1	0	0	0	0	4
	Alliances (1)	0	0	0	0	0	3	1	0	0	0	0	4
	Alliances (2)	0	0	0	0	0	0	0	0	0	0	0	0
	Total alliances	0	0	0	0	0	3	1	0	0	0	0	4
Integra Life Sciences	Patents	0	0	0	0	0	1	1	2	0	0	0	4
	Alliances (1)	0	0	0	0	1	0	0	0	0	0	0	1
	Alliances (2)	0	0	0	0	0	0	0	0	0	0	0	0
	Total alliances	0	0	0	0	1	0	0	0	0	0	0	1
Lifecell	Patents	0	0	0	2	0	0	0	0	0	0	1	3
	Alliances (1)	0	0	0	1	0	0	0	0	0	0	0	1
	Alliances (2)	0	0	0	0	0	1	1	0	1	0	0	3
	Total alliances	0	0	0	1	0	1	1	0	1	0	0	4
Stemcells	Patents	0	0	0	0	0	2	5	14	5	2	0	28
	Alliances (1)	0	1	1	0	1	0	1	0	0	0	0	4
	Alliances (2)	0	1	1	0	1	0	0	0	0	0	0	3
	Total alliances	0	2	3	0	2	0	0	0	0	0	0	7

bring out a key distinction between the alliances formed by these companies in the following discussion.

At a finer level of analysis, we divided the alliances into two categories. The first category included those where there was an inflow of technical knowledge. These might include inward licensing arrangements, joint R&D projects and so on. The second category included alliances where there was no direct inflow of technical knowledge. These included outward licensing arrangements, funding alliances, and marketing and distribution agreements, among others. Hypothesis 2 predicted that the first category of alliances would have a more significant impact on patent count than the second category of alliances. Again, we observe some support for hypothesis. Though Advanced Tissues Sciences and Organogenesis have similar levels of alliance formation, Advanced Tissue has been involved in many more (8 versus 4 or less for any of the other firms) technology acquisition and learning alliances and hence exhibits a higher patent count. On the other hand, Organogenesis has been involved in several outward-licensing arrangements, which might have led to greater financial viability but not to increased patent count. In fact it was involved in only 2 technology acquisition or learning alliances, exactly identical to Osiris, which again might explain the similar levels of patent filing for these two firms. Thus, the critical variable seems to be the number of technology and learning alliances rather than the aggregate number of alliances. It is also apparent from Table 2 (columns 4 and 6) there is a close correspondence between samples firms' rank in the formation of learning alliances and the no of tissue engineering patents.

From Table 3, there appears to be a lagged effect of alliances on patent filing. For instance the impact of the technology acquisition or learning alliances formed by Advanced Tissues Sciences between 1992 and 1994 is apparent (in terms of increased patent counts) in 1995 and subsequent years. A similar pattern can be observed with Stemcells (Cytotherapeutics) where the lag appears to be 2-3 years. Organogenesis seems to be benefiting from an alliance formed in 1996 in 1998 and 1999. The lagged effect is quite consistent with expectations since the absorption of new knowledge as well as capitalizing on it would need time. In addition, there is a lag in filing for patent and obtaining approval.

5.   Summary and Conclusions

In summary, we find some support for both the hypotheses. External linkages appear to spur innovation. Technology acquisition or learning alliances seem to have a greater and more direct impact on the innovative output of firms. There seems to be a lagged effect of alliances on innovation—that is alliances lead to greater innovation only a few years after their formation.

This paper is a report of ongoing research work on the issues of innovation and business performance in the tissue engineering sector. While this paper has provided some initial results, much remains to be done. In the next phase of research, we intend to broaden the sample to include many more firms, incorporate business performance (including sales growth, profitability etc.) as an outcome variable and also include variables such as age of the company, financial resources raised from the market (including public offerings as well as private placements).

In the next phase of research, we also intend to analyse companies outside the US and also do a comparison between US-based versus Europe-based firms. We believe that firms outside the US will play an increasingly important role in the industry. We also believe that there are some key differences between the broader environment of the US versus Europe which will make the comparison informative. Sixteen out of the reviewed 73 companies in the survey of Lysaght et al are listed outside the US. Especially European start-ups have become aggressive and entrepreneurial. Isotis in the Netherlands; Modex in Switzerland; and Codon and Biotissue in Germany have been able to go public in 2001on national stock exchanges during a time in which IPO window was effectively closed for US-based companies. In addition, one advantage enjoyed by European companies is a regulatory environment that allows product sales far sooner than in the United States.

One limitation of the present study is that it focuses solely on patents as a measure of innovation output. While they have their limitations (not all patents lead to useful new products), patents are objective and unbiased indicators. Since they are approved by a regulatory authority (government), patents meet the test of newness - that is they are substantially different from anything that exists currently.

6.   References

1. Clark, K.B. and Fujimoto, T. (1991) *Product Development Performance*, Harvard Business School Press, Boston.
2. Brown, W. and Eisenhardt, K. (1995) Product development: Past research, present findings and future directions, *Academy of Management Review*, 20:343-378.
3. Lysaght, M.J. and Reyes, J. (2001). The growth of tissue engineering. *Tissue Engineering*, 7(5), 485-493.Business Week (1993). Flops, August 16th, 76-82.
4. Cooper, R.G. (1993) *Winning at new products: Accelerating the process from idea to launch*, 2nd edition, Addison Wesley , Reading, Mass:.
5. Whittaker, E. and Bower, D.J. (1994) A shift to external alliances for product development in the pharmaceutical industry, *R&D Management*, 24(3), 249-260.
6. Schoonhoven, C.B., Eisenhardt, K.M. and Lyman, K. (1990) Speeding products to the market: Waiting time to first product introduction in new firms, *Administrative Science Quarterly*, 35, 177-207.
7. Business Week (1993). Flops, August 16th, 76-82.
8. Zirger, B.J. and Maidique, M.A. (1990). A model of new product development: An empirical test, *Management Science*, 36(7), 867-883.Dodgson, M. (1993) Learning, trust and technological collaboration, *Human Relations*, 46, 77-95.
9. Teece, D.J. (1992) Competition, cooperation and economic behaviour, *Journal of Economic Behaviour and Organization*, 18, 1-25.

418

10. Robertson, T.S. and Gatignon, H. (1998) Technology development mode: A transaction cost conceptualization, *Strategic Management Journal*, 19, 515-531.
11. Gemuden, H.G., Heydebreck, P. and and Herden, R. (1992). Technological interweavement: A means of achieving innovation success, *R&D Management*, 22(4), 359-376.
12. Bidault, F, and Cummings, T. (1994) Innovating through alliances: Expectations and limitations, *R&D Management*, 24(1), 33-45.
13. Dyer, J.H. and Singh, H. (1998) The relational view: Cooperative strategy and sources of organizational advantage, *Academy of Management Review*, 23(4), 660-679.
14. Mowery, D.C. (1988) *International collaborative ventures in US manufacturing*, Ballinger, Cambridge, MA.Hagedoorn, J. (1993). Understanding the rationale of strategic technology partnering: Interorganizational modes of cooperation and sector differences, Strategic Management Journal, 14(5), 371-385.
15. Dodgson, M. (1993) Learning, trust and technological collaboration, *Human Relations*, 46, 77-95.
16. Hagedoorn, J. (1995). A note on international market leaders and networks of strategic technology partnering, Strategic Management Journal, 16, 241-250.
17. Hagedoorn, J. (1993). Understanding the rationale of strategic technology partnering: Interorganizational modes of cooperation and sector differences, Strategic Management Journal, 14(5), 371-385.
18. Hamel, G., Doz, Y. and Prahalad, C.K. (1989). Collaborate with your Competitors and Win, *Harvard Business Review*, 67(1), 133-39.

INDEX